Case Studies in Environmental Archaeology

Second Edition

INTERDISCIPLINARY CONTRIBUTIONS TO ARCHAEOLOGY

Series Editor: Michael Jochim, *University of California, Santa Barbara, California*
Founding Editor: Roy S. Dickens, Jr. *Late of University of North Carolina, Chapel Hill, North Carolina*

A Continuation Order Plan is available for this series. A continuation order will bring delivery of each new volume immediately upon publication. Volumes are billed only upon actual shipment. For further information please contact the publisher.

Case Studies in Environmental Archaeology

Second Edition

Edited by

ELIZABETH J. REITZ

University of Georgia
Athens, Georgia

and

C. MARGARET SCARRY

University of North Carolina
Chapel Hill, North Carolina

and

SYLVIA J. SCUDDER

University of Florida
Gainesville, Florida

 Springer

Editors:
Elizabeth J. Reitz
Georgia Museum of Natural History
University of Georgia, Athens, USA
Athens, GA 30602
ereitz@uga.edu

C. Margaret Scarry
Department of Anthropology
University of North Carolina
Chapel Hill, NC 27599-3115
Margie_Scarry@unc.edu

Sylvia J. Scudder
Florida Museum of Natural History
University of Florida
Gainesville, FL 32611
Scudder@flmnh.ufl.edu

ISBN-13: 978-0-387-71302-1   e-ISBN-13: 978-0-387-71303-8

Library of Congress Control Number: 2007929503

For students, the future of environmental archaeology

Foreword to the Second Edition

The first edition of this volume exhibited environmental archaeology of the 1990s, an intensely interdisciplinary field that arose half-a-century ago from widespread concern with environmental issues, both past and present. Environmental archaeology had clearly come of age by the last decade of the 20th century, but was still centered upon methods, techniques, and basic epistemological matters: what kinds of knowledge claims can be made by environmental archaeologists, and how can those claims be most effectively tested and justified? In the second edition of *Case Studies in Environmental Archaeology*, greater maturity of the sub-discipline is highlighted by a shift in emphasis to results, and to the relevance of those results for major themes and research problems within Americanist archaeology and anthropology.

Hence, the second edition differs significantly from edition number one in scope, organization, substantive content, and emphasis. Most of the first-edition authors are also included in this second edition, having updated, revised, or rewritten their chapters, but there are several new authors and half-a-dozen new chapters. Both revised chapters and new ones have been accorded the same careful attention from the editors as was the case with the first edition, so that the volume as a whole is well written, well organized, and well integrated. Anyone seriously interested in environmental and ecological archaeology will want to own both books.

The first (1996) edition was a surprise gift from students and colleagues to Elizabeth Wing when she was presented by the Society for American Archaeology with the Fryxell Award for distinguished contributions to interdisciplinary research in archaeology. Now, ten years later, having accrued another award of high distinction (election to the National Academy of Sciences in 2006), but no longer this volume's honoree, Wing has enhanced the second edition by contributing her own chapter on people, animals, and their reciprocal relations in Caribbean pre- and proto-history. Although this second edition of *Case Studies in Environmental Archaeology* is no longer overtly dedicated to her, this is still Liz Wing's book, because the contributors — whether discussing sediments or landscapes, past and present; flora or fauna, ancient or modern, domestic or wild; climatic regimes, synchronic and diachronic; or past human subsistence patterns, and human susceptibility to endoparasites — are all individuals inspired and guided by her exemplary record as a team member, leader, and all-around practitioner of environmental archaeology. Once again Elizabeth Wing and her group have clearly demonstrated the diverse, highly significant knowledge gains possible for those who follow ecological and environmental pathways to understanding the human past.

Patty Jo Watson
Washington University-St. Louis
University of Montana-Missoula
St. Louis, Missouri

Preface to the Second Edition

A few years ago, one of us was approached by a student who asked if we really *knew* anything about human behavior from environmental archaeology. This student was taking an environmental archaeology course in which taphonomy and the biases of the techniques and methods used by environmental archaeologists figured prominently. The goal was to train students to be informed users of environmental data. But the student, quite rightly, wanted to know if beyond taphonomy, recovery biases, and analytical biases, there was anything environmental archaeologists could say about the human condition. Her question was the stimulus for the first edition of this volume and continued to guide us as we edited this second edition.

Our purpose is to show students and scholars, through a series of case studies, that there are things environmental archaeologists do know about the human-environmental relationship; many of which defy conventional archaeological expectations. To do so, however, we must break with the traditional organization of environmental archaeology along disciplinary lines. Environmental archaeology is intrinsically interdisciplinary. The demands of each discipline often mean, however, that specialists become further specialized in a very limited technical focus. While almost every environmental technique and method is represented in this edition, they are used as tools, not as ends in themselves.

To enhance the focus on interpretation, each author has reduced or eliminated discussions of methodology and technique. Thus, each had to balance the methodological responsibility required by the scientific method with a focus on theory and interpretations. The studies focus on insights obtained from methods, rather than on the methods themselves. Readers are referred to other literature for discussions of biases and methodological limitations.

Environmental archaeology is a broad and diverse field united by ecological and anthropological theories and interests that rarely are made explicit. It is not possible, in a few case studies, to do justice to the entire field. Some themes, however, are consistently embraced by environmental archaeology. These themes include, but are not limited to: the systemic relationships between humans and the physical, chemical, and biological world in which they live; human nutrition and health; and complex human behaviors associated with acquiring resources. These behaviors include strategies for accommodating variations in temporal and spatial availability of resources, such as settlement patterns, domestication, and exchange systems; and the emergence of social complexity. These interlaced topics are addressed to varying degrees in this volume.

The case studies focus on research problems using data from the Americas; however, a quick review of these studies shows that the subject matter of each study has a global perspective. We hope that by demonstrating the power of environmental archaeology to explore interesting aspects of human behavior using examples from a hemisphere not known for its strength in environmental archaeology that we will draw students to these topics.

Preparing these studies was a challenge to us all. The scientific method requires that the validity of scientific finds be assessed first by evaluating the methods used to derive the results. But, we also believe it is important to look up from the microscope occasionally and evaluate what we have learned about culture. We ask forbearance from our colleagues for down-playing how we know what we know in order to focus on broader questions. We hope that students and colleagues alike will appreciate the joy of discovery that motivates us to persevere.

The first edition was dedicated to Elizabeth S. Wing, who is a friend, colleague, and mentor to many of the authors in this volume. Without her inspiration, guidance, and encouragement, this volume would not have been conceived. The first edition was intended to be a surprise for her on the occasion of the Society for American Archaeology's Fryxell Award in Interdisciplinary Studies. Thus, we did not ask her to contribute a case study. We are delighted to include her among the authors of this second edition.

Many others made substantial contributions either to the first edition or to the second edition. In particular, we thank Gisela Weis-Gresham and Tara Odorizzi, who drafted most of the figures. Robin Cook, Ervan G. Garrison, H. Stephen Hale, Lee A. Newsom, Rochelle A. Marrinan, Bonnie G. McEwan, Deborah M. Pearsall, Irvy R. Quitmyer, Herman J. Reitz, Michael J. Rodeffer, Donna L. Ruhl, Clara JoAnn Scarry, John F. Scarry, Izumi Shimada, Myrna Sulsona, Eliot Warner, Timothy S. Young, and the staff of Plenum Press provided invaluable assistance on the first edition. To this list we add Teresa Krauss, Katie Chabalko, and the staff of Springer for the second edition. Through the generosity of Kitty Emery, the Florida Museum of Natural History provided essential support during the final preparation of the second edition. Finally, we thank our home institutions, students, colleagues, families, and the authors of these studies for their patience, cooperation, and support.

Contents

PART III: HUMAN ECOLOGY

PART IV. SOCIAL AND ECONOMIC STRATEGIES

Contributors

Karen R. Adams, Crow Canyon Archaeological Center, Cortez, CO

C. Fred T. Andrus, Department of Geological Sciences, University of Alabama, Tuscaloosa, AL

Susan R. Arter, San Diego, CA

Richard Cooke, Smithsonian Tropical Research Institute (STRI), Ancón, Panamá

Kathleen A. Deagan, Florida Museum of Natural History, University of Florida, Gainesville, FL

Catherine S. Fowler, Department of Anthropology, University of Nevada at Reno, Reno, NE

Gayle J. Fritz, Department of Anthropology, Washington University in St. Louis, St. Louis, MO

Jochen Gerber, Division of Invertebrates, Field Museum of Natural History, Chicago, IL

Robert J. Hard, Department of Anthropology, University of Texas at San Antonio, San Antonio, TX

Brian E. Hemphill, Department of Sociology and Anthropology, California State University, Bakersfield, CA

Dale L. Hutchinson, Department of Anthropology, University of North Carolina at Chapel Hill, Chapel Hill, NC

Máximo Jiménez, Smithsonian Tropical Research Institute (STRI), Ancón, Panamá

Robert L. Kelly, Department of Anthropology, University of Wyoming, Laramie, WY

David B. Landon, Department of Anthropology, University of Massachusetts, Boston, MA

Clark Spencer Larsen, Department of Anthropology, The Ohio State University, Columbus, OH

Sarah W. Neusius, Department of Anthropology, Indiana University of Pennsylvania, Indiana, PA

Lee A. Newsom, Department of Anthropology, The Pennsylvania State University, University Park, PA

Evan Peacock, Cobb Institute of Archaeology, Mississippi State, University, MI

Deborah M. Pearsall, Department of Anthropology, University of Missouri, Columbia, MO

Irvy R. Quitmyer, Florida Museum of Natural History, University of Florida, Gainesville, FL

Anthony J. Ranere, Anthropology Department, Temple University, Philadelphia, PA

Karl J. Reinhard, School of Natural Resources, University of Nebraska-Lincoln, Lincoln, NE

Elizabeth J. Reitz, Georgia Museum of Natural History, University of Georgia, Athens, GA

John R. Roney, Colinas Cultural Resource Consulting, Albuquerque, NM

Christopher B. Ruff, Center for Functional Anatomy and Evolution, The Johns Hopkins University School of Medicine, Baltimore, MD

Michael Russo, National Park Service, Southeast Archeological Center, Tallahassee, FL

Daniel H. Sandweiss, Department of Anthropology/Climate Change Institute, University of Maine, Orono, ME

C. Margaret Scarry, Department of Anthropology, University of North Carolina at Chapel Hill, Chapel Hill, NC

Kari M. Schmidt, Department of Anthropology, University of New Mexico, Albuquerque, NM

Margaret J. Schoeninger, Department of Anthropology, University of California at San Diego, La Jolla, CA

Elizabeth M. Scott, Department of Sociology and Anthropology, Illinois State University, Normal, IL

Sylvia J. Scudder, Florida Museum of Natural History, University of Florida, Gainesville, FL

Kristin D. Sobolik, Department of Anthropology, University of Maine, Orono, ME

Julie K. Stein, Burke Museum of Natural History and Culture, University of Washington, Seattle, WA

Gail E. Wagner, Department of Anthropology, University of South Carolina, Columbia, SC

Patty Jo Watson, Department of Anthropology, Washington University in St. Louis, St. Louis, MO

Elizabeth S. Wing, Florida Museum of Natural History, University of Florida, Gainesville, FL

Melinda A. Zeder, Department of Anthropology, National Museum of Natural History, Smithsonian Institution, Washington, DC

Part I

Introduction

Chapter 1

Introduction to Environmental Archaeology

ELIZABETH J. REITZ, LEE A. NEWSOM, SYLVIA J. SCUDDER,
AND C. MARGARET SCARRY

Environmental archaeology is an eclectic field directed toward understanding the ecology of human communities (Albarella 2001; Bintliff et al. 1988; Branch et al. 2005; Butzer 1971, 1982:5; Dimbleby 1965, Evans 1978, 2003; Evans and O'Connor 2001; Luff and Rowley-Conwy 1994; O'Connor 1998; Shackley 1981; Wilkinson and Stevens 2003). Environmental researchers apply information and techniques from the earth and biological sciences to study relationships among peoples and their environments using organic and inorganic evidence from archaeological sites. Environmental archaeology encompasses many interests, such as those subsumed by the terms paleoecology, paleoenvironment, paleo-economy, and paleogeography; in practice it often eludes such intellectual boundaries (see discussions in Albarella 2001). Cultural and non-cultural environments are dynamic factors in the formation of archaeological sites. Although much environmental research traces ecological relationships at a site or within a region, at its best environmental archaeology interprets human behavior set in an environmental framework that includes broad social, spatial, temporal, physical, and biotic parameters.

The study of organic and inorganic remains from archaeological sites has a long history. In its early years, the field was a collection of biological, physical, and chemical techniques applied with little theoretical framework beyond, perhaps, environmental reconstruction or the evolution of domestic plants and animals. The earliest investigations generally produced lists of identifications with limited, if any interpretations, and little reference to the processes of deposition, the fragmentary nature of the archaeological record, recovery techniques, or methodologies (Bryant 1989; Butzer 1982; Cowan and Watson 1992; Ford 1979; Luff and Rowley-Conwy 1994; Miksicek 1987; Pearsall 2000:3–6; Smith 1976; Watson 1997). It was not until recovery techniques were improved and evolutionary, ecological, and interdisciplinary frameworks adopted that environmental archaeology became the holistic and diverse field that it is today. Most current practitioners endeavor to apply insights obtained from the environmental sciences to questions concerning relationships among people, cultural systems, and ecosystems. In addition, they explore the biases of the archaeological record, evaluate the methods used to obtain environmental data, and interpret the data themselves.

Environmental scientists, some of whom are neither archaeologists nor anthropologists, often conduct their research as members of multidisciplinary teams. In such settings, good research designs are fundamental to maximizing the potential of environmental data. If environmental data are peripheral to the major research questions of a project, they are likely to appear in highly descriptive appendices. Such reports lack the linking structures needed for the insights produced by environmental archaeologists to be communicated to other members of the team and incorporated into the interpretive fabric. Only in the context of problem-oriented archaeology in which concepts about ecological relationships are pertinent and articulated with the overall research agenda is the potential of environmental research realized. Such contributions are strengthened by evidence from multiple sources of data.

It is not the case, however, that environmental archaeologists are dependent upon questions generated by other members of a research team. Theoretical perspectives rather than methodological concerns are present in much environmental research, albeit often as unstated assumptions. These perspectives include, but are not limited to, the systemic relationships between people and the physical and biological world in which they live; human nutrition and health; the complex human behaviors associated with acquiring food and other resources; strategies for accommodating seasonal and spatial variability in resources, including settlement patterns, domestication, and exchange systems; and the emergence of social complexity.

Concepts about human behavior and cultural systems are central to work encompassed under the broad umbrella of environmental archaeology. Implicit in most environmental research are assumptions about the relationship between inheritance and learning. To what extent do cultural systems reflect a genetic basis and to what extent are they the product of learned behaviors? In what ways are the variables of time and space reflected in archaeological data? How do the structure and function of cultural systems relate to environmental processes? Are there biological and physical constraints on cultural innovation? If so, how are these reflected in archaeological data? What are the relationships between natural systems and cultural ones? To what extent has human behavior been responsive to, or even caused by, environmental variables and to what extent are people responsible for environmental changes? These are basic questions that are as central to environmental archaeology as they are to any other discipline.

Environmental archaeology is divided for convenience, if not in practice, into four subfields: the earth sciences, archaeobotany, zooarchaeology, and bioarchaeology. These are roughly defined by methods and data appropriate to one subfield or another. The field's multi-disciplinary nature is reflected in the backgrounds of environmental archaeologists, who may be zoologists, veterinarians, botanists, paleontologists, human biologists, agricultural scientists, chemists, geographers, geneticists, ecologists, anatomists, forest managers, geologists, and nutritionists, as well as anthropologists and archaeologists.

EARTH SCIENCES

Geological or earth sciences emphasize the non-cultural aspects of sites and their environments. These studies are variously referred to as archaeogeology, archaeometry, geoarchaeology, and archaeological geology (Brown 1997; French 2003; Garrison 2003; Gladfelter 1981; Herz and Garrison 1998; Holliday 1992; Taylor and Payen 1979; Waters 1992). The distribution of major land masses, volcanic and earthquake zones, topography, and inorganic raw materials such as rocks and minerals; as well as chemical, physical, and stratigraphic properties of soils and sediments all influence human behavior. Study of these phenomena yields climatic, chronological, geological, and cultural information critical to reconstructing and understanding human activities.

Geomorphologists place sites within natural landscapes such as river valleys, alluvial terraces, coastal plains, or mountain ranges and investigate structural features that dictate surface geomorphology; effects of relief, slope, and hydrology; and the edaphic relationships of soils and vegetation. A settlement's location may suggest an arrangement of landscape elements which no longer exist. Combining modern regional geomorphology with paleoenvironmental indicators and an understanding of the dynamics of landscape evolution allows the geomorphologist to reconstruct those conditions which favored settlement in a specific area.

Sedimentologists examine the relationships among size class distributions and physical conditions of so-called "clastic" particles from fine clay to gravel and cobbles in loose sediments and soils as well as lithified forms. Grain-size distributions of sediments typically reflect the agent of deposition (wind, water, gravity) as well as the energy needed for transport. Particle size analyses can be used to identify processes such as erosion by wind or water, freeze-thaw cycles, dust and ash falls, landslides, and floods. Stratigraphic relationships among sediment types define sequences of depositional environments (e.g., marine-nearshore-continental) which can be used to infer changes in sea level, climate, river courses, and other events on a geologic time scale. Disruptions in natural processes of sediment transport and accumulation brought about by human activities which alter local and regional landscapes can be identified by episodes of increased erosion or deposition. The correlation of sediment types and sequences with microfauna identified within those sequences contributes additional data to interpretations of climate and environmental change (Scudder 2001, 2003).

Archaeopedologists characterize the chemical and physical attributes of soils from archaeological sites and compare these with local native soils not impacted by human activity (Collins et al. 1995). Major elements such as phosphorus, calcium, and magnesium as well as trace minerals which accumulate in anthropogenic (human-generated) soils are compared with residual levels of those same elements in native soils to evaluate the intensity and nature of human impact on the soil environment. Patterns of chemical element accumulation within the soil can aid in locating and interpreting the function of intrasite features such as hearths, food preparation and storage areas, and burials. Altered physical characteristics such as changes in soil texture (ratio of sand, silt, and clay), thickness or arrangement of horizons (layers), or micromorphology, are used to examine the effects of human habitation on pedogenesis, or soil formation.

Archaeometrists use techniques that measure physical and chemical properties for site prospecting, chronometric (absolute) dating, physical and chemical artifact analysis, and sourcing raw materials and artifacts (Taylor and Payen 1979). Geophysical methods of remote sensing such as magnetometry, resistivity, and ground-penetrating radar are used to explore for sites and to "map" buried architectural features. Geochemical prospecting, such as soil phosphate testing, gives a three-dimensional map of subsurface element distribution. These methods, combined with chronometric dating techniques, such as ^{14}C, potassium-argon, or dendrochronology, help archaeologists decide whether to excavate a site more extensively, where to locate excavation units, how to orient them, and to what depths to dig. Provenance of artifacts can be estimated from their chemical and physical properties and provides information about trade relationships and exchange routes as well as manufacturing processes.

Archaeometry also provides information about paleodiets and paleotemperatures through stable isotopes (Herz 1990; Price 1989). Elements such as carbon and nitrogen fractionate as they pass through the food chain. Hence it is possible to relate carbon ($^{13}C/^{12}C$) or nitrogen ($^{15}N/^{14}N$) ratios in bone to diet. Ratios such as these are widely applied as researchers attempt to define, for example, the relative proportions of plants and animals or of terrestrial and aquatic foods in diets. This research is enhanced by examining trace elements such as strontium, zinc, or copper. Oxygen ($^{18}O/^{16}O$) isotopes fractionate in response to temperature and offer information about paleotemperatures and seasonal patterns in resource use and residential choices.

ARCHAEOBOTANY

Archaeobotanists, or paleoethnobotanists, study carbonized, subfossil, waterlogged, or desiccated plant remains (Ford 1979; Fritz 1994; Pearsall 2000; Piperno 2006; Watson 1997). The array of plant materials studied by archaeobotanists includes mature seeds, pollen, and spores; whole organs such as fruits, flowers, leaves, stems, and roots; the conductive and supportive tissue systems, such as wood, bark, and epidermis; isolated components of plant tissues such as fibers, stomata, starch grains, and phytoliths; chemical compounds such as resins, tannins, lignin, lipids; and more. Archaeobotanists tend to specialize in particular classes of plant materials such as wood anatomy, seeds and/or other macroremains; pollen, starch, or phytolith remains (plant microremains); or cryptic data such as chemical residues.

Wood remains provide a variety of information. Growth-rings in wood are used to construct floating and long-term relative dating sequences (dendrochronology), to obtain climatic information (dendroclimatology), and to examine forest structure as well as the impact of people on forests (dendroecology). Wood also provides details about uses of plants for as tools, baskets, building materials, ornaments, or fuels, as well as environmental information (Pearsall 2000; Smart and Hoffman 1988).

Seeds, fruits, roots, and other edible structures provide basic information about the plants people used for food and medicine. Besides providing evidence about what people ate, such macroremains often are used to investigate subsistence strategies (e.g., foraging and/or farming); seasonality and scheduling decisions; domestication processes; social and ethnic variations in foodways; and a host of other questions. Analyses of macroremains complement environmental information gleaned from microremains and wood. For example the types and quantities of "weed" seeds can yield clues to farming practices or anthropogenic landscapes.

Palynologists study spores, isolated stomates, and the relative quantities of carbonized particulate matter as constituents of sediment cores, but they primarily work with fossil and subfossil pollen produced by seed-bearing plants. Deposits of wind-borne pollen provide regional profiles of vegetation, which researchers use to reconstruct floristic environments and track landscape evolution. Pollen distributed by animals, water, or self-pollinating plants provides information about rare plants, about environments closer to the site, about the site itself, and about the use of wild and domesticated plants.

Phytoliths form in epidermal and other tissues of plants and provide information similar to that obtained from pollen. They may yield evidence for thatch, bedding, mats, and containers, and other plant materials that often do not survive in other forms. Phytoliths from peat and animal dung offer indirect evidence of husbandry techniques and/or the use of these materials as fuel and fertilizer. Phytoliths on tool surfaces may indicate the function of such objects and signal the presence of economic plant taxa. Calcium oxalate crystals and stone cells from tubers and fruits may document the use of plants which are otherwise elusive because they are often pulverized, boiled, or ingested as whole, uncooked food. Starch grains analysis is particularly useful in the identification of such remains (Loy 1994).

Cryptic data come from chemical residues or DNA analysis. Compositional analysis of residues on pottery and other artifacts offers insights about plants (and animals) that were processed, consumed or used as oils, medicines, or perfumes (Fankhauser 1994; Reber 2005). Isozyme (protein) and DNA analyses provide information about genetic relatedness. Such studies, for example, can compare the DNA of wild and domesticated plant varieties to identify the origin and antiquity of particular domesticated species (Decker-Walters et al. 1993; Matthews and Terauchi 1994; Renfrew and Boyle 2000; Zeder et al. 2006).

A few archaeobotanists specialize in the identification of organisms such as seaweeds, algae, and fungi (Evans 1978; Shackley 1981). Some of these organisms were consumed or used as medicines and others provide environmental information or indirect evidence for the use of host organisms or the presence of diseases with which they are commonly associated. Diatoms and other microscopic unicellular or colonial organisms, which live in freshwater and marine habitats, are typical of specific depositional environments, providing information about sources of clays used in ceramics as well as aquatic and vegetational histories. They may serve as sensitive microclimatic indicators. Fungi play an important role in the relationship between people and their environments sometimes as disease agents and sometimes as food or medicines. Ferns and mosses are additional sources of information.

ZOOARCHAEOLOGY

Zooarchaeologists, or archaeozoologists, study subfossil, and sometimes fossil, non-human animal remains from archaeological sites (Davis 1987; O'Connor 2000; Reitz and Wing 2007). The most commonly studied tissues are non-human bones and teeth, mollusk shells, crustaceans, and echinoderms; but hair, insect exocarps, DNA, and other tissues are included in this discipline. The identity of the animals and parts of animals present in an assemblage, as well as cryptic data contained in these remains, provide information on paleoenvironments and relationships among people and animals. Zooarchaeological analysis offers insights about faunal succession; husbandry and capture decisions related to the habits, habitats, seasonal characteristics, age, sex, and size of animal resources; as well as butchering, transportation, and redistribution decisions. Technologies used to capture, butcher, or tend animals; the relative frequency with which species were used based on the number of individuals and/or the amount of meat animals contributed; and the health of animal species can be inferred from such data. The relationships may reflect faunal changes in a region, including those bearing upon plant and animal domestication. Animal remains provide insights into local exchange systems and more extensive trade routes. Mollusks, crustaceans, and sea urchins not only were important foods in many places, but were often important trade items and status markers.

Several other groups of animals are important, but less frequently studied (Evans 1978; Shackley 1981). Insects such as beetles, flies, ants, wasps, spiders, and mites, tell us about climate, regional and local environment, and economies. Some insects, such as grasshoppers, termites, and locusts, were used as food. Important evidence for the use of insects as food is found in paleofeces or coprolites. Ectoparasites and endoparasites provide information about sanitation; the level of infestation of plant, animal, and human populations; and indicate habitats frequented by people, plants, and animals. Many of these parasites are associated with food species and provide indirect evidence that these were used. Parasitic arthropods such as fleas, lice, and ticks provide information about health, sanitation, and environment. Foraminifera, small single-celled Protozoa, are found in sediments of marine and freshwater origin and are important climatic indicators.

BIOARCHAEOLOGY

In some research traditions, bioarchaeology is an umbrella term for the study of plant, animal, and human remains (Albarella 2001; Wilkinson and Stevens 2003:17). In this volume, bioarchaeology refers to study of the biological aspects of human behavior from the perspective of human osteology (Larsen 1997). Bioarchaeology grows out of traditional anthropological interests in human skeletal remains from archaeological contexts. With increased awareness of relationships among health, morphology, and the environment, bioarchaeology is critical to environmental archaeologists. Particularly important are dietary reconstructions made via the analysis of stable isotopes and trace elements; assessments of health and well-being made via the study of pathological conditions such as growth arrest markers and infectious diseases; and behavioral reconstructions made via the study of bone structure.

Human remains provide information about nutrition, diet, demography, paleopathology, and population movements (Armelagos 1994; Buikstra and Mielke 1985; Cox and Mays 2000; Huss-Ashmore et al. 1982; Larsen 1997; Price 1989; Reed 2005). Analyses of tooth eruption and wear sequences; growth curves of long bones; estimates of adult stature; sexual dimorphism; and bone chemistry; combined with evidence for porotic hyperostosis, osteoporosis, hypoplasias, traumas, and other pathological conditions provide invaluable information about growth, disease processes, nutrition, and environmental relationships. Activity levels are inferred from variations in skeletal joints, bone form, and bone function as these reflect responses of bone to mechanical stresses. Although some pathologies, such as deficiency diseases (e.g., rickets), have specific origins, most of the information available to human biologists provides general evidence of health and biobehavioral characteristics of human/environmental relationships.

THEMES IN ENVIRONMENTAL ARCHAEOLOGY

The systemic relationship among people and their environments is studied from many perspectives. For example, in some regions, the impact of food production strategies is very important; in other regions, the impact of climate on cultural fluorescence is a major issue. As diverse as the methods and questions of environmental archaeology are, however, three research themes are common to them: concern about site formation processes and methodological issues; geological and biological discoveries; and anthropological interpretations.

Site Formation Processes and Methodological Issues

By considering the depositional setting and the principal agents of deposition, environmental archaeologists distinguish between site formation processes that are the products of peoples' activities and those resulting from other forces (Lyman 1994; Weigelt 1989:62). Much environmental research focuses on taphonomy and other site formation processes both as sources of bias and as interesting research areas in their own right (Gifford 1981; Gordon and Buikstra 1981; Lyman 1994; Miksicek 1987). Ethnoarchaeology and other actualistic studies are useful ways to explore how people and other agents create and alter the archaeological record.

Methodological concerns encompass recovery techniques, identification procedures, analytical measures, and the challenges these offer to environmental and behavioral interpretations (Ford 1979; Grayson 1984; Hastorf and Popper 1988; Larsen 1997; Lyman 1994; Monks 1981; Pearsall 2000; Piperno 2006). Data are substantially altered by procedural decisions, especially those related to field work and quantification. Our ability to explore significant questions is influenced by the confidence we have that the material was competently recovered and accurately identified. Deceptively simple decisions made in the field or laboratory impact subsequent interpretations.

Geological and Biological Discoveries

Geological and biological discoveries encompass phytogeographical and zoogeographical distributions, morphological characteristics of organic and inorganic remains, population structures, paleoenvironmental conditions, and ecological relationships of extant organisms for which archaeological materials provide historical background. Many of these studies focus on regions rather than on specific sites. Much of this research is concerned with paleoenvironmental reconstruction, a rendering of the biological and physical surroundings, resource base, and climatic conditions that defined ecosystems in the past. Documenting sea level fluctuations, landscape evolution, and climate change are fundamental to such reconstructions. Although many of these studies ignore humans, one major area of research focuses on the question of causality in human/environment relationships. To what extent have people altered the landscape and to what extent have people responded to natural environmental change? What were the consequences for people of change, or stasis, in the environment?

Anthropological Interpretations

Broad-scale anthropological questions include an array of topics embedding human institutions in their environments. Among other issues, environmental archaeologists interested in anthropological research study intra-site spatial organization, local physiography and site placement, patterns of plant and animal use, and human health. Non-food uses of organic and inorganic materials include tools and craft production, housing and shelter, clothing, containers, toys, firewoods and fuels, ornaments, medicines, labor, pets, or social and religious symbols.

Often anthropologically oriented research explores nutrition, diet, and subsistence strategies through combinations of ecological theory, environmental data, human biology, and anthropological concepts. Examining biological remains for evidence of subsistence strategies may require defining predator-prey relationships related to periodic schedules of resource use; niche width, optimal foraging behavior, site catchment areas, and settlement patterns; nutritional contribution and needs; taphonomy; activity areas within archaeological sites; collection and processing technologies; exchange, division of labor, and other aspects of economic systems; social stratification; political institutions; belief systems; human and non-human population levels; and other elements of social organization. The topics subsumed under subsistence provide an important window on such systems because acquisition of foods, medicines, and raw materials is fundamental to the survival of the community and impinges upon many cultural institutions.

The importance of asking good questions is relevant to all environmental studies; however, the importance of applying appropriate recovery and laboratory techniques, and being open to unexpected answers, is particularly clear in subsistence analysis. Environmental archaeologists often find that the organization and timing of cultural innovations in subsistence strategies and their consequences are considerably different from perceptions held by scholars who ignore environmental evidence. Many common models for cultural change that are uninformed by environmental archaeology are contradicted by environmental evidence. For example, during most of the Holocene, people were primarily gatherers, casual gardeners, fishers, trappers, and hunters. In large part, anthropologists have presumed an association between residential mobility, seasonal scarcity in resources, and low population sizes for these people. One of the most important contributions environmental archaeologists make to this debate is the compelling evidence that foragers made choices among the resources they used. Under no circumstances did they make use of every resource available to them. Nor is there a single, monolithic, mobile, foraging strategy. In many areas, sedentism on the part of all or a portion of a community was possible without relying on domesticated plants. Yet many archaeologists continue to associate reliance on domesticated plants with sedentism, increased population size, monumental construction, ceramic technology, and social complexity in spite of evidence that these traits are not necessarily linked. In terms of food selection, food preference, nutritional quality, residential patterns, diseases, and demography, neither foraging communities nor those that made use of domestic plants and animals fit into the concept of evolutionary stages advancing toward more complex social relationships based on a transition from mollusk use to hunting deer to plant cultivation. As often as not, the evidence of environmental archaeology suggests long periods of subsistence stability or trajectories at odds with simplistic linear models.

In addition to information about paleoenvironments, health, and subsistence strategies, environmental data provide insights about political institutions, divisions of labor, ethnicity, social identity, and other elements of social organization. Anthropological questions include an array of topics connecting social institutions to the environments in which they are found. Where people live; how they acquire their foods, medicines, and other raw materials; how, when, where, and by whom these resources are used; and which ailments afflict people are all reflected in environmental data. The uses of environmental resources are important ways to symbolize and affirm social categories and to enhance social and political stability. They are closely related to the structure and function of a local community and the region within which the community lives.

In many respects the question of whether, and on what scale, ecological, taphonomic, methodological, geological, biological, or anthropological issues should be emphasized reflects the wide variety of exciting information available from inorganic and organic materials recovered from archaeological contexts. Environmental archaeologists address many inter-related issues drawing upon many different methodologies. Depending upon the specialists' training and interests, the research objectives of the project, and the funding available, analyses may include studies of bone, seeds, enzymes, otoliths, pollen, wood, isotopes, feces, soil, phytoliths, egg shell, or DNA. Any one, or many, of these may be central to a study, occasionally examined, or ignored altogether. Successional changes of plants and animals in response to human predation, nutritional consequences of resource intensification, socioeconomic parameters involved in redistribution, or feasting are all topics studied by environmental archaeologists from a variety of perspectives.

Most environmental archaeologists do not consider taphonomy, methodology, biology, geology, or anthropology mutually exclusive domains. It is their unique combination that makes the field interesting. Environmental archaeologists are not strictly earth scientists, botanists, zoologists, or anthropologists but combine the earth and biological sciences with anthropology to study the interrelationship of people with their environment: the dynamics of human ecology. The interchange among these disciplines is probably the most significant source of novelty for the field, which is strengthened and enlivened by the diverse interests of its practitioners.

ORGANIZATION OF THE VOLUME

Environmental archaeology is intrinsically interdisciplinary, a characteristic amply demonstrated by the case studies in this volume. Following three introductory chapters, the studies are grouped into three topics: the physical environment and environmental change, human ecology, and social and economic strategies. Because each case study uses many lines of evidence, the sections are not mutually exclusive. Each study focuses on an anthropological or environmental question with interpretations supported by multiple lines of evidence generated by diverse methods and disciplines. The results demonstrate that research questions have an influence on the generation of new knowledge equal to or greater than that of new methods and techniques. None of the conclusions in these case studies is final, because we continue to seek verification and amplification of our results through further research. A comparison of the case studies in this second edition with those by the same authors in the first edition amply demonstrates that the authors' understanding of their materials changed as new data became available.

Deagan (Chapter 2) shows that environmental questions are as germane to the recent past as to other time periods and cultural systems. Her review of "environmental historical archaeology" demonstrates the importance of two elements in environmental archaeology: interdisciplinary teamwork and asking good questions. Studies of social and market variables such as social stratification and complexity, group affiliations, global markets, ethnicity and gender, acculturation, labor requirements, consumer patterns, colonialism, and urban land use typify environmental historical archaeology. The documentary record strengthens and improves environmental research, but does not make it unnecessary. Many of the topics surveyed by Deagan are relevant to colonial expansion whenever and where ever it occurred. In many respects, the cultural responses to the colonial expansions of the last 500 years are not so very different from ones that marked ethnic migrations, political expansions, urbanization, and similar phenomena throughout human history.

In Chapter 3, Fowler's study of Timbisha Shoshone foragers reveals that they groomed and manipulated their environment to enhance the yield of valued resources. Her study i an example of the insights that ethnographic studies offer archaeologists. It provides a useful caution against assuming that environmental data from foraging populations represent a "natural," unmodified landscape. It also suggests habits of resource management that might have led to domestication of plant resource where foraging and tending plants are closely connected. This case study should be considered carefully by conservation biologists who presume that people substantially impacted the environment only after the industrial revolution.

Part II focuses on the physical environment and paleoenvironments. The relationships among people and soils, local climatic and topographic conditions, resource distributions, and regional landscapes yield bioclimatological, geological, and cultural information critical to the study of human communities. Paleoenvironmental and cultural indicators, combined with a knowledge of landscape evolution, can be superimposed on modern regional geomorphology to reconstruct the conditions that favored settlement in a specific area. Because the choice of settlement location, foraging area, or economic pursuits may have been based on environmental factors that no longer exist, such reconstructions are vital to understanding human ecology. Conversely, people influence soil formation processes, landscape evolution, and distribution of native vegetation. The premise that a relationship exists between the physical environment and material culture is fundamental to all archaeological studies. In Chapter 4, Stein shows that postdepositional site formation processes may alter deposits within a shell midden, producing stratigraphy that is unrelated to material culture. Scudder's study (Chapter 5) demonstrates that soil morphology as well as chemical and physical soil characteristics testify to the construction history of a sand burial mound.

The next three case studies in Part II examine complex relationships among changes in the landscape, resource use, residential patterns, and political alliances. These studies address the extent to which people altered the landscape and the extent to which human behavior responded to environmental change or stasis. The study by Cooke, Jiménez, and Ranere (Chapter 6) shows the importance of considering the archaeological, artistic, and documentary record to obtain a more complete picture of paleoenvironments. Peacock and Gerber (Chapter 7) evaluate an historical landscape using archaeological mollusks to query environmental factors underlying the initiation and spread of farming in the American Southeast. Andrus, Sandweiss, and Reitz (Chapter 8) draw upon evidence from many different sources to characterize El Niño/Southern Oscillation and its relationship to the development of social complexity in Peru. These three studies have serious implications for restoration ecology and conservation biology in so far as many ecologists presume environmental conditions that prevailed over the last few centuries characterize the entire Holocene.

The case studies in Part III explore human nutrition, health, mobility, sedentism, domestication, diet; and trade, encapsulated under the term human ecology. Environmental variables and cultural systems are closely related.

Implicitly or explicitly, the quality of that relationship is assessed in terms of human health and nutrition. Human physical responses to the stresses of survival are present in the form of disease, chronic wear, or acute injury; or as robust health. Combined studies of bone morphology, parasites, bacteria, viruses, and stable isotopes provide the evidence for some of the physical challenges and responses, contributing to bioarchaeological studies of human health.

The first two case studies in Part III focus on non-linear relationships among nutrition, activity levels, health, and settlement patterns. Larsen, Kelly, Ruff, Schoeninger, Hutchinson, and Hemphill (Chapter 9) demonstrate that hunter-gatherers were adequately nourished and not subject to overbearing amounts of work, although mobility had consequences. Sedentism has consequences too, one of which is increased exposure to certain types of parasitic infections. Reinhard (Chapter 10) uses parasite ecology to identify human behaviors and environmental conditions that may promote or reduce disease. Both studies indicate that seasonal periodicity in resource acquisition, plant domestication, and population size are not the only significant variables in either mobility or sedentism.

The remaining studies in Part III explore subsistence strategies and resource availability. The decisions of each population impacted land-use patterns, residential decisions, diet, and demography. Seasonal availability of resources is one of the chief explanations for residential mobility, as is demonstrated by Sobolik in Chapter 11 for a population in the northern Chihuahuan Desert. Russo and Quitmyer (Chapter 12) explore the origin of sedentism and social complexity in the absence of a domestic food source on the southwest coast of the Gulf of Mexico. Evidence for very early coastal sedentism in Chapters 8 and 12 demonstrate that sedentism by foragers in coastal locations was not unusual.

On the other hand, farming involved carefully scheduled activities and delicate nutritional balances which might be met by residential mobility. Pearsall demonstrates (Chapter 13) that plant domestication does not insulate human societies from environmental disasters such as volcanic eruptions. She argues that maintaining a broad subsistence base was highly adaptive in a setting subjected to catastrophic events. The more crops and other resources used, the more likely some would survive. Wagner (Chapter 14) shows that farmers may be residentially mobile, at least in part, and that access to meat and food storage are important elements in such patterns. She contrasts responses to predictable shortages with responses to unpredictable shortages, arguing that the latter pressures lead to political and economic changes. Neusius (Chapter 15) argues that garden-hunting is used to overcome scheduling conflicts between tending crops and capturing animals in temperate horticultural systems. Hard, Adams, Roney, Schmidt, and Fritz (Chapter 16) explore why farming was adopted at all, since farming may require more work and result in a decline in health compared with most hunting and gathering strategies. They show that highly variable combinations of population size, sedentism, and settlement patterns are associated with plant cultivation. In combination, the case studies in Part III demonstrate that models based on a simple switch from hunting to gardening are incomplete.

The case studies in Part IV examine intra-community social relations, rural/urban relationships, and ethnic identity. These remind us of the importance of long-term studies of environmental data recovered during extensive site excavations and the importance of regional synthesis. Zeder and Arter (Chapter 17) explore foodways within a community to detect differential access to resources and patterns of resource sharing among households. Their study is possible because the site was almost completely excavated. Scott (Chapter 18) demonstrates that food remains provide evidence of socioeconomic position as well as the ethnic identity of a site's occupants at Fort Michilimackinac. In Chapter 19, Landon investigates urban meat distribution systems by comparing seasonal slaughter patterns of domestic animals at urban and rural sites from the mid-17th century through the end of the 18th century. Scarry (Chapter 20) examines how native farmers raised and tended indigenous crops and the implications of the inferred husbandry practices for the subsequent adoption of maize. Wing (Chapter 21) concludes the case studies by drawing upon evidence for animal domestication in the Caribbean to suggest that the process of domestication was more complex than usually described and domestic animals were more widespread in the Americas than previously thought. These studies remind us that plants and animals are more than nutrition; they are among the most fundamental components of human life and form part of the ideological fabric that binds communities together.

CONCLUSION

The primary objectives of environmental archaeology should be interpretations rather than identifications, measurements, and lists. At the same time, methodological concerns require a great deal of consideration and experimentation in themselves. Environmental archaeologists must balance the demands of their procedures against the potential they have to contribute to our understanding of the human/environmental relationship. The relationship between that interaction and the conditions under which biological and geological materials are present at archaeological sites must be considered. Resource managers and conservation biologists who use environmental data should familiarize themselves with these biases and methodologies.

However, environmental archaeologists must not become so focused on the rigors of their discipline that they neglect the broader implications of their research. Balancing methodology and interpretation is difficult, but these case studies show that achieving this goal is possible and worth the effort. It is not the aim of these chapters to describe environmental materials or critique the biases and methodologies of environmental archaeology. Ample literature is available for studies traditionally encompassed within environmental archaeology and these are cited where appropriate. The purpose of these studies is to explore larger anthropological issues, hoping to encourage students and colleagues alike to look beyond the technical details of each specialty.

As the following studies illustrate, the importance of long-term team research directed by good questions and communication networks cannot be over-emphasized. If these studies have no other unifying characteristic, it is that they arose out of such research environments.

REFERENCES

Albarella, U. (ed.), 2001, *Environmental Archaeology: Meaning and Purpose*, Kluwer, Dordrecht, The Netherlands.

Armelagos, G. J., 1994, You Are What You Eat, in: *Paleonutrition: The Diet and Health of Prehistoric Americans* (K. D. Sobolik, ed.), *Center for Archaeological Investigations, Occasional Paper* 22, Southern Illinois University at Carbondale, Carbondale, pp. 235–244.

Bintliff, J. L., Davidson, D. A., and Grant, E. G. (eds.), 1988, *Conceptual Issues in Environmental Archaeology*, Edinburgh University Press, Edinburgh.

Branch, N., Canti, M., Clark, P., and Turney, C., 2005, *Environmental Archaeology: Theoretical and Practical Approaches*, Hodder Arnold, London.

Brown, A. G., 1997, *Alluvial Geoarchaeology: Floodplain Archaeology and Environmental Change*, Cambridge University Press, Cambridge, England.

Bryant, V. M., Jr., 1989, Pollen: Nature's Fingerprints of Plants, in: 1990 Yearbook of Science and the Future, *Encyclopedia Britannica*, Chicago, Illinois, pp. 92–111.

Buikstra, J. E., and Mielke, J. H., 1985, Demography, Diet, and Health, in: *The Analysis of Prehistoric Diets* (R. L. Gilbert, Jr., and J. H. Mielke, eds.), Academic Press, Orlando, Florida, pp. 359–422.

Butzer, K. W., 1971, *Environment and Archeology: An Ecological Approach to Prehistory*, Aldine, Chicago, Illinois.

Butzer, K. W., 1982, *Archaeology as Human Ecology: Method and Theory for a Contextual Approach*, Cambridge University Press, Cambridge, England.

Collins, M. E., Carter, B. J., Hart, J. T., and Foss, J. E. (eds.), 1995, *Pedological Perspectives in Archaeological Research*, Soil Science Society of America Special Publication 44, Madison, Wisconsin.

Cowan, C. W., and Watson, P. J., 1992, Introduction, in: *The Origins of Agriculture: An International Perspective* (C. W. Cowan and P. J. Watson, eds.), Smithsonian Institution Press, Washington, DC, pp. 1–6.

Cox, M., and Mays, S. (eds.) 2000, *Human Osteology in Archaeology and Forensic Science*, Greenwich Medical Media, London.

Davis, S. J. M., 1987, *The Archaeology of Animals*, Yale University Press, New Haven.

Decker-Walters, D. S., Walters, T. W., Cowan, C. W., and Smith, B. D., 1993, Isozymatic Characterization of Wild Populations of *Cucurbita pepo*, *Journal of Ethnobiology* 13(1):55–72.

Dimbleby, G. W., 1965, Environmental Studies and Archaeology, *Bulletin of the Institute of Archaeology* 14:1–12, University of London, London.

Evans, J. G., 1978, *An Introduction to Environmental Archaeology*, Cornell University Press, Ithaca, New York.

Evans, J. G., 2003, *Environmental Archaeology and the Social Order*, Routledge, London.

Evans, J., and O'Connor, T., 2001, *Environmental Archaeology: Principles and Methods*, Sutton, Stroud, England.

Fankhauser, B., 1994, Protein and Lipid Analysis of Food Residues, in: *Tropical Archaeobotany: Applications and New Developments* (J. G. Hather, ed.), Routledge, London, pp. 227–250.

Ford, R. I., 1979, Paleoethnobotany in American Archaeology, in: *Advances in Archaeological Method and Theory*, Volume 2 (M. B. Schiffer, ed.), Academic Press, New York, pp. 285–336.

French, C., 2003, *Geoarchaeology in Action: Studies in Soil Micromorphology and Landscape Evolution*, Routledge, London.

Fritz, G. J., 1994, The Value of Archaeological Plant Remains for Paleodietary Reconstruction, in: *Paleonutrition: The Diet and Health of Prehistoric Americans* (K. D. Sobolik, ed.), *Center for Archaeological Investigations, Occasional Paper* 22, Southern Illinois University at Carbondale, Carbondale, pp. 21–33.

Garrison, E. G., 2003, *Techniques in Archaeological Geology*, Springer-Verlag, Berlin.

Gifford, D. P., 1981, Taphonomy and Paleoecology: A Critical Review of Archaeology's Sister Disciplines, in: *Advances in Archaeological Method and Theory*, Volume 4 (M. B. Schiffer, ed.), Academic Press, New York, pp. 365–438.

Gladfelter, B. G., 1981, Developments and Directions in Geoarchaeology, in: *Advances in Archaeological Method and Theory*, Volume 4 (M. B. Schiffer, ed.), Academic Press, New York, pp. 344–364.

Grayson, D. K., 1984, *Quantitative Zooarchaeology: Topics in the Analysis of Archaeological Faunas*, Academic Press, Orlando, Florida.

Gordon, C. C., and Buikstra, J. E., 1981, Soil pH, Bone Preservation, and Sampling Bias at Mortuary Sites, *American Antiquity* 46(3):566–571.

Hastorf, C. A., and Popper, V. S. (eds.), 1988, *Current Paleoethnobotany: Analytical Methods and Cultural Interpretations of Archaeological Remains*, University of Chicago Press, Chicago, Illinois.

Herz, N., 1990, Stable Isotope Geochemistry Applied to Archaeology, in: *Archaeological Geology of North America* (N. P. Lasca and J. Donahue, eds.), *Geological Society of America, Centennial Special,* Volume 4, Boulder, Colorado, pp. 585–595.

Herz, N., and Garrison, E. G., 1998, *Geological Methods in Archaeology*, Oxford University Press, Oxford, England.

Holliday, V. T. (ed.), 1992, *Soils in Archaeology, Landscape Evolution and Human Occupation*, Smithsonian Institution Press, Washington, DC.

Huss-Ashmore, R., Goodman, A. H., and Armelagos, G. J., 1982, Nutritional Inference from Paleopathology, in: *Advances in Archaeological Method and Theory*, Volume 5 (M. B. Schiffer, ed.), Academic Press, New York, pp. 395–474.

Larsen, C. S., 1997, *Bioarchaeology: Interpreting Behavior from the Human Skeleton*, Cambridge University Press, Cambridge.

Loy, T. H., 1994, Methods in the Analysis of Starch Residues on Prehistoric Stone Tools, in: *Tropical Archaeobotany: Applications and New Developments* (J. G. Hather, ed.), Routledge, London, pp. 36–114.

Luff, R., and Rowley-Conwy, P. (eds.), 1994, *Whither Environmental Archaeology? Oxbow Monograph* 38, Oxford, England.

Lyman, R. L., 1994, *Vertebrate Taphonomy*, Cambridge University Press, Cambridge, England.

Matthews, P., and Terauchi, R., 1994, The Genetics of Agriculture: DNA Variation in Taro and Yam, in: *Tropical Archaeobotany: Applications and New Developments* (J. G. Hather, ed.), Routledge, London, pp. 251–262.

Miksicek, C. H., 1987, Formation Processes of the Archaeobotanical Record, in: *Advances in Archaeological Method and Theory*, Volume 10 (M. B. Schiffer, ed.), Academic Press, San Diego, California, pp. 211–247.

Monks, G. G., 1981, Seasonality Studies, in: *Advances in Archaeological Method and Theory*, Volume 4 (M. B. Schiffer, ed.), Academic Press, New York, pp. 177–240.

O'Connor, T., 1998, Environmental Archaeology: A Matter of Definition, *Environmental Archaeology* 2:1–6.

O'Connor, T., 2000, *The Archaeology of Animal Bones*, Sutton, Stroud, England.

Pearsall, D. M., 2000, *Paleoethnobotany: A Handbook of Procedures*, 2nd ed., Academic Press, San Diego, California.

Piperno, D. R., 2006, *Phytoliths: A Comprehensive Guide for Archaeologists and Paleoecologists*, Altamira Press, Lanham, Maryland.

Price, T. D. (ed.), 1989, *The Chemistry of Prehistoric Human Bone*, Cambridge University Press, Cambridge, England.

Reber, E., 2005, What Did They Eat? in: *Annual Editions: Archaeology 04/05* (L. L. Hasten, ed.), McGraw-Hill/Dushkin, Dubuque, Iowa, pp. 132–133.

Reed, D. M. (ed.), 2005, *Biomolecular Archaeology: Genetic Approaches to the Past, Center for Archaeological Investigations, Occasional Paper 32*, Southern Illinois University at Carbondale, Carbondale.

Reitz, E. J., and Wing, E. S., 2007, *Zooarchaeology*, 2nd ed., Cambridge University Press, Cambridge, England.

Renfrew, C., and Boyle, K. (eds.), 2000, *Archaeogenetics: DNA and the Population of Prehistory Europe*, McDonald Institute for Archaeological Research, Cambridge, England.

Scudder, S. J., 2001, Evidence of Sea Level Rise at the Early Ostionan Coralie Site (GT-3), c. AD 700, Grand Turk, Turks and Caicos Islands, *Journal of Archaeological Science* 28:1221–1233.

Scudder, S. J., 2003, Sea Level Rise or Shallow-water Midden Deposition? Archaeopedology at the Seminole Rest Archaeological Site, Coastal East-Central Florida, *Journal of Archaeological Science* 30:1551–1557.

Shackley, M., 1981, *Environmental Archaeology*, George Allen & Unwin, London, England.

Smart, T. L., and Hoffman, E. S., 1988, Environmental Interpretations of Archaeological Charcoal, in: *Current Paleoethnobotany: Analytical Methods and Cultural Interpretation of Archaeological Plant Remains* (C. A. Hastorf and V. S. Popper, eds.), University of Chicago Press, Chicago, Illinois, pp. 167–205.

Smith, B. D., 1976, "Twitching": A Minor Ailment Affecting Human Paleoecological Research, in: *Cultural Changes and Continuity* (C. E. Cleland, ed.), Academic Press, New York, pp. 275–292.

Taylor, R. E., and Payen, L. A., 1979, The Role of Archaeometry in American Archaeology: Approaches to the Evaluation of the Antiquity of *Homo sapiens* in California, in: *Advances in Archaeological Method and Theory*, Volume 2 (M. B. Schiffer, ed.), Academic Press, New York, pp. 239–283.

Waters, M. R., 1992, *Principles of Geoarchaeology: A North American Perspective*, University of Arizona Press, Tucson.

Watson, P. J., 1997, The Shaping of Modern Paleoethnobotany, in: *People, Plants, and Landscape: Studies in Paleoethnobotany* (K. J. Gremillion, ed.), University of Alabama Press, Tuscaloosa.

Weigelt, J., 1989, *Recent Vertebrate Carcasses and Their Paleobiological Implications* (J. Schaefer, trans.), University of Chicago Press, Chicago, Illinois.

Wilkinson, K., and Stevens, C., 2003, *Environmental Archaeology: Approaches, Techniques, and Applications*, Tempus, Stroud, England.

Zeder, M. A., Bradley, D. G., Emshwiller, E., and Smith, B. D. (eds.), 2006, *Documenting Domestication: New Genetics and Archaeological Paradigms*, University of California Press, Berkeley, California.

Chapter 2

Environmental Archaeology and Historical Archaeology

KATHLEEN A. DEAGAN

Although environmental archaeology and historical archaeology share a common bond with anthropology, they originated as distinct fields with close ties respectively to biology and history. Their trajectories only connected as practitioners in both fields discovered their common questions and mutual interdisciplinary needs. The two disciplines have, however, become increasingly integrated in both practice and training since their formative years, suggesting that "environmental historical archaeology" is a legitimate subdisciplinary distinction in the study of the past.

In it's broadest sense in the Americas, environmental historical archaeology is concerned with understanding the endlessly recursive relationships between people and their environments during the "historic" periods of the past. "Historic" in this sense refers to the period after Eurasians and Africans arrived in the Americas and began to write about the peoples and environments of the region. When the historic period begins obviously varies among regions, but the critical distinguishing methodological issue is the presence of documentary sources that must be integrated with biological and archaeological data in the investigation of the past.

Since the initial publication of this paper in 1996 (Deagan 1996), the theoretical foundation of Americanist historical archaeology (and environmental historical archaeology) has become both more refined and more diverse. Environmental historical archaeologists are able to articulate the early methodological advances of the field with contemporary social theory, and, through

this, to address anthropological and historical questions that are of broad intellectual importance (see discussions in Landon 2006; Mrozowski 2006a).

This chapter considers some of these developments from the perspective of
a field archaeologist who both collaborates with and depends on the insights
of environmental archaeologists in our mutual attempts to understand the
past. The discussion attempts to convey the nature of contemporary environmental historical archaeology within reluctantly self-imposed limits. One of
these is the recognition that it would be impossible, given space constraints, to
fully consider or do justice to the many important historical archaeological
studies concerned with social constructions of landscape at various scales,
how people give meaning to landscapes and space, and the ways in which
space and landscape shape social meaning (e.g., Pauls 2006; Yemin and
Metheny 1996; Young 2000). Although environmental archaeology obviously
contributes significantly to the general issues of landscape archaeology, I chose
here to focus on investigations of human-environmental interplay through the
integration of excavated archaeological environmental data sets (animal, plant,
geological, and chemical) with archaeological context and written information. Few such studies, as Kealhofer (1999) points out, are truly integrated
with historic landscape archaeology (for an exception see Ruhl 2003a).

My examples are furthermore drawn primarily from the Americas, owing to
limits of space and focus (Figure 2-1).

Nonetheless, one of the most significant developments in environmental
historical archaeology is the globalization of the field. Researchers on every
continent study the relationships between historic-period peoples and

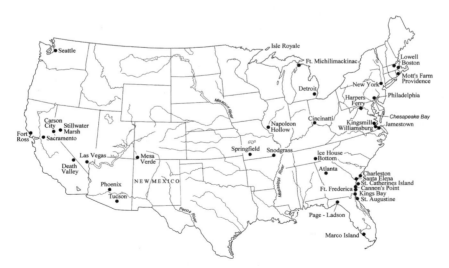

Figure 2-1. Map showing North American sites mentioned in this chapter and their
location relative to other sites discussed in this volume.

environments in contexts and sites that are relevant to their own historical and national concerns. These studies are gradually being synthesized to inform studies of the "modern world" that are inherently global in scope, and of interest to Americanist historical archaeologists. They include colonialism; slavery and labor systems; the rise and spread of capitalism with its associated class structure, urbanization, and industrialism; and environmental degradation (see Hardesty 1999; Mrozowski 2006a; Orser 1996). Discussions of environmental and historical archaeology in Europe, Africa, and Australia can be found elsewhere (de Moulins and Wier 1997; Giorgi 1999; Hall 2000; Hardesty and Fowler 2001; Jack and Cremin 1994; Shrire 1995, among others).

ORIGINS OF ENVIRONMENTAL HISTORICAL ARCHAEOLOGY

When American historical archaeology first emerged as a formal discipline in 1967, most environmental archaeologists (usually zooarchaeologists and less commonly archaeobotanists, human biologists, or geoarchaeologists) were concerned with understanding the articulations between pre-Columbian environments and cultures through analyses of the plant and animal resources available to individual groups. These concerns resulted in both theoretical principles and tools grounded on the assumptions of pre-industrial economic systems and based on regional biota. By these means, environmental archaeologists investigated such critical issues as animal and plant domestication, patterns of settlement and seasonality, procurement strategies, trade networks among resource zones, resource change and depletion, and a myriad of other topics.

In contrast, historical archaeologists, who were at that time primarily concerned with articulating the questions most appropriate to the complex, and multi-evidentiary data of historical archaeology, and in developing methodologies with which to approach those questions. Most American historical archaeologists since that time have been concerned with examining the culturally pluralistic and ethnically complex post-Columbian American experience.

Because the relationships between the peoples and environments of post-Columbian America were, in many instances, fundamentally different from those of pre-Columbian America, new and different requirements were placed on historical archaeologists in their studies of the past. Underlying many of these requirements is the need for an interdisciplinary effort that includes not only the traditional disciplines of archaeology, anthropology, geology, and the biological sciences, but also history, literature, art, architecture, and folklore.

Americanist historical archaeologists generally also assume a global interaction sphere driven by a market economy, which implies a profoundly different relationship between people and their environments than the locally-focused, non-market economies presumed for many pre-Columbian cultures. For the most part, concepts important to pre-Columbian archaeologists, such as seasonality, procurement strategies, and resource zones, could not be used in the same ways in studies of colonial, post-colonial, or urban sites. Issues of frontier expansion, the social consequences of multi-cultural encounter and exchange, mercantile economy, urbanism, consumer behavior, critical analysis of documentary sources, and the classification of worldwide artifactual and technological categories concerned historical archaeologists more than did many of the issues of interest to pre-Columbian environmental archaeologists.

Articulating these questions with those of environmental archaeologists within a coherent theoretical framework was furthermore often made problematical by the respective explanatory foundations of the two fields. While social theory prevailed among historical archaeologists, evolutionary biology models were emphasized by environmental archaeologists, who necessarily trained at least partly in biological science.

This divergence of concerns tended to inhibit collaboration between environmental archaeologists and historical archaeologists in the early years of the discipline. When collaboration did take place, it was evident that the primary database of much pre-Columbian environmental archaeology, the biotic communities of a given region, was not sufficient to address the questions asked by historical archaeologists. The detailed documentary record available to historical archaeologists emphasizes that locally available resources, while always important, are not necessarily central to understanding dietary, subsistence, or settlement strategies of historic period societies. From the very beginning of the post-Columbian era in the Americas, the environmental variables on which people depended, and to which they adapted, included plants, animals, and diseases from many parts of the globe in addition to those existent locally. This variety obtained particularly for urban settings and sites occupied after the 15th century A.D., when market economies, mass production methods, and advances in shipping and preservation led to widespread availability of material objects and foodstuffs from throughout the world.

WRITTEN TEXTS AND ENVIRONMENTAL HISTORICAL ARCHAEOLOGY

One of the most difficult methodological issues in both the development of historical archaeology and its integration with environmental archaeology was articulating the role of written text in archaeological inquiry. Documentary data provide a context and platform for the study of cultural and environmental relationships during the historic period. It can also, when used appropriately in combination with archaeological environmental remains, permit interpretations not accessible in any other way. A concern with avoiding investigations that simply affirm, repeat, or refute information already known through documentary sources is, in fact, central in all aspects of historical archaeology. For environmental historical archaeologists, this requires identifying questions that can be informed only through the integration of environmental, textual, iconic, and contextual information.

The development of such appropriate questions was the topic of the Society for Historical Archaeology's plenary session in 1987 on "questions that count" (Honerkamp 1988). Although many of these questions implicitly incorporated environmental archaeology, only one participant (Mrozowski 1988) made explicit note of faunal, botanical, and soil studies. The presence of the immense documentary database clearly overwhelmed the discussion in historical archaeology. Consequently less attention was devoted to the creative incorporation of environmental elements until quite recently.

The documentary record relevant to the concerns of historical environmental archaeologists might include both traditional and nontraditional historical sources, such as narrative accounts, letters, shipping records, tax records, wills and inventories, census data, cookbooks, animal husbandry treatises, newspaper ads, menus, and paintings (see Beaudry 1988; Yentsch 1988). McEwan, for example, in her effort to assess Spanish dietary adaptations to the early Caribbean colonies, used travelers' accounts, household fiscal records, cookbooks, hospital purchase records, and monastery records, as well as army subsistence and ships' crew rations to characterize the consistencies and variations in 16th-century Spanish foodways, for which no reliable archaeological database existed. This model was used as a baseline against which to measure the divergence of 16th-century Spanish-American foodways (for which there are no reliable documentary data) from traditional Spanish patterns (McEwan 1988).

Environmental historical archaeologists increasingly construct models and define principles that diverge from those traditionally used in pre-Columbian environmental archaeology, but are necessary for the study of post-Columbian human-environmental interactions. One basic principle is that social environment and market variables are often more directly relevant to understanding subsistence strategies than are local environmental variables and their scheduling. Bowen (1988), for example, drew upon domestic animal sizes, exchange networks along a rural-urban continuum, and climatic concerns of meat preservation to define a model that is more useful than traditional ones for the study of Euro-American societies (see also Landon 1993). Another uniquely historical measure of seasonality may be historically-documented holiday foods, such as turkeys (*Meleagris gallopavo*), in an otherwise non-local environmental assemblage (Davidson 1982; Wilson and Rathje 2001). Rothschild (1989) and Henry (1987a, 1991) have shown that consumer behavior, standardized food production, and market distribution tend to lower faunal diversity in urban settings. Diversity measures cannot be understood without incorporating these social factors.

The tendency to assume that all animal remains in archaeological sites represent food also has been addressed and corrected. Environmental historical archaeologists have used archaeological, biological, documentary, and ethnographic data to develop more accurate interpretive methods for this question (Reitz 1986b, 1991, 1992; Reitz and McEwan 1995), including practices for integrating documentary accounts with archaeological remains (Bowen 1975, 1999) and the development of zooarchaeological methods for recognizing the processing of animals as market commodities (Whittaker 1999). The discovery of a monkey (*Cebus* spp.) in a 19th-century New York City privy was identified through associated textual information as a working animal (in this case with an organ grinder) rather than as an exotic pet or food import (Crabtree and Milne 1997).

One of the most compelling principles underscored by interdisciplinary historical environmental integration is the duality of human-perceived diet (and by extension, nature) and biological diet. Perhaps no other focus in archaeology is as directly pertinent to this concern as environmental historical archaeology, which provides an incontrovertible nexus for physical biological elements and cultural constructions. Reitz and Scarry (1985) were among the first researchers to explore this. They showed that when documentary accounts were articulated with biological data from St. Augustine, Florida, it was clear that the colonists' perceptions of appropriate food, as contrasted with the physical dietary remains, led to arguably false claims of starvation. These perceptions were in themselves an important element in the overall Spanish colonial subsistence strategy in this subsidized military town. Although the dialectical relationship between culturally-constructed and physically constituted categories of nature are widely acknowledged and accepted by archaeologists in general, it is rarely as directly viewable as it can be in environmental historical archaeology.

SOCIAL VARIABLES AND PATTERNS IN ENVIRONMENTAL HISTORICAL ARCHAEOLOGY

The obvious advantage of site-associated documentary data is the potential to recognize many of the social variables that identify individuals and groups. Whether self-expressed or externally attributed, documentation of a study subject's income, race, gender, occupation or other group affiliation can often refine our understanding of the ways in which socially-dictated and environmentally-dictated choices are managed by people in order to construct diet, subsistence strategies and landscapes. It is not surprising, therefore, that the most common research undertaking in the early practice of environmental historical archaeology was the attempt to correlate dietary patterns with specific ethnic, national, economic, or social affiliations.

Such correlations were pursued for a variety of reasons, including the construction of middle-range theory in the interpretation of social variability in archaeological remains; understanding the nature of subsistence strategies during the historic period; and contributing to the historical ethnology of past societies (see Cusick 1993; Jolley 1983).

Some of the earliest studies (e.g., Cleland 1970; Mudar 1978; Otto 1984) incorporated all these goals, and demonstrated that there were, in fact, distinctive patterns in the use of animals among people with different economic and class affiliations in their study sites. Others employed a cross-cultural comparative approach to investigate dietary distinctions among national or ethnic groups living in the same regions (Cumbaa 1975; Reitz and Cumbaa 1983; Reitz and Scarry 1985; Schulz and Gust 1983). Scott (see Scott, this volume) correlated general differences in diet with national distinctions, in the sense that certain preferred animals or butchering and preparation techniques could be identified with practices in native homelands. In other cases, national identity was correlated with a measurable difference in the relative proportions of species utilized. In general, however, similarities among the diets of distinct ethnic or national groups in a given region were often as pronounced (albeit through parameters such as dominant domestic species) as the differences (Reitz 1985; Reitz and Honerkamp 1983).

Building on this work, some subsequent investigations such as Cusick's multivariate analysis of diet, documents, and material culture for Minorcans and Spaniards of known income in late 18th-century St. Augustine (Cusick 1993) and Henry's analysis of 19th-century Phoenix (Henry 1987a, 1987b), were not as enthusiastic about correlating ethnic identity and subsistence patterns. The studies reluctantly concluded that although consistent correlations between dietary pattern and socioeconomic status (class) can be demonstrated, the relationships between diet and ethnicity, family composition or occupation are less clear, and can confound one another (see Cantwell and Wall [2001:222–223] for a mitigating view). These assessments support the widely-held assertions of contemporary post–processualist archaeologists that patterns of variability in the human use of environmental elements can be characterized only in terms of the communities in which they occur; are affected by the function of the site itself; and reflect the multiple influences of gender, ethnic affiliation, economic status, ideology, social position, and local environment (Cusick 1993; deFrance 2003; Martin 1991; Reitz 1987). Environmental archaeologists underscore the difficulty of isolating the roles played by specific social factors in shaping site-specific subsistence patterns, pointing out that taphonomic, environmental, temporal, and sampling factors are responsible for many observed patterns (Reitz et al. 2006; Rothschild 1990; Rothschild and Balkwill 1993; Schmitt and Zeier 1993).

Because of these factors, much environmental historical archaeology has not gone beyond simply demonstrating correlations between social variables and nonrandom patterns of environmental remains. Although this has added detail and enhanced understanding of local histories, it has not provided general principles to explain the shape and direction of variance or to suggest why and under what circumstances variance can be expected to occur.

I should note that in the study of pre-Columbian societies, empirically grounded, descriptive correlations of social groups with environmental adaptive behavior are critically important, in no small measure because they are the only source of information about those societies. On many historic sites, however, such simple correlations often serve primarily to reinforce or verify the social patterning already made evident in considerably greater detail in the documentary record. In other words, the predictive goals of scientific archaeology in understanding the nature of human behavior remain largely unmet by this approach.

CONTEMPORARY CONTEXTS

The failure of many studies, environmental or otherwise, to produce insights that transcended the particular historic group or community being studied led many archaeologists after 1980 to reject deductive science as an appropriate methodology for the study of the past. They objected to reductionist approaches designed to provide a generalized expression or description of cultural behavior such as statistical correlations reflecting group characteristics. They especially decried efforts to arrive at general statements or law-like generalizations about human behavior, arguing that such representation not only masked the true internal diversity of a society, but also inevitably privileged the interests and values of those with power over those without power, and buried the role of resistance (see Leone and Potter 1988; Little 1994; McGuire and Paynter 1991). By the end of the 20th century, individual agency (that is, how individuals within specific local historical contexts engage with and manipulate their social and physical environments to assert, resist, control or change) was seen as more important than either group norms or environmental variables for understanding social dynamics and their archaeological imprints. Archeologists have generally come instead to focus on how individuals or small aggregate groups construct and manipulate social and material life in local settings, as well as the ways in which social and material conditions in turn alter the ideational constructs and systems of meaning for individuals.

The role of environment in human affairs is, of course, central to environmental archaeology, and remains so in the post-processual era. Only rarely, however, is environment invoked by 21st-century archaeologists as a causal explanation for human activities. Some recent exceptions to this come from historical environmental archaeology, however, and underscore the important role of biological science in the concerns of historical archaeology. For example, tree-ring analysis has established that the colonization of Jamestown, Virginia (1607) coincided with the worst drought the region had experienced in more than eight centuries (Blanton 2000; Stahle et al. 1998). These conditions made water supply a concern and farming problematic, contributed to disease, and may have contributed to tension between the indigenous people of the region and the English intruders. Climatic conditions offer a compelling association for the "starving time" experienced by the Jamestown colonists in

1609–1610, and possibly with the late 16th-century struggles and abandonment of the Spanish colony of Santa Elena (1567–1587) (Anderson et al. 1995). As Mrozowski (2006a) points out, however, environmental circumstances contributing to colonial difficulties were exacerbated and, in some cases, perhaps even created by different ecological knowledge systems among indigenous and colonizing people (see also Hassan 2002).

Far more commonly today, however, the research of environmental historical archaeologists in the Americas is embedded in questions about the "modern world" (Orser 1996). The mechanics and consequences of European colonial expansion in the Americas, and particularly those related to the rise of capitalism (e.g., labor organization, class, capitalism, industrialization, environmental change), are among the dominant areas of inquiry for historical archaeology (see, for example, papers in Cusick 1998; Honerkamp 1988; Leone and Potter 1988; Orser 1996). Because the meeting of Europe and America involved encounter and exchange among literate and nonliterate peoples, the unique biological-archival-archaeological perspective of environmental historical archaeology is essential to understanding both environmental and social change in the post-Columbian Americas. Although a larger understanding of such issues ultimately demands a global scale of analysis, in practice such work is generally carried out at local or regional levels with an emphasis on placing local experience in a global context, and most attempt to address questions of general concern by studying the relationships between people and environments in specific settings. The remainder of this chapter will identify some of these issues and offer examples of some successful approaches.

ANIMALS, PLANTS, AND SOCIAL IDENTITY

One of the questions related to American colonialism that is most frequently explored by environmental archaeologists is that of how traditional European foodways and environmental understanding changed or persisted when European colonists were required to adapt to new American conditions. These studies contribute significantly to the current concerns with identity, self representation, and the processes of creolization in the modern Americas.

Cumbaa (1975), who assessed archaeologically-based dietary reconstructions in the light of documentary-derived models from Spain, was one of the first to study this issue. This methodology was subsequently extended and refined by Reitz and Scarry (1985) and McEwan (1988). Drawing on articulation of documentary, environmental, and archaeological data, historical environmental archaeologists have also addressed alterations and continuities in traditional foodways transferred to the Americas by English colonists (Anderson 1971; Cheek 1999; Honerkamp and Reitz 1983; Reitz and Honerkamp 1983; Walsh et al. 1997), French colonists (Blanchette 1981; Martin 1991), and Dutch colonists (Huey 1990; Janowitz 1993). All of these studies suggest that European colonists in America, when possible, attempted to retain the familiar traditional food elements of their homelands. All, however, found it necessary to make changes to varying degrees, depending on the suitability of the local environment for preferred plants and animals. Moreover, zooarchaeological analyses in English colonial sites suggest that adjustments to local resources coalesced into distinct regional Anglo-American cuisine patterns within less than a century after initial colonization (Cheek 1999).

Studies of adjustment in traditional European foodways when transplanted to America indicate that the colonists' choices in changing their foodways were not made predictably along ethnic or national lines, but rather were made in response to the presence of environmental conditions suitable for European domestic animals. In other words, different kinds of domestic mammals were dominant in the colonial diet depending on their suitability for a particular region, regardless of their traditional uses in a European homeland. On the southeastern Atlantic coastal plain, for example, cattle (*Bos taurus*) were more abundant than pigs (*Sus scrofa*) in both English and Spanish sites, and caprines (Caprinae, goats [*Capra hircus*], sheep [*Ovis aries*]) were rare, reflecting varying degrees of suitability of these species to regional environmental conditions (Kelso 1984; Reitz 1986a; Reitz and Honerkamp 1983; Reitz and Zierden 1991). In contrast, caprines, particularly sheep, dominated the domestic mammal assemblages from Spanish colonial New Mexico (Rothschild 2003:140–144). In other regions, pigs dominated the diet regardless of the national origin of the sites' occupants (Blanchette 1981; Bowen 1975; Faulkner and Faulkner 1987:222–224; Stewart-Abernathy and Ruff 1989). A similar pattern existed with the transfer of traditional plant foods such as wheat (*Triticum aestivum*) and wine grapes (*Vitis vinifera*) (Reitz and Scarry 1985; Ruhl 1990, 2003b; Trigg 2004).

Although these conclusions may seem intuitively obvious, they challenge the common assumption that colonial settlers in the Americas arrived with a complex plant and animal domestication technology and were immune to the need to adapt significantly to new natural environments. Some propose a model for colonial adjustment that places the requirements of European plants and animals in a primary causal role in predicting the viability of colonial ventures (Reitz 1992; Reitz and McEwan 1995; Reitz and Scarry 1985). In Spanish-American colonies of the Caribbean, as well as in the South American Andes, for example, it can be argued that the consistent incorporation of indigenous material and social elements was most pronounced in those areas in which European domestic animals and plants did not flourish (deFrance 2003; Ewen 1991; Newsom and Quitmyer 1992; Reitz and McEwan 1995; Reitz and Scarry 1985; Scarry and Reitz 1990). Conversely, both English and Dutch colonists were dependent on white-tailed deer (*Odocoileus virginianus*) as a source of meat during the initial years of colonization, but reverted to preferred domestic mammals and European food preparation traditions as herds were able to establish themselves in the new landscape (Bowen 1999; Cantwell and Wall 2001:179–180). These environmentally-derived propositions offer intriguing directions for archaeologists interested in the ways in which American social, material, and biological elements were incorporated into European colonial societies.

Within specific cultural-geographic regions, the extent to which preferred dietary composition could be retained at a household scale appears to be related to elite status and high income. Elite colonists in the Bolivian Andes, for example, went to considerable effort and expense to import animal products that were both unavailable in and inappropriate to the region, although the material culture used to prepare and consume them was locally-produced (deFrance 2003). Elite colonists in colonial New Mexico asserted their identity in part through the transfer and use of wheat, rather than the better-adapted indigenous maize (*Zea mays*) (Trigg 2004).

Environmental historical archaeologists have until recently paid less attention to the consequences of colonization among non-European groups in the Americas than to those that affected the European newcomers. Studies oriented toward non-Europeans, however, have provided surprising insights. Assessments of Native American subsistence patterns in villages (Reitz 1991) and missions (Newsom 1987; Newsom and Quitmyer 1992; Reitz 1990, 1993; Ruhl 1990; Scarry 1993), for example, indicate that relatively little significant change can be documented in post-contact Native American subsistence as a consequence of European influence in the southeastern United States. A similar disinterest in (or exclusion from) European plants and animals is documented among 18th-century Mohawk groups in contact with the Dutch colonists of New York, Pueblo peoples of New Mexico (Rothschild 2003:140–144), and post-Columbian Taino Indians on the Caribbean island of Hispaniola (Deagan 2004). Instead, these studies suggest that somewhat more adjustment and alteration was made by the European settlers in these regions through indigenous influence.

Excavations at the Russian site of Fort Ross in California have revealed a somewhat different practice. There, the Kashaya people in contact with Russians did, in fact, extensively adopt European mammals in their diets. They processed and prepared the animals, however, in distinctively native ways, consistent with pre-contact treatment of elk (*Cervus elaphus*), deer, and sea mammals (Lightfoot 2005:167–168). Environmental archaeologists have also shown that the introductions of certain European economic institutions–the fur trade, for example–often had a more apparent impact on indigenous American subsistence and life ways than did introduced plants and animals. Through analysis of deer remains at Native American sites, Lapham (2004) has shown that economic intensification occurred among native people of the Chesapeake region shortly after European contact, suggesting a shift from subsistence to commercial deerskin hunting. More prime-age (heavy fur bearing) male animals were killed, hunting became year-round rather than seasonal, and deer-processing artifacts became more frequent. Stress on the deer population was inevitable.

Other compelling characterizations of American colonialism have come from studies at sites of African slavery. Efforts to understand African-American diet, food procurement strategies and other uses of the environment under conditions of slavery reveal an active and conscious use of animals and plants from Europe, America, and sometimes Africa, and their recombination into a distinctly African-American, creole subsistence pattern (Armstrong 1989:273–276; Crader 1984; McKee 1999; Otto 1984; Reitz et al. 1985). Environmental archaeologists have consistently found that slaves supplemented rations of European plants and animals by hunting, fishing, and gathering and cultivating plants of Native American, Eurasian, and African origin (Reitz 1986a, 1994). In the Bahamas, Wilkie and Farnsworth (2005) found that African slaves subsisted principally on fish and shellfish and packaged foods, while at the same time raising hogs and fowl. Both of the latter animals, however, seem to have been commodities plantation slaves raised to sell in local markets (Wilkie and Farnsworth 2005:237–240).

Analysis of animal remains from sites of slavery underscore the undeniable misery of slavery and the global economic factors that sustained it. One of the most compelling of these is Klippel's groundbreaking study of slave diets on the Caribbean island of St. Kitts in the 18th century, where the highly lucrative sugar monoculture economy made it more profitable to import food for slaves than it was to produce it (Klippel 2001). Using stable isotope analyses, Klippel showed that although sheep and goats were raised on the island, beef came from as far away as Europe and North America, apparently imported as barreled meat. Element analysis furthermore revealed beef marrow bones, which are normally excluded from barreled beef, and may indicate that low-quality, cheap products were sought and imported as slave provisions.

These insights into change, persistence and creative adaptation in patterns of resource use are primarily relevant to initial and colonial occupation of the Americas, prior to the emergence of industrialism, global markets, and large industrial/market cities. The study of dietary adaptations of more recent immigrants offers another important area and a different kind of challenge through which to address questions of change and persistence in traditional foodways (e.g., Cantwell and Wall 2001:179–180; Langenwalter 1979; Longnecker and Stapp 1993; Mudar 1978).

ENVIRONMENTAL HISTORICAL ARCHAEOLOGY IN CITIES

A number of specific methodologies appropriate to the study of life in cities have developed in urban archaeology (Cantwell and Wall 2001; Mrozowski 2006b; Rothschild 1990; Spencer-Wood 1987; Staski 1987; Zierden and Calhoun 1986). Until recently, less attention was given to environmental archaeology in cities than to methods for recognizing consumer behavior, status, group identities, and land use patterns through artifact and documentary sources. This reluctance was due in large part to the very dense and complex archaeological record in cities, the immense documentary record usually associated with cities, and the intensively destructive taphonomic factors typical of cities (Kelso 1993; Rothschild and Balkwill 1993; Schávelson 1999).

Urban economies and subsistence systems based almost exclusively on markets controlled by production and distribution specialists also complicates traditional archaeological dietary analyses. Recognizing this, some environmental archaeologists have instead turned to a consumer behavior framework that emphasizes the linkages between environmental artifacts and consumer choice (Holt 1991; Huelsbeck 1991; Spencer-Wood 1987). This approach can potentially articulate factors shaping local consumption, such as geography, gender, class, technology, ideology, and identity creation, with global considerations of markets – price, income, marketing, advertising, transportation, preservation technologies, and food distribution systems (e.g., Bowen 1999; Landon 1996; Shackel 2000:149–66).

Some work in urban settings, such as colonial Boston, focuses on how animal bones from urban contexts can be used to investigate cultural practices of food distribution and exchange in cities (Landon 1996; see Landon, this volume). Other research is concerned with issues of urban land use and vegetation history, as well as with urban health and sanitation (e.g., Geismar and Janowitz 1993; Mrozowski 2006b). Shackel has assessed urban uses of plants, animals, and domestic landscapes in late 19th-century Harpers Ferry in the light of late Victorian rules of etiquette and household management to challenge traditional notions of community history (Shackel 2000). These studies are more successful than simple dietary reconstructions in providing insights into urban life, largely owing to the integrated, multifaceted environmental approach they employ, i.e., analytical attention to plants, animals, soils, documents, and material culture.

One of the most interesting of these urban environmental efforts is the long-term study of New England's urban landscapes conducted by Mrozowski, Beaudry, and their colleagues (Mrozowski 2006b; Mrozowski et al. 1996; Reinhard et al. 1986). Combined analyses of pollen, plant macrofossils, archaeological features, and archival data from several New England settlements resulted in a characterization of emergent and established urban landscapes. Early urban centers contained residential household gardens characterized by dry, disturbed soils, weedy plant species, exotic weeds, and edible plants. As cities grew and land use became more intensive, residential areas were segregated in suburbs with households and yards not used for food gardens and other economic activities.

Patterns of health could be correlated with economic levels through the analysis of coprolites from urban households (Reinhard et al. 1986). Coprolites from documented lower economic artisan households in Providence, Rhode Island, yielded parasites thought to have originated in human feces used as fertilizer in household gardens. These parasites were not present at the more affluent merchant household, where it was thought that such supplemental economic activity as gardening did not take place. The merchant's household, however, yielded evidence of the West African hookworm (*Necator americans*), introduced to the Americas through the African slave-trade diaspora, and may indicate the presence of Africans in this affluent 18th-century household (Mrozowski 2006b:40–48; Reinhard et al. 1986). A contrast to the correlation of parasites and low economic status in Providence was provided by data from late 19th-century Harper's Ferry (Shackel 2000:166–169). Analysis of privy contents showed that parasites were considerably more prevalent in the coprolites from a well-to–do merchants household than they were from a lower-class boarding house. Archaeobotanical and documentary analysis from the same sites also reveal that the merchant ate a much broader variety of fruits than did the boarding house residents. The higher rate of parasites in the high status household may be due to the merchant's consumption of locally-available, garden-grown produce fertilized with night soil, while the boarding house residents ate more imported, market-distributed foods.

Closely related to the archaeology of cities is the historical archaeology of industrialization (Hardesty 1999). Environmental historical archaeology in cities is increasingly contributing to the understanding of the broader impacts of industrialization, labor, and environmental degradation. Health consequences of industrialization, for example, can be seen at the Boott boarding house for factory workers in Lowell, Massachusetts, where soil analysis showed dangerous levels of lead in the soil, and rodent infestations were documented (Mrozowski 2006a:34).

The intensification of consumption and disposal in the industrial era causes it's own set of problems that are particularly appropriate to environmental historical archaeology (e.g., Steinberg 2002). These are addressed most visibly by Rathje, whose studies of modern garbage reveal rates of decomposition of food and other remains in landfills; sources and levels of pollution, and changing dietary and consumption patterns in cities (Rathje and Murphy 2001).

POST-COLUMBIAN ENVIRONMENTAL CHANGES

Environmental archaeology not only informs our understanding of human engagement with new environments, but is increasingly providing insights into the responses of both imported and indigenous plants and animals to human settlement and exploitation (e.g., Hales and Reitz 1992). Mrozowski (1987), for example, documented through archaeology that the tree-of-heaven (*Ailanthus altissima*), which was used to nourish silkworms, was established in New England by 1806, just a few decades after its introduction.

Reitz found that 16th-century Spanish cattle on the Caribbean island of Hispaniola rapidly become larger, heavier, and possibly more fertile than their

European counterparts (Reitz 1986b; Reitz and McEwan 1995; Reitz and Ruff 1994). This fluorescence was due to the absence of competing herbivores and indigenous bovine disease in the Caribbean, as well as to the presence of fertile grassland areas. Cattle in the piney woods and scrub regions of peninsular Florida, however, followed an opposite trend, resulting in the modern *criollo* animal. Similarly, the European domesticates in colonial New York were smaller than their Dutch counterparts, a development attributed to preferential choice by the colonists because the smaller animals required less care (Cantwell and Wall 2001:140).

Archaeobotanical and zooarchaeological analyses of post-Columbian deposits are increasingly contributing to a better understanding of environmental change and degradation as a consequence of colonial expansion and subsequent intensification of development (Hardesty 1999; Hardesty and Fowler 2001). New and often more intensive forms of land use placed restrictions on the ways people could articulate with or use the natural environment (see Bowen 1999). Deforestation for farmland, introduction of potentially destructive and weedy species (e.g., pigs or kudzu [*Pueraria montana*]), diversion of natural water systems, fencing, and urbanization are all well-known examples of such restrictions (see Cronin 1983; White 1991).

Intensification of hunting, fishing, herding, and farming after the arrival of Europeans resulted in dramatic changes in native American plant and animal communities. Reitz, for example, has demonstrated that overharvesting of fishes created a negative impact on fish species in colonial-period Florida, and that the modern fisheries ecology characterization of "fishing down the food web" occurred in Florida as early as the 18th century (Reitz 2004). The commercially-packed Atlantic cod analyzed from the Bertrand shipwreck similarly show evidence of commercial intensification even before the mid-19th century (Klippel and Falk 2002).

Macrobotanical and pollen analyses in colonial contexts, although not yet widely employed, have considerable potential for elucidating the consequences and timing of land clearing through identification of emergence of secondary growth patterns in some regions (Kelso et al. 1993, 1995; Newsom 1993; Ruhl 2003a) and the identification of managed landscapes such as gardens and pathways (Miller 1989; Miller et al. 1990; Newsom 1993). The "lagomorph index" (the ratio of cottontail rabbits (*Sylvilagus* spp.) or jackrabbits (*Lepus* spp.) to all rabbits (Lagomorpha) is used in faunal assemblages in the southwestern United States to suggest the periodicity and impacts of land clearing through farming and herding (Rothschild 2003:146–147). In some areas, such as on the Caribbean islands, these processes greatly exacerbated the rate of demise during the early colonial period of native species such as the Greater Antillean hutia (*Isolobodon portoricensis*) and rice rat (*Oryzomys* spp.) (Wing 1989; Woods 1989), and doubtless many more cases remain to be documented.

CONCLUSION

Environmental historical archaeology is perhaps unique in American archaeology today as a focus for the convergence of biological data and cultural meaning, expressed both intentionally (principally through texts) and non-intentionally (principally through archaeological remains). Over the past decades, American environmental historical archaeologists have greatly expanded and diversified the kinds of questions they ask, and have successfully integrated their analyses into broader questions of anthropological theory and general intellectual interest. Many studies address the social and ecological consequences of European colonial expansion, including the adaptive strategies of both people and animals, cultural creolization, the construction and representation of identity in pluralistic societies, and slavery and labor dynamics. Many others have addressed issues related to the impacts of early industrial capitalism and global markets, including the expression of class differentiation, urban land use and diet, health issues, changes in American consumption patterns, and, perhaps most importantly for the future, the nature, rates and consequences of environmental degradation and landscape change.

Much of the success of environmental archaeology in American historic-period sites appears to depend on the identification of those aspects of globally relevant research questions that can be uniquely informed by the integration of environmental and behavioral information from documents, literary and artistic sources, plants, animals, artifacts, and soils. It has also depended on the development and incorporation of methodologies that explicitly recognize and incorporate the differences in human uses of the environment in the "modern" world of America from those of pre-Columbian societies. These areas are by no means the only areas of environmental historical archaeology that are producing, or have the potential to produce, significant results, but they are examples of the kinds of information that we can learn in no other way.

REFERENCES

Anderson, D. G., Stahle, W., and Cleveland, M. K., 1995, Paleoclimate and Potential Food Reserves of the Mississippian Societies: A Case Study from the Savannah River Valley, *American Antiquity* 60 (2):258–286.

Anderson, J. A., 1971, "*A Solid Sufficiency" An Ethnography of Yeoman Foodways in Stuart England*, Ph.D. dissertation, Department of Folklore and Folklife, University of Pennsylvania, Philadelphia, University Microfilms, Ann Arbor Michigan.

Armstrong, D., 1989, *The Old Village at Drax Hall Plantation: An Archaeological Study of Afro-Jamaican Settlement*, University of Illinois Press, Chicago.

Beaudry, M. (ed.), 1988, *Documentary Archaeology in the New World*, Cambridge University Press, Cambridge, England.

Blanchette, J. F., 1981, *The Role of Foodways in New France 1720–60*, History and Archaeology 52, Parks Canada, Ottowa.

Blanton, D., 2000, Drought as a Factor in the Jamestown Colony, *Historical Archaeology* 34(2):74–81.

Bowen, J., 1975, Probate Inventories: An Evaluation from the Perspective of Zooarchaeology and Agricultural History at Mott Farm, *Historical Archaeology* 9:11–25.

Bowen, J., 1988, Seasonality: An Agricultural Construct, in: *Documentary Archaeology in the New World* (M. Beaudry, ed.), Cambridge University Press, Cambridge, England, pp. 161–171.

Bowen, J., 1999, The Chesapeake Landscape and the Ecology of Animal Husbandry, in: *Old and New Worlds* (G. Egan and R. Michael, eds.), Oxbow Books, Oxford, England, pp. 358–367.

Cantwell, A., and Wall, D. Z., 2001, *Unearthing Gotham: The Archaeology of New York City*, Yale University Press, New Haven.

Cheek, C., 1999, An Evaluation of Regional Differences in Colonial English Foodways, in: *Old and New Worlds* (G. Egan and R. Michael, eds.), Oxbow Books, Oxford, England, pp. 349–357.

Cleland, C., 1970, Comparison of Faunal Remains from French and British Refuse Pits at Ft. Michilimackinac: A Study in Changing Subsistence Patterns, *Canadian Historic Sites Occasional Papers in Archaeology and History* 3:8–23.

Crabtree, P., and Milne, C., 1997, Monkey in the Privy! *Archaeology* (March–April):49.

Crader, D. C., 1984, The Zooarchaeology of the Storehouse and the Dry Well at Monticello, *American Antiquity* 49(3):542–559.

Cronin, W., 1983, *Changes in the Land*, Hill and Wang, New York.

Cumbaa, S. L., 1975, *Patterns of Resource Use and Cross-cultural Dietary Change in the Spanish Colonial Period*, Ph.D. dissertation, Department of Anthropology, University of Florida, Gainesville, University Microfilms, Ann Arbor, Michigan.

Cusick, J. G., 1993, *Ethnic Groups and Class in an Emerging Market Economy: Spaniards and Minorcans in Late Colonial St. Augustine*, Ph.D. dissertation, Department of Anthropology, University of Florida, Gainesville.

Cusick, J. G., (ed.), 1998, *Studies in Culture Contact: Interaction, Culture Change, and Archaeology, Center for Archaeological Investigations, Occasional Paper 25*, Southern Illinois University at Carbondale, Carbondale.

Davidson, P., 1982, Patterns in Urban Foodways: An Example from Early Twentieth Century Atlanta, in: *Archaeology of Urban America: The Search for Pattern and Process* (R. Dickens, Jr., ed.), Academic Press, New York, pp. 381–398.

Deagan, K. A., 1996, Environmental Archaeology and Historical Archaeology, in: *Case Studies in Environmental Archaeology* (E. J. Reitz, L. A. Newsom, and S. J. Scudder, eds.), Plenum Press, New York, pp. 339–358.

Deagan, K., 2004, Reconsidering Taíno Social Dynamics after Spanish Conquest: Gender and Class in Culture Contact Studies, *American Antiquity* 69(4):597–626.

de Moulins, D., and Wier, D., 1997, The Potential and Use of Environmental Techniques in Gardens, *Journal of Garden History* 17:40–46.

deFrance, S. D., 2003, Diet and Provisioning in the High Andes: A Spanish Colonial Settlement on the Outskirts of Potosí, Bolivia, *International Journal of Historical Archaeology* 7(2):99–126.

Ewen, C., 1991, *From Spaniard to Creole: The Archaeology of Hispanic American Cultural Formation at Puerto Real, Haiti*, University of Alabama Press, Tuscaloosa.

Faulkner, A., and Faulkner, G., 1987, *The French at Pentagoet 1635–1674, Occasional Publications in Maine Archaeology 5* and *Special Publications of the New Brunswick Museum*.

Geismar, J., and Janowitz, M. F. (eds.), 1993, Health, Sanitation and Foodways in Historical Archaeology, *Historical Archaeology* 27(2):1–111.

Giorgi, J., 1999, Archaeobotanical Evidence from London on Aspects of Post-Medieval Economies, in: *Old and New Worlds* (G. Egan and R. Michael, eds.), Oxbow Books, Oxford, England, pp. 342–348.

Hales, L. S., Jr., and Reitz, E. J., 1992, Historical Changes in Age and Growth of Atlantic Croaker, *Micropogonias Undulatus* (Perciformes: Sciaenidae), *Journal of Archaeological Science* 19(1):73–99.

Hall, M., 2000, *Archaeology and the Modern World: Colonial Transcripts in South Africa and the Chesapeake*, Routledge Press, London.

Hardesty, D., 1999, Historical Archaeology in the Next Millennium: A Forum, *Historical Archaeology* 33(3):51–58.

Hardesty, D. L., and Fowler, D., 2001, Archaeology and Environmental Change, in: *New Directions in Anthropology and Environment: Intersections* (C. Crumley, A. E. van Deveneter and J. J. Fletcher, eds.), Altamira Press, Walnut Creek, California, pp. 72–89.

Hassan, F., 2002, Environmental Perception and Human Responses in History and Prehistory, in: *The Way the Wind Blows: Climate, History and Human Action* (R. J. McIntosh, J. A. Tainter and S. K. McIntosh, eds.), Columbia University Press, New York, pp. 121–140.

Henry, S., 1987a, Factors Influencing Consumer Behavior in Turn-of-the-century Phoenix, Arizona, in: *Consumer Choice in Historical Archaeology* (S. Spencer-Wood, ed.), Plenum Press, New York, pp. 359–369.

Henry, S., 1987b, A Chicken in Every Pot: The Urban Subsistence Pattern of Turn-of-the-century Phoenix, Arizona, in: *Living in Cities: Current Research in Urban Archaeology* (E. Staski, ed.), *Society for Historical Archaeology Special Publication* 5:19–28.

Henry, S., 1991, Consumer, Commodities, and Choices: A General Model of Consumer Behavior, *Historical Archaeology* 25(2):3–14.

Holt, C., 1991, Plants, Humans and Culture: An Edible Model of Consumer Behavior, *Historical Archaeology* 25(2):46–61.

Honerkamp, N. (ed.), 1988, Questions of Substance, Questions That Count, *Historical Archaeology* 22(1):2–25.

Honerkamp, N., and Reitz, E. J., 1983, Social Status as Reflected by Faunal Remains from an Eighteenth Century British Colonial Site: The Faunal Evidence, in: *Forgotten Places and Things: Archaeological Perspectives on American History* (A. Ward, ed.), *Contributions to Anthropological Studies* 3, Center for Anthropological Research, Albuquerque, New Mexico, pp. 335–340.

Huelsbeck, D. R., 1991, Faunal Remains and Consumer Behavior: What is Being Measured?, *Historical Archaeology* 25(2):62–76.

Huey, P., 1990, The Dutch at Fort Orange, in: *Historical Archaeology in Global Perspective* (L. Falk, ed.), Smithsonian Institution Press, Washington, DC, pp. 21–68.

Jack, I. R., and Cremin, A., 1994, *Australia's Age of Iron: History and Archaeology*, Oxford University Press, Oxford, England.

Janowitz, M. F., 1993, Indian Corn and Dutch Pots: Seventeenth Century Foodways in New Amsterdam, *Historical Archaeology* 27(2):6–24.

Jolley, R., 1983, North American Historic Sites Zooarchaeology, *Historical Archaeology* 17(2):64–79.

Kealhofer, L., 1999, Adding Content to Structure: Integrating Environment and Landscape, in: *Old and New Worlds* (G. Egan and R. L. Michaels, eds.), Oxbow Books, Oxford, England, pp. 378–389.

Kelso, G. K., 1993, Pollen-record Formation Processes, Interdisciplinary Archaeology, and Land Use by Millworkers and Managers: The Boott Mill Corporation, Lowell, Massachusetts, 1836–1942, *Historical Archaeology* 27(1):70–94.

Kelso, G. K., Karish, J. F., and Smith, C., 1993, Pollen Analysis in Historic Landscape Studies: Fort Necessity, Pennsylvania, *Park Science* (1993):8–10.

Kelso, G. K., Mrozowski, S. A., Currie, D., Edwards, A. C., Brown, III, M. R., Horning, A. J., Brown, G. J., and Dandoy, J. R., 1995, Differential Pollen Preservation in a Seventeenth-Century Refuse Pit, Jamestown Island, Virginia, *Historical Archaeology* 29(2):43–54.

Kelso, W., 1984, *Kingsmill Plantations, 1619–1800: Archaeology of Country Life in Virginia*, Academic Press, Orlando, Florida.

Klippel, W. E., 2001, Sugar Monoculture, Bovid Skeletal Part Frequency, and Stable Carbon Isotopes: Interpreting Enslaved African Diet at Brimstone Hill, St. Kitts, West Indies, *Journal of Archaeological Science* 28:1191–1198.

Klippel, W., and Falk, C. R., 2002, Atlantic Cod in the Missouri River: *Gadus morhua* from the Steamboat Bertrand, *Archaeofauna* 11:23–44.

Landon, D. B., 1993, Testing a Seasonal Slaughter Model for Colonial New England Using Tooth Cementum Increment Analysis, *Journal of Archaeological Science* 20(4):439–457.

Landon, D. B., 1996, Feeding Colonial Boston: A Zooarchaeological Study, *Historical Archaeology* 30(1):1–153.

Landon, D. B., 2006, Zooarchaeology and Historical Archaeology: Progress and Prospects, *Journal of Archaeological Method and Theory* 12(1):1–36.

Langenwalter, P., 1979, 19th Century Chinese Subsistence, in: *Archaeological Perspectives on Ethnicity in America* (R. Schuyler, ed.), Baywood, Farmingdale, New Jersey, pp. 80–90.

Lapham, H., 2004, "Their Complement of Deer-Skins and Furs" Changing Patterns of White-Tailed Deer Exploitation in the Seventeenth Century Southern Chesapeake and Virginia Hinterlands, in: *Indian and European Contact in Context. The Mid-Atlantic Region* (D. Blanton and J. King, eds.), University Press of Florida, Gainesville, pp. 172–192.

Leone, M., and Potter, P., 1988, Introduction: Issues in Historical Archaeology, in: *The Recovery of Meaning in the Historical Archaeology of the Eastern United States* (M. Leone and P. Potter, eds.), Smithsonian Institution Press, Washington, DC, pp. 1–22.

Lightfoot, K. G., 2005, *Indians, Missionaries and Merchants. The Legacy of Colonial Encounters on the California Frontiers*, University of California Press, Berkeley.

Little, B., 1994, People with History: An Update on Historical Archaeology in the United States, *Journal of Archaeological Method and Theory* 1:5–40.

Longnecker, J., and Stapp, D., 1993, The Study of Faunal Remains from an Overseas Chinese Mining Camp in Northern Idaho, in: *Hidden Heritage: The Archaeology of the Overseas Chinese* (P. Wegars, ed.), Baywood, Amityville, New York, pp. 97–123.

Martin, T. J., 1991, An Archaeological Perspective on Animal Exploitation Patterns at French Colonial Sites in the Illinois Country, in: *French Colonial Archaeology: The Illinois Country and the Western Great Lakes* (J. Walthall, ed.), University of Illinois Press, Urbana, pp. 189–200.

McEwan, B. G., 1988, *An Archaeological Perspective of Sixteenth Century Spanish Life in the Old World and the Americas*, Ph.D. dissertation, Department of Anthropology, University of Florida, Gainesville.

McGuire, R. H., and Paynter, R. (eds.), 1991, *The Archaeology of Inequality*, Basil Blackwood, Oxford, England.

McKee, L., 1999, Food Supply and Plantation Social Order, in: *I, Too, Am America: Archaeological Studies of African American Life* (T. Singleton, ed.), University of Virginia Press, Charlottesville, pp. 218–239.

Miller, N. F., 1989, What Mean These Seeds: A Comparative Approach to Archaeological Seed Analysis, *Historical Archaeology* 23(2):50–59.

Miller, N. F., Yentsch, A. E., Piperno, D., and Paca, B., 1990, Two Centuries of Landscape Change at Morven, Princeton, New Jersey, in: *Earth Patterns: Essays in Landscape Archaeology* (W. M. Kelso and R. Most, eds.), University Press of Virginia, Charlottesville, pp. 257–275.

Mrozowski, S., 1987, Exploring New England's Evolving Urban Landscape, in: *Living in Cities: Current Research in Urban Archaeology* (E. Staski, ed.), *Society for Historical Archaeology Special Publication* 5:1–9.

Mrozowski, S., 1988, Historical Archaeology as Anthropology, *Historical Archaeology* 22(1):18–24.

Mrozowski, S. A., 2006a, Environments of History: Biological Dimensions of Historical Archaeology, in: *Historical Archaeology* (M. Hall and S. W. Silliman, eds.), Blackwell Publishing, Oxford, England, pp. 23–41.

Mrozowski, S.A., 2006b, *The Archaeology of Class in Urban America*, Cambridge University Press, Cambridge, England.

Mrozowski, S., Ziesing, G., and Beaudry, M., 1996, *Living on the Boott: Historical Archaeology at the Boott Mills Boarding House, Lowell, Massachusetts*, University of Massachusetts Press, Amherst.

Mudar, K., 1978, Effects of Socio-cultural Variables on Food Preferences in Early 19th Century Detroit, *Conference on Historic Sites Archaeology Papers, 1977* 12:323–391.

Newsom, L. A., 1987, Analysis of Botanical Remains from Hontoon Island (8Vo202), Florida: 1980–1985 Excavations, *Florida Anthropologist* 40(1):47–84.

Newsom, L. A., 1993, *Plant Use Among Saladoid and Later People of the Caribbean*, Ph.D. dissertation, Department of Anthropology, University of Florida, Gainesville.

Newsom, L. A., and Quitmyer, I. R., 1992, Appendix E: Archaeobotanical and Faunal Remains from Fig Springs Mission (8Co1), in: *Excavations on the Franciscan Frontier: Archaeology at the Fig Springs Mission* (B. Weisman, ed.), University Press of Florida, Gainesville, pp. 206–233.

Orser, C., 1996, *A Historical Archaeology of the Modern World*, Plenum Press, New York.

Otto, J., 1984, *Cannon's Point Plantation, 1794–1860*, Academic Press, New York.

Pauls, E. P., 2006, The Place of Space: Architecture, Landscape and Social Life, in: *Historical Archaeology* (M. Hall and S. W. Silliman, eds.), Blackwell Publishing, Oxford, England, pp. 65–83.

Rathje, W. J., and Murphy, C., 2001, *Rubbish! The Archaeology of Garbage*, 2nd ed., University of Arizona Press, Tucson.

Reinhard, K., Mrozowski, S., and Orloski, K., 1986, Privies, Pollen, Parasites and Seeds: A Biological Nexus in Historical Archaeology, *MASCA* 4(1):31–36.

Reitz, E. J., 1985, Comparison of Spanish and Aboriginal Subsistence on the Atlantic Coastal Plain, *Southeastern Archaeology* 4(1):41–50.

Reitz, E. J., 1986a, Urban/Rural Contrasts in Vertebrate Fauna from the Southern Coastal Plain, *Historical Archaeology* 20(2):47–58.

Reitz, E. J., 1986b, Vertebrate Fauna from Locus 39, Puerto Real, Haiti, *Journal of Field Archaeology* 13(3):317–328.

Reitz, E. J., 1987, Vertebrate Fauna and Socio-economic Status, in: *Consumer Choice in Historical Archaeology* (S. Spencer-Wood, ed.), Plenum Press, New York, pp. 101–119.

Reitz, E. J., 1990, Zooarchaeological Evidence for Subsistence at La Florida Missions, in: *Columbian Consequences: Volume 2, Archaeological and Historical Perspectives on the Spanish Borderlands East* (D. H. Thomas, ed.), Smithsonian Institution Press, Washington, DC, pp. 543–554.

Reitz, E. J., 1991, Animal Use and Culture Change in Spanish Florida, in: *Animal Use and Culture Change* (P. J. Crabtree and J. Ryan, eds.), *MASCA Research Papers in Science and Archaeology* 8, University Museum, University of Pennsylvania, Philadelphia, pp. 62–77.

Reitz, E. J., 1992, The Spanish Colonial Experience and Domestic Animals, *Historical Archaeology* 26(1):84–91.

Reitz, E. J., 1993, Evidence for Animal Use at the Missions of Spanish Florida, in: *The Spanish Missions of La Florida* (B. G. McEwan, ed.), University Presses of Florida, Gainesville, pp. 376–398.

Reitz, E. J., 1994, Zooarchaeological Analysis of a Free African Community: Gracia Real de Santa Teresa de Mose, *Historical Archaeology* 28(1):23–40.

Reitz, E. J., 2004, Fishing Down the Food Web: A Case Study from St. Augustine, Florida, U.S.A. *American Antiquity* 69(1):63–83.

Reitz, E. J., and Cumbaa, S. L., 1983, Diet and Foodways of Eighteenth Century Spanish St. Augustine, in: *Spanish St. Augustine: The Archaeology of a Colonial Creole Community* (K. A. Deagan, ed.), Academic Press, New York, pp. 147–181.

Reitz, E. J., Gibbs, T., and Rathbun, T., 1985, Archaeological Evidence for Subsistence on Coastal Plantations, in: *The Archaeology of Slavery and Plantation Life* (T. Singleton, ed.), Academic Press, New York, pp. 163–194.

Reitz, E. J., and Honerkamp, N., 1983, British Colonial Subsistence Strategy on the Southeastern Coastal Plain, *Historical Archaeology* 17(2):4–26.

Reitz, E. J., and McEwan, B. G., 1995, Animals, Environment and the Spanish Diet at Puerto Real, in: *Puerto Real: The Archaeology of a Sixteenth-century Spanish Town in Hispaniola* (K. A. Deagan, ed.), University Presses of Florida, Gainesville, pp. 287–334.

Reitz, E. J., and Ruff, B., 1994, Morphometric Data for Cattle from North America and the Caribbean Prior to the 1950s, *Journal of Archaeological Science* 21(5):699–713.

Reitz, E. J., Ruff, B., and Zierden, M., 2006, Pigs, Time Period, Status, and Site Function in Charleston, South Carolina, *Historical Archaeology*:40(4):104–124.

Reitz, E. J., and Scarry, C. M., 1985, *Reconstructing Historic Subsistence with an Example from Sixteenth-Century Spanish Florida, Society for Historical Archaeology Special Publication* 3:1–150.

Reitz, E. J., and Zierden, M., 1991, Cattle Bones and Status from Charleston, South Carolina, in: *Beamers, Bobwhites and Blue-points: Tributes to the Career of Paul W. Parmalee* (J. Purdue, W. Klippel, and B. Styles, eds.), Illinois State Museum, Springfield, pp. 395–407.

Rothschild, N. A., 1989, The Effect of Urbanization on Faunal Diversity: A Comparison between New York and St. Augustine, Florida, in the Sixteenth to Eighteenth Centuries, in: *Quantifying Diversity in Archaeology* (R. D. Leonard and G. T. Jones, eds.), Cambridge University Press, Cambridge, England, pp. 92–99.

Rothschild, N. A., 1990, *New York City Neighborhoods: The Eighteenth Century*, Academic Press, San Diego.

Rothschild, N. A., 2003, *Colonial Encounters in a Native American Landscape, The Spanish and Dutch in North America*, Smithsonian Institution Press, Washington, DC.

Rothschild, N. A., and Balkwill, D., 1993, The Meaning of Change in Urban Faunal Deposits, *Historical Archaeology* 27(2):71–89.

Ruhl, D. L., 1990, Spanish Mission Paleoethnobotany and Culture Change: A Survey of the Archaeological Data and Some Speculations on Aboriginal and Spanish Agrarian Interactions in Spanish Florida, in: *Columbian Consequences,* Volume 2, *Archaeological and Historical Perspectives on the Spanish Borderlands East* (D. H. Thomas, ed.), Smithsonian Institution Press, Washington, DC, pp. 555–580.

Ruhl, D. L., 2003a, Imagining Sixteenth and Seventeenth Native American and Hispanic Transformations of the Georgia Bight Landscapes, *Bulletin of the Florida Museum of Natural History* 44(1):183–198.

Ruhl, D. L., 2003b, Archaeobotanical Remains, in: *Presidio Santa Maria de Galve: A Struggle for Survival in Early 18th-Century Spanish Colonial Pensacola* (J. Bense, ed.), University Press of Florida, Gainesville, pp. 229—256.

Scarry, C. M., 1993, Plant Production and Procurement in Apalachee Province, in: *The Spanish Missions of La Florida* (B. G. McEwan, ed.), University Presses of Florida, Gainesville, pp. 357–376.

Scarry, C. M., and Reitz, E. J., 1990, Herbs, Fish, Scum, and Vermin: Subsistence Strategies in Sixteenth-Century Spanish Florida, in: *Columbian Consequences,* Volume 2, *Archaeological and Historical Perspectives on the Spanish Borderlands East* (D. H. Thomas, ed.), Smithsonian Institution Press, Washington, DC, pp. 343–354.

Schávelson, D., 1999, *The Historical Archaeology of Buenos Aires: A City at the End of the World*, Plenum Press, New York.

Schmitt, D. N., and Zeier, C. D., 1993, Not by Bones Alone: Exploring Household Composition and Socioeconomic Status in an Isolated Historic Mining Community, *Historical Archaeology* 27(4):20–38.

Schulz, P. D., and Gust, S. M., 1983, Faunal Remains and Social Status in 19th Century Sacramento, *Historical Archaeology* 17(1):44–53.

Shackel, P., 2000, *Archaeology and Created Memory. Public History in a National Park*, Springer, New York.

Shrire, C., 1995, *Digging Through Darkness. Chronicles of an Archaeologist*, University of Virginia Press, Charlottesville.

Spencer-Wood, S. (ed.), 1987, *Consumer Choice in Historical Archaeology*, Plenum Press, New York.

Staski, E. (ed.), 1987, *Living in Cities: Current Research in Urban Archaeology, Society for Historical Archaeology Special Publication* 5:1–101.

Stahle, D. W., Cleveland, M. K., Blanton, D. B., Therell M. D., and Gay, D. A., 1998, The Jamestown and Lost Colony Droughts, *Science* 280:564–567.

Steinberg, T., 2002, *Down to Earth: Nature's Role in American History*, Oxford University Press, Oxford, England.

Stewart-Abernathy, L., and Ruff, B., 1989, A Good Man in Israel: Zooarchaeology and Assimilation in Antebellum Washington, Arkansas, *Historical Archaeology* 23(2):96–112.

Trigg, H., 2004, Food Choice and Social Identity in Early Colonial New Mexico, *Journal of the Southwest* 46(2):223–252.

Walsh, L. S., Martin, A. S., and Bowen, J., 1997, Tidewater Towns: Provisioning Early American Towns, (Research Project Supported by the National Endowment for the Humanities), Colonial Williamsburg Foundation, Williamsburg, Virginia,
http://research.history.org/Archaeological_Research/Research_Articles/ThemeZooarch/Provisioning.cfm.

White, R., 1991, *"It's Your Misfortune and None of My Own": A History of the American West*, University of Oklahoma Press, Norman.

Whittaker, W., 1999, Production of Animal Commodities at Plum Grove, Iowa City, *Historical Archaeology* 33(4):44–57.

Wilkie, L., and Farnsworth, P., 2005, *Sampling Many Pots. An Archaeology of Memory and Tradition at a Bahamian Plantation*, University Press of Florida, Gainesville.

Wilson, D. C., and Rathje, W. J., 2001, Garbage and the Modern American Feast, in: *Feasts: Archaeological and Ethnographic Perspectives on Food, Politics, and Power* (M. Dietler and B. Hayden, eds.), Smithsonian Institution Press, Washington, DC, pp. 404–422.

Wing, E. S., 1989, Evidences for the Impact of Traditional Spanish Animal Uses in Parts of the New World, in: *The Walking Larder* (J. Clutton-Brock, ed.), Orwin Hyman, London, pp. 72–79.

Woods, C. A., 1989, The Biogeography of West Indian Rodents, in: *Biogeography of the West Indies* (C. Woods, ed.), Sandhill Crane Press, Gainesville, Florida, pp. 741–798.

Yemin, R., and Metheny, K. B. (eds.), 1996, *Landscape Archaeology: Reading and Interpretation of American Historical Landscapes*, University of Tennessee Press, Knoxville.

Yentsch, A., 1988, Farming, Fishing, Whaling, Trading: Land and Sea as Resource on Eighteenth-century Cape Cod, in: *Documentary Archaeology in the New World* (M. Beaudry, ed.), Cambridge University Press, Cambridge, England, pp. 138–160.

Young, A. (ed.), 2000, *Archaeology of Southern Urban Landscapes*, University of Alabama Press, Tuscaloosa.

Zierden, M. A., and Calhoun, J. A., 1986, Urban Adaptation in Charleston, South Carolina, 1730–1820, *Historical Archaeology* 20(1):29–43.

Chapter 3

Historical Perspectives on Timbisha Shoshone Land Management Practices, Death Valley, California

CATHERINE S. FOWLER

During the past two decades, scholars have paid increasing attention to the role of indigenous peoples and communities, both at present and in the past, in land and resource management and in other forms of environmental manipulation. Few anthropologists would deny that hunter–gatherers and subsistence farmers know a considerable amount about their resources and environments, but most would see an important difference between the knowledge systems of these peoples and their resulting influences on landscapes. Agricultural people were known to practice techniques such as slash and burn, various forms of soil disturbance and manipulation, water diversion, and other techniques that resulted in altered landscapes. By implication, these activities altered habitats and even genetics for plants and animals.

Hunter-gatherers, however, were viewed as more passive in their effects on the landscape, with occasional attempts at manipulation such as burning tracts of land or broadcast sowing of wild seeds, but little else (Blackburn and Anderson 1993; Harris and Hillman 1989). Now, with new field studies, the differences in environmental knowledge and procedures among such groups are less obvious, and most see instead a real continuum in principles of

environmental manipulation and resource management among peoples practicing various modes of primary subsistence. Hunter-gatherers, and former
hunter-gatherers, hold some very sound ideas about these topics, most of
which are well worth exploring and assessing, not only for their contributions
to ongoing discussions about the ultimate courses and processes of domestication, but also for a broader understanding of past and present environments
and distributions (see Harris 1989; Ucko 1989).

This case study reviews the preliminary results of field studies among the
Timbisha Shoshone people of Death Valley National Monument (now in Death
Valley National Park) in southern California (Figure 3-1).

The Timbisha people in the 1840s were hunter-gatherers in what is often
seen as one of the harshest of environments. They were closely related in culture as well as language to various other Great Basin Indian people, classically
profiled by Steward (1938, 1955) as representing "band level" societies. Due to
federal land use regulations, it is no longer possible to directly observe
Timbisha land and resource management systems in action in Death Valley.
Death Valley National Monument was established in 1933. In the 1940s, monument staff began implementing federal regulations on the use of monument
lands and resources. Bans were instituted against killing or collecting animals
and gathering some types of plants. Camping outside approved campgrounds,
building fires, and cutting wood were also banned.

The indigenous peoples still remember enough about former activities to
outline some of what used to occur. Through an approach that involved extensive interviews of elders of the tribe, as well as on-site visits to former camps
and other locations, a picture of these practices begins to emerge. It is clearer
for activities related to plants than it is for animals, possibly only because
today more women than men recall former times. Gathering plant resources,
classically women's work, seems to have persisted longer than game animal
hunting, classically men's work.

BACKGROUND

When sustained contact with Euro-Americans began in the 1840s, small kin-
based groups of Shoshone people lived in and around Death, Panamint, Saline,
and Eureka valleys and in the surrounding and intervening mountain ranges
in the northern Mojave Desert. In the older literature, they are often referred
to as "Panamint" or "Panamint Shoshone" or "Koso Shoshone," although more
recently the people in Death Valley have come to be known as Timbisha
Shoshone after their place name for Death Valley (Dayley 1989; Thomas et al.
1986). In former times, these peoples lived by hunting and gathering diverse
botanical and faunal resources, following a seasonal round that took them
from the hot valley floors (44–54 °C maximum) to the cooler mountains
(–18 to –1 °C minimum) in appropriate seasons. Plant resources particularly

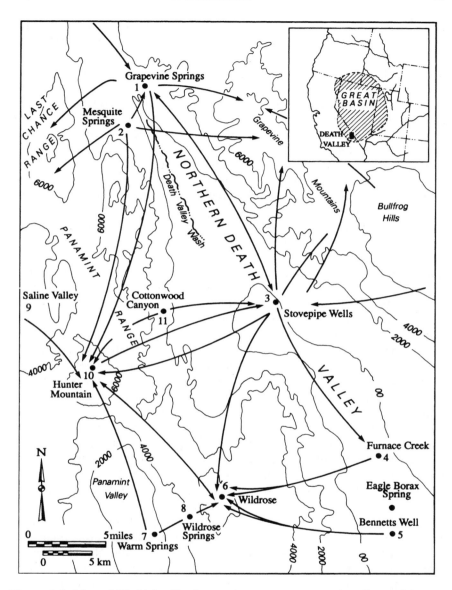

Figure 3-1. Map of Timbisha Shoshone winter camps in Death Valley, California. Arrows indicate seasonal movements. (1) Grapevine Springs; (2) Mesquite Springs; (3) Stovepipe Wells; (4) Furnace Creek; (5) Bennetts Well; (6) Wildrose; (7) Warm Springs; (8) Wildrose Springs; (9) Saline Valley; (10) Hunter Mountain; (11) Cottonwood Canyon.

favored for food were honey mesquite (*Prosopis glandulosa*) and single-leaf pinyon (*Pinus monophylla*), both producers of large and abundant fruits ("beans" and "nuts"); several genera and species of plants with edible leaves, fruits, bulbs, or corms; and many small-seeding annuals and perennials (Fowler 1986; Irwin 1980; Steward 1938). Some of these resources were carefully managed using such practices as burning, cleaning, clearing, pruning, and coppicing, along with minimal transplanting. Others seem not to have been the focus of these activities, but rather to have reproduced without human intervention.

From ethnographic work in the early 1930s by Steward (1938, 1941), it is known that in the 1840s there were four main areas on the floor of Death Valley and at least one in Panamint Valley that served as common winter village sites. At these sites, people usually built from three to ten conical brush houses, each serving a nuclear or extended family. The houses were often within mesquite groves, although in good weather people lived under the trees without shelters. The mesquite groves sheltered a range of small game animals and birds to be hunted. In spring and early summer, they also provided mesquite beans, an exceedingly important and storable food crop.

From these sites, people moved out to the high country on all sides, especially in summer and fall, for other plant resources—including seeds, roots, berries, and pinyon nuts, as well as for large and small game, including bighorn sheep (*Ovis canadensis*), mule deer (*Odocoileus hemionus*), yellow-bellied marmot (*Marmota flaviventris*), black-tailed jackrabbit (*Lepus californicus*), and chuckwalla (*Sauromalus obesus*). They established base camps in the pinyon-juniper woodlands of the Grapevine, Panamint, and Cottonwood mountains in late summer and fall and remained there until snows sent them back to the valley floor camps for the winter. In most seasons, subsistence was decidedly a mixture of plant and animal resources, although during specific harvest times, a single food might be eaten almost exclusively for a short period. People lived almost entirely on cached vegetable products during some periods of winter.

Kinship connected the people of Death Valley with those of Panamint Valley and, in turn, these two groups with all other speakers of the Panamint Shoshone language in southern California and Nevada (Dayley 1989). Families from these areas also sometimes met in the high country of the Panamint, Cottonwood, or Grapevine mountains or at other points to the west or east in the summer and fall. The number who habitually camped near each other in these areas was probably most often under 50 persons. During good years for pinyon nut harvests, however, camps of 100 people or more were reported (Dutcher 1893). Material culture for these groups in former times stressed the light and portable, as befitting quite mobile strategies (Driver 1937; Steward 1941). Social and religious systems likewise evidenced "conspicuous environmental conditioning" (Steward 1938:1).

MANAGEMENT PRACTICES

The environmental management practices for plants that can be documented for the Timbisha Shoshone people within memory are these: the use of fire; clearing, pruning, and coppicing; and transplanting and cultivation. Some of these practices follow general principles stating that the land and its plants should be generally tended; others have to do with activities that are more species-specific, producing a desired result or end product. Another practice, cleaning and clearing springs, generally benefits wildlife as well as people.

The Use of Fire

Timbisha people today recall that fire was used as a management tool for at least three purposes in earlier times: to encourage the growth of tobacco (*Nicotiana attenuata*); to clear riparian and marshy areas of dense growth of willows (*Salix* spp.), emergents, and fringing grasses; and to promote the growth of certain types of seeds, particularly white-stemmed blazing star (*Mentzelia albicaulis*). They tend to associate each use with specific places they have heard discussed in the past, but in all likelihood these activities took place in other locations as well. They also say more generally that fire is good and useful to clean the country; this they were told by the Old People. The best fires were natural ones caused by lightning, which were thought to be always beneficial to the land.

Hunter Mountain, at the south end of the Cottonwood Mountains, is particularly well known for its good tobacco. In June 1993, a considerable amount of tobacco was growing in disturbed conditions along the roadway leading up to the mountain. Several individual plants also were associated with pinyon nut harvesting camps on the summit. In former times, the people who habitually used Hunter Mountain for winter camps and seasonal camps (Saline and Panamint Valley Timbisha) apparently burned certain areas on the mountain to encourage tobacco growth. Timbisha people are no longer certain where these places are, but other people in the region may recall them.

Both Steward (1941:281) and Driver (1937:84) were told by the people they interviewed that burning for tobacco was formerly a practice of the Death Valley, Saline Valley, and Koso people. Tobacco was known to be a fire follower by many native California and Great Basin peoples, who fired a few acres in the fall or spring, usually in areas dominated by big sagebrush (*Artemisia tridentata*) or by sagebrush and juniper (*Juniperus* spp.) (Fowler 1986; Kroeber 1941; Steward 1938).

The use of fire in and around marshes to remove unwanted growth of emergent vegetation and fringing grasses is best documented for Warm Springs in Panamint Valley, but is likely to have taken place in Death Valley as well. According to what Timbisha people have heard, a Panamint Valley resident used to tell his relatives that it was time to burn the marsh at Warm Springs when the vegetation was dry (late fall, winter). This burning would clear the area and provide open water for waterfowl. Although what they heard probably relates to the 1890s or later, there is historical evidence that this site was burned much earlier in the past to encourage the growth of grasses for horses (*Equus caballus*). In the 1930s, a Saline Valley man provided a tale involving a horse raid into southern California, probably prior to 1850. In order to get ready, Warm Springs was fired by the men so there would be grass when they returned with the horses (Irwin 1980:69).

Cattail (*Typha* spp.), especially, can become a weedy problem if conditions are right and it is not controlled (Morton 1975). Bulrush (*Scirpus* sp.) is less troublesome, although three-square (*S. americanus*) is known to be invasive. Some managers today advocate deep dredging to remove cattails, or cutting below the waterline two to three times during the growing season (Morton 1975:13). Burning also is probably effective, however, especially if water levels are slightly lower so that the fire reaches part of the rhizome. It certainly clears the area of old matted vegetation, which itself can facilitate the spread of cattail.

Timbisha people have noticed particularly how open water is lost to cattail (and probably three-square) at the pond at Eagle Borax, a site in south central Death Valley. In the 1930s, there was considerable open water at this location, and duck hunting occurred regularly. Two men who camped here kept the growth of cattails and an unidentified bunchgrass in check by letting their horses and burros (*Equus asinus*) graze the area each fall. Otherwise, the animals were kept in corrals. It is possible that the men used fire before acquiring livestock.

Firing riparian areas, potential sites for cultivation and plant collecting, is best documented for Hunter Mountain and Cottonwood Canyon, but it probably also occurred elsewhere as part of general management. Dried willows, grasses, and other annual and perennial growth was fired in the fall to clear areas for next year's planting and harvesting. Burning was felt to keep the willow under control and keep it from invading other areas of moist ground. Two sites that were fired and then planted from the 1890s to the 1930s were visited in 1994; both are now dense willow thickets with little evidence of cleared ground.

A number of annual plants formerly harvested for their seeds are known to be fire followers today, and it is quite clear that people knew their habits in the past. Timbisha people today remember primarily this characteristic for white-stemmed blazing star, and they recall looking for it in places where natural fires had occurred. But both Steward (1941:281) and Driver (1937:65) record that all the individuals they interviewed in the 1930s said that areas were purposefully burned to encourage the growth of several types of seeds. Also in the 1930s, one man reported that while women were collecting blazing star seeds, the men burned in the same region during rabbit drives (Irwin 1980:15).

Documented in the ethnographic literature for the larger Panamint territory is the use of fire in deer and rabbit drives, which secondarily could have achieved some of the same ends as purposeful firing for seed growth. The deer drives apparently involved firing around the base of a small hill so that the flames would drive animals, but especially deer, past waiting hunters (Irwin 1980:22–23). Firing the brush for rabbits involved setting several linear fires one at a time and one ahead of the other at roughly 150 to 200 m intervals. In this way, the rabbits were driven in one direction until they became exhausted and overheated, at which time they were said to turn back to seek refuge under unburned brush. There, the hunters shot them with bow and arrow, taking as many as they needed (Irwin 1980:24).

Clearing, Pruning, and Coppicing

An additional set of related management tools used in several ways by the Timbisha people involved clearing undergrowth as well as pruning and coppicing perennial plants and trees. These techniques were used extensively in pinyon nut harvesting areas and in mesquite groves, but coppicing willow also occurred near springs or seeps, or along stream banks where good willows for basketry were known to grow.

All areas in the pinyon-juniper forests where people habitually camped or collected pinyon nuts were carefully cleaned of underbrush as part of routine and necessary maintenance. Lower branches of trees were cut close to the trunk to allow freer access underneath. Timbisha people were taught by their elders that these activities were an important part of learning to care for the land properly—to keep it clean and litter-free. Secondarily, such care also provided wood for fires and made the collection of pinyon nuts easier, as those that fell from the cones could be seen more easily on cleared ground. Even beyond these purposes, cleaned and cleared areas would not catch fire, so that in the unlikely event that a campfire got away, the fire could be put out quickly.

In the pinyon nut camps visited in 1992 and 1993 in the Wildrose district and on Hunter Mountain, evidence of this clearing and cleaning process was still present. There was very little big sagebrush or other undergrowth under the trees, especially those immediately surrounding the camps. The lower branches to many of the trees were cut close to the trunk, up to about 2 m or so. The Hunter Mountain camps, which are outside the monument boundary, are still kept clean and clear by the Panamint and Saline Valley people who use them. The Wildrose sites have not been thoroughly cleaned since the 1940s, when people were prevented from camping there by Monument regulations. The axe-scarred trees are the primary reminders of these former activities.

This same type of activity also was required in the mesquite groves, particularly those near Furnace Creek in central Death Valley. Timbisha people say their elders told them that when the people formerly camped in the middle of a mesquite clone, they kept the area clean and clear of undergrowth, dead limbs, and lower branches. This cleaning made it easier for people to make paths through the trees while avoiding the thorns, but also it was easier to collect beans from properly trimmed and tended trees. The wood and other debris were used for fires. Wood was always a scarce commodity on the valley floor. People did not kill living trees for firewood; they depended on this cleaning process for fuel.

In the case of the mesquites, there was yet another benefit to this clearing process, especially in areas where there were dunes. Today, a number of the mesquites on the floor of Death Valley, at Furnace Creek, and along the western edge of the valley are being overtaken by blowing dune sand. Although this process is probably natural to some degree (it was noted as early as 1891 by Frederick Coville of the U.S. Biological Survey [Coville 1891]), many Timbisha people feel that it is much more common today than in former times because the trees are not being properly tended. Deadwood is left to accumulate without cutting, and lower branches are not trimmed well above ground level; now, the blowing sand is stopped, and eventually the mesquite is totally engulfed (Figure 3-2).

The trees are still alive under these hummocks of sand, but they do not seem to be flowering or fruiting properly. Although it is unlikely that Timbisha people formerly kept all mesquites clear of dune sand, they certainly were responsible for liberating a much higher number than at present.

Clearing and cleaning mesquite groves may have fostered seedling development. According to authorities (Mooney et al. 1977), mesquite seedlings are not good competitors with grasses and other types of undergrowth. They also require sunlight, which could have been provided by trimming the trees to open up the groves. The seedlings do best if somehow the seeds are pressed into the ground even slightly, as might occur as a by-product of people walking among the trees. When mesquite pods were processed for food, an activity that normally was done within the grove, a wooden mortar and a long stone pestle were used to grind into meal the spongy mesocarp as well as part of the

Figure 3-2. Mesquite (*Prosopis glandulosa*) engulfed by dune sand in Death Valley, California.

stronger exocarp and endocarp. The hard seeds were normally discarded as inedible after the pounding process, itself an excellent form of scarification. Pestle scarification might have given seeds an extra impetus to break dormancy, further solidifying the relationship of people to mesquite. It is interesting to note that mesquites are doing well and spreading today in the southwestern United States only where cattle (*Bos taurus*) are grazed (Fisher 1977). Cattle appear to produce several of the same conditions as native peoples did for these plants in the past: reduced competition, trampling, cleaning, and clearing. Today, the mesquite groves of Death Valley are contracting rather than expanding. Lack of water is one suspected cause; changes in management may well be another.

Although trimming pinyons and mesquites can be a form of pruning, additional techniques that seem to have fostered more cone production were used on pinyons in former times. These techniques were whipping the trees as well as pinching or breaking the growth tips. Whipping was done with the long harvesting poles used to remove cones in the fall. After the harvest, the poles were used again to whip the trees vigorously to remove any dead cones. In addition to removing the dead cones, the whipping seemingly broke the ends of the branch tips, thus bringing about the production of one or two growth buds. Pinching or breaking the growth buds by hand accomplished the same thing, although this was usually done only on the lower branches. People say that both processes were "good for the trees," and the trees responded by producing more pinyon nuts. Although these processes have not been thoroughly studied, it is known that new cones will be produced on the faster-growing branches of the tree (Lanner 1981:79). Perhaps the pruning activity stimulates this growth. Whipping trees to stimulate production is known in other world areas, a case in point being 17th-century Europe, where walnut trees were whipped (Eugene Anderson 1993, personal communication; see also McCarthy 1993). Today, no one is whipping or pruning the pinyon pines of Death Valley National Park, but a few people still do on Hunter Mountain and in the Grapevine Mountains. A number of people feel that pinyon nut crops have suffered in recent years because no one is tending the trees.

People also pruned plants as part of routine maintenance. Two species in particular were routinely pruned as part of the food-collecting process in the spring, these being the prince's plumes: Panamint prince's plume (*Stanleya elata*) and desert prince's plume (*S. pinnata*). As the new growth was removed to be used as a green, people broke off last year's flower stalks and any dead leaves on these perennials, thus cleaning them up to make ready for continued new growth (Figure 3-3).

These plants absorb increasing amounts of toxic selenium during the growth cycle. One must boil even young plants and discard the water. People therefore harvested only young and tender leaves in the early spring, giving the plant ample time to put on additional leaves to carry it through the late spring to early summer bloom. The pruning and cleaning promoted healthy growth for next year, according to what people were taught by their elders.

Figure 3-3. Pruned prince's plume (*Stanleya elata*). Dried stems and leaves have been removed and new leaves will be harvested.

Coppicing of willow was widely practiced throughout this region to produce straight stems for basketry. Each winter, after last year's stems were harvested, those remaining in a willow patch were cut to the ground. The large root then responded by sending up new and vigorous sprouts or canes (known as rods in Europe). These canes were straight and without side branches. This type of first-year growth was most prized for basketry, especially for splitting strands for wefts. Side branches interfere with the splitting process, as do any insect borings or scales. They weaken the stem so that the splits hinge off, and the person doing the splitting cannot maintain three even strands throughout the length. Thus, it was very important to tend willow patches in this fashion each year, or else a weaver would be unable to get materials that were workable or worth her time. Cut willows could be refreshed by burying them in the damp sand, so that taking them in quantity was not wasteful. People knew that coppicing helped keep insect infestations under control.

Nothing is known of tending deer grass (*Muhlenbergia rigens*), a plant used for warp in basketry. DeDecker (1984:89) lists it as occurring only west of the Coso Range on the slopes of the Sierra Nevada along the western edge of the Great Basin. However, it was identified as a basketry plant by Coville (1892), seemingly east of the Coso Range. In several areas of southern California, deer grass clumps were burned to stimulate new growth (Anderson 1993).

Transplanting and Cultivation

Although the history of agriculture is not fully documented for the Timbisha Shoshone, it is suggested that some people may have cultivated traditional crops (maize [*Zea mays*], beans [*Phaseolus* spp.], and gourd/squash [*Cucurbita* spp.]) for some time before Death Valley was settled by Euro-Americans. Driver (1937:113) was told by a consultant that his great-grandfather had visited the Mohave people on the lower Colorado River, south of Death Valley, and brought back seeds of various domesticated plants. Driver interpreted the date of this acquisition to be about 1840. Included were maize, beans, and unidentified varieties of yellow squash. The man's great-grandfather apparently planted these seeds at a ranch in the Panamint Range at a well-known gardening spot. Jaeger (1941:284) remarks, presumably about the same individual, that he visited Fort Mohave "about 80 years ago" and obtained seeds of devil's claw (*Proboscidea parviflora*), used in basketry, which he planted at the same gardening spot. Other accounts from the 1870s speak of gardens in what is probably this location, as well as in Grapevine Canyon and at Furnace Creek (Wallace 1980).

Whatever the sources and timing of the introduction of agriculture among the Timbisha people (for a discussion, see Wallace 1980), they were familiar with the ideas of planting, tending, and irrigating crops for quite some time. Thus, either from this knowledge or from even earlier sources, people had ideas about planting and transplanting that they put to use in several instances. There is some indication that moving plants too far from their native situations may have met with disapproval.

In Wildrose Canyon, one individual transplanted willows into a location near a spring so that his sisters would have a source of supply for their basketry. He seems to have done so in the 1920s or 1930s, while the camp in the area was still being heavily used. Another member of the same family planted at the same site a plum tree (*Prunus* sp.) that reached maturity and fruited. Other members of the Timbisha community were against these activities, as the willows were known to be invasive and would probably ruin the spring, and the plum tree "did not belong there." The willow, indeed, has taken over quite an area around the spring. A few years ago, when a Timbisha tribal member attempted to remove some of its growth to increase the water flow, a monument ranger stopped the activity as destructive, probably not realizing that the willow had originally been transplanted there. Other Native Californians transplanted small shrubs such as ceanothus (*Ceanothus* sp.) and manzanita (*Arctostaphylos* spp.) (Shipek 1989:380).

Cleaning Water Sources

Timbisha people also were taught by their elders to care for other types of resources, such as those that provide water. Springs and tanks or potholes were routinely cleaned when people arrived at a site to camp, this procedure being part of the general cleaning process, one that specifically aided people and wildlife. Springs choked with willow, such as the one mentioned above, were cleared and dug out so that the water could accumulate better. Potholes that had filled with debris since the last visit were similarly cleared. Fresh water would then collect in future rains, and thus aid all animals in the vicinity. Water sources were never to be fouled, or overused, as such action would jeopardize all life forms that depended upon them.

DISCUSSION

It is equally erroneous to view the landscape as unmanipulated wilderness and to romanticize native environmental control and management. Common ground and common sense dictate, however, that native management principles and procedures at Death Valley National Park, as well as elsewhere, be further investigated and explored. Although the practices described herein have been suppressed within the monument since the 1940s, there are still elders who would be willing to cooperate with resource managers toward solutions that might lead to healthier mesquite groves, pinyon-juniper forests, and marshes. The ethic of caring for the environment is still very much alive, and the people see themselves as having something to contribute on this level. They are willing to listen to others with different ideas, as long as they can see some results.

Seeing results perhaps is one important general feature of Timbisha management principles that needs to be stressed. For the Timbisha people, it appears that management should show, thereby creating, in certain circumstances, habitats that appear to be tended as opposed to what they feel is unkempt. This is particularly true of areas they formerly pruned and cleared, such as mesquite groves, camps, and favorite gathering areas in pinyon-juniper woodlands, and patches of willow and prince's plume. Elders explain that plants need to "feel" the presence of people; they are used to it. This is how plant–human interrelationships are in part maintained. Plants are not destroyed when people harvest or care for them in proper ways; they are actually enhanced.

Ucko (1989:xii–xiii) remarked, with reference to the Australian Aborigines, that they "domesticated' the environment, *including plants, not* by practicing agriculture but by developing a complex system of mental categorizations which gave them control over their plants (and animals)." Although the Timbisha people would probably not agree with the term "control," they would certainly understand the concept of a "domesticated environment," based on some that they have known. This is not to say that they have necessarily created anthropogenic environments of the types recognized in parts of southern California (e.g., chaparral (Lewis 1993a:67), northern California (coastal prairie) (Blackburn and Anderson 1993:22), and the Southwest (Sonoran oases) (Nabhan et al. 1982). They altered conditions to fit their standards of a managed landscape. It remains to be determined whether there are any negative or additional positive benefits beyond those suggested.

Certainly one of the most effective and widely used technologies for creating true anthropogenic landscapes or for encouraging succession is fire, and its effects are chronicled in Australia, Africa, Canada, and the United States (for a discussion, see Lewis 1993b). Fire opens grasslands, cleans out forest litter, promotes the germination and growth of species that have coevolved with it, provides mulch, and much more. Timbisha uses of fire fit quite well into this picture, even though the details of its use are no longer clear. Likewise, their activities in coppicing willow fit previously known patterns for the region (Anderson 1991, 1993; Fowler 1986). Pruning and cleaning prince's plume, pruning mesquite, and whipping pinyon are new records, but parallel principles are documented for California and elsewhere (McCarthy 1993; Shipek 1989).

Considerable attention has been given to questioning why native cultivation of maize, beans, and squash seemingly did not spread from the Southwest into the southern California deserts (and well beyond) prior to the 17th century (for a review, see Bean and Lawton 1973). Although Timbisha agriculture also appears to be relatively late, the reason it was not used does not seem to be that the Timbisha peoples lacked knowledge of how to manipulate plants. As with the broader question of the factors involved with the transition from hunter-gatherers to farmers, explanations other than simplistic ones are generally required (Harris and Hillman 1989; Rindos 1984). Continued exploration of human-plant relationships among hunter–gatherers, and former hunter-gathers such as the Timbisha Shoshone, may provide additional clues.

In 2000, the Timbisha Shoshone Tribe, through the Timbisha Shoshone Homeland Act (PL 106–423), was given the right to enter into traditional use and co-management agreements with the National Park Service and the Bureau of Land Management within the homeland. In 2001, a small pilot project was initiated to reintroduce indigenous management techniques (pruning, coppicing, cleaning, clearing) in small areas of the Furnace Creek mesquite grove and the Wildrose Canyon pinyon forest. Although it may be some time before the results of this project can be fully assessed, the Timbisha people feel very positive about this small step toward again caring for the trees (Fowler et al. 2003).

ACKNOWLEDGMENTS

Research for this chapter was conducted under the auspices of the Applied Ethnography Program of the National Park Service, Washington DC, through a cooperative agreement with the University of Nevada, Las Vegas (CAO No. 8011–92–9003). The Timbisha Shoshone Tribe's Historic Preservation Committee worked hard on all aspects of this and the larger land use project of which this is a part. The committee's cooperation and expertise are gratefully acknowledged, as are the funding and advice of the National Park Service. Molly Dufort and Mary Rusco also participated in all phases of the work.

REFERENCES

Anderson, M. K., 1991, California Indian Horticulture: Management and Use of Redbud by the Southern Sierra Miwok, *Journal of Ethnobiology* 11:145–157.

Anderson, M. K., 1993, Native Californians as Ancient and Contemporary Cultivators, in: *Before the Wilderness: Environmental Management by Native Californians* (T. C. Blackburn and K. Anderson, comps. and eds.), Ballena Press, Menlo Park, California, pp. 151–174.

Bean, L. J., and Lawton, H. W., 1973, Some Explanations for the Rise of Cultural Complexity in Native California with Comments on Proto-Agriculture and Agriculture, *Ballena Press Anthropological Papers* 1, Socorro, New Mexico.

Blackburn, T. C., and Anderson, K. (comps. and eds.), 1993, *Before the Wilderness: Environmental Management by Native Californians*, Ballena Press, Menlo Park, California.

Coville, F. V., 1891, *Death Valley Expedition Itinerary*, unpublished manuscript, U.S. Fish and Wildlife Service, 1860–1961, Smithsonian Institution Archives, Washington, DC.

Coville, E. V., 1892, The Panamint Indians of California, *American Anthropologist* (o.s.) 5:351–356.

Dayley, J. P., 1989, *Tümpisa (Panamint) Shoshone Grammar,* University of California Publications in Linguistics 116, Berkeley.

DeDecker, M., 1984, *Flora of the Northern Mojave Desert, California*, California Native Plant Society, Special Publication 7, Berkeley.

Driver, H. E., 1937, Culture Element Distributions. VI. Southern Sierra Nevada, *University of California Anthropological Records,* Berkeley, 1:53–154.

Dutcher, B. H., 1893, Piñon Gathering among the Panamint Indians, *American Anthropologist* (o.s.) 6:377–380.

Fisher, C. E., 1977, Mesquite and Modern Man in Southwestern North America, in: *Mesquite: Its Biology in Two Desert Ecosystems* (B. B. Simpson, ed.), Dowden, Hutchinson and Ross, Stroudsburg, Pennsylvania, pp. 177–188.

Fowler, C. S., 1986, Subsistence, in: *Handbook of North American Indians, Great Basin*, Volume 11 (W. C. Sturtevant, general ed.), (W. L. d'Azevedo, vol. ed.), Smithsonian Institution Press, Washington, DC, pp. 64–97.

Fowler, C. S., Esteves, P., Goad, G., Helmer, W., and Watterson, K., 2003, Caring for the Trees: Restoring Timbisha Shoshone Land Management Practices in Death Valley National Park, *Ecological Restoration* 21:203–306.

Harris, D. R., 1989, An Evolutionary Continuum of People-Plant Interaction, in: *Foraging and Farming: The Evolution of Plant Exploitation* (D. Harris and G. C. Hillman, eds.), Unwin Hyman, London, pp. 11–26.

Harris, D. R., and Hillman, G. C. (eds.), 1989, *Foraging and Farming: The Evolution of Plant Exploitation*, Unwin Hyman, London.

Irwin, C. (ed.), 1980, The Shoshone Indians of Inyo County, California: The Kerr Manuscript, *Ballena Press Publications in Archaeology, Ethnology and History* 15, Socorro, New Mexico.

Jaeger, E. C., 1941, *Desert Wild Flowers*, Stanford University Press, Palo Alto, California.

Kroeber, A. L., 1941, Culture Element Distributions. XV. Salt, Dogs, Tobacco, *University of California Anthropological Records,* Berkeley, 6:1–20.

Lanner, R. M., 1981, *The Pinyon Pine: A Natural and Cultural History*, University of Nevada Press, Reno.

Lewis, H. T., 1993a, Patterns of Indian Burning in California: Ecology and Ethnohistory, in: *Before the Wilderness: Environmental Management by Native Californians* (T. C. Blackburn and K. Anderson, comps. and eds.), Ballena Press, Menlo Park, California, pp. 55–116.

Lewis, H. T., 1993b, In Retrospect, in: *Before the Wilderness: Environmental Management by Native Californians* (T. C. Blackburn and K. Anderson, comps. and eds.), Ballena Press, Menlo Park, California, pp. 389–400.

McCarthy, H., 1993, Managing Oaks and the Acorn Crop, in: *Before the Wilderness: Environmental Management by Native Californians* (T. C. Blackburn and K. Anderson, comps. and eds.), Ballena Press, Menlo Park, California, pp. 213–228.

Mooney, H. A., Simpson, B. B., and Solbrig, O. T., 1977, Phenology, Morphology, Physiology, in: *Mesquite: Its Biology in Two Desert Ecosystems* (B. B. Simpson, ed.), Dowden, Hutchinson and Ross, Stroudsburg, Pennsylvania, pp. 26–43.

Morton, J., 1975, Cattails (*Typha* spp.)—Weed Problem or Potential Crop?, *Economic Botany* 29:7–29.

Nabhan, G. P., Rea, A. M., Reichhardt, K. L., Mellink, E., and Hutchinson, C. F., 1982, Papago Influences on Habitat and Biotic Diversity: Quitovac Oasis Ethnoecology, *Journal of Ethnobiology* 2:124–143.

Rindos, D., 1984, *The Origins of Agriculture: An Evolutionary Perspective*, Academic Press, San Diego, California.

Shipek, F. C., 1989, An Example of Intensive Plant Husbandry: The Kumeyaay of Southern California, in: *Foraging and Farming: The Evolution of Plant Exploitation* (D. Harris and G. C. Hillman, eds.), Unwin Hyman, London, pp. 159–170.

Steward, J. H., 1938, Basin-Plateau Aboriginal Sociopolitical Groups, *Bureau of American Ethnology Bulletin* 120, Washington, DC.

Steward, J. H., 1941, Cultural Element Distributions. XIII. Nevada Shoshone, *University of California Anthropological Records,* Berkeley, 4:209–360.

Steward, J. H., 1955, *Theory of Culture Change*, University of Illinois Press, Urbana.

Thomas, D. H., Pendleton, L. S. A., and Cappannari, S. C., 1986, Western Shoshone, in: *Handbook of North American Indians, Great Basin*, Volume 11 (W. C. Sturtevant, general ed.), (W. L. d'Azevedo, vol. ed.), Smithsonian Institution Press, Washington, DC, pp. 262–283.

Ucko, P., 1989, Foreword, in: *Foraging and Farming: The Evolution of Plant Exploitation* (D. Harris and G. C. Hillman, eds.), Unwin Hyman, London, pp. ix–viii.

Wallace, W. J., 1980, Death Valley Indian Farming, *Journal of California and Great Basin Anthropology* 2:269–272.

Part II

The Physical Environment and Environmental Change

Chapter *4*

Geoarchaeology and Archaeostratigraphy: View from a Northwest Coast Shell Midden

JULIE K. STEIN

Geoarchaeology is a fundamental part of archaeology because artifacts, by their very nature, are found in the ground and require analysis of more than just their manufacture, use, and age. Approaching the archaeological record with questions other than those about an artifact's technological or chronological aspects has been popular since Butzer (1964) introduced the environmental approach and Schiffer (1972) introduced archaeologists to site formation processes. From these leaders, archaeologists now know that artifacts experience changes in the ground and that these changes affect interpretations of technology and chronology.

The following study of a northwest coast shell midden provides an example of how geoarchaeology (Butzer 1982; Goldberg et al. 2001; Rapp and Hill 1998; Schiffer 1987; Waters 1992) and a modern stratigraphic approach (Stein 1987, 1990, 1993, 2000b), combined with artifact analysis, improves our understanding of the past. The midden displayed a two-layered structure that has been used to group artifacts and to contrast changes in artifacts over time (Stein 1992b, 2000a). Stratigraphic analysis suggested that this two-layered stratification should not be used to group artifacts because it was created by weathering processes that occurred after the deposits were laid down. In this case study, I describe the two-layered stratification observed at this coastal shell midden, provide data that explain how archaeological stratigraphy was used to dissect the complex formation history of the site, and discuss the implications for using geoarchaeology and stratigraphy.

THE PROBLEM

Throughout the coastal regions of the Americas, archaeological deposits containing shell dominate the archaeological record (Claassen 1998). Such shell middens have larger than average volumes compared to archaeological sites without shell. The presence of shell increases the volume of each deposit, yet the number of artifacts per deposit remains the same. These characteristics force archaeologists to excavate in large units and to use large stratigraphic layers if they want to recover enough artifacts to use in constructing chronologies (Stein 1992a).

Shell is not the best attribute to use when creating and defining contrasting layers within shell middens because shell is everywhere, often in frequencies that change only slightly from layer to layer. A more common attribute used is the color of the matrix (fine-grained fraction) surrounding and supporting the shell (Stein 1992e). On the Pacific Northwest Coast, shell midden matrices often present sharp contrasts between dark deposits at the base of the midden and light deposits at the surface (Figure 4-1).

Figure 4-1. At the excavation of the English Camp shell midden the dark/light boundary can be seen in the profile of unit 310/300 behind the students. The light, shell-bearing layers lie over the dark, shell-bearing layers. The dark/light distinction dips to the right (toward the water), which is related to the flow of groundwater through the shell deposit (Photo by J. K. Stein).

Such a site would be excavated so as to keep the artifacts found within these two large layers separate. The frequencies of artifacts found in each layer would then be compared and, if found to change, would be used to define chronologies.

For example, the Marpole phase (approximately 2500–1500 years ago) is in part defined by the presence of chipped stone tools and high frequencies of debitage. In the later San Juan phase (approximately 1500–200 years ago), such stone tools are found in low frequencies (Burley 1980; Carlson 1960; Mitchell 1971, 1990). In a site with a deeper dark layer and an upper light layer, the deeper, presumably older, layer containing numerous lithics would be tentatively assigned to the Marpole phase, and the shallower layer with fewer lithics would be tentatively assigned to the San Juan phase. Assignments would then be checked with radiocarbon dates of appropriate material in each layer. Finally, the lithological attributes of the layers (how they looked) would be used to define the groups into which artifacts were summed and compared. The artifact groupings are used to define the phases and thus the chronology. The question remains: Was the boundary of the deposits defined by the color of the matrix an appropriate boundary to use for the definition of cultural phases based on the artifacts?

THE STRATIFICATION PROCESS

Although the light/dark stratigraphic distinction is noted at many Northwest Coast sites (e.g., Mitchell 1971:88; Sullivan 1993), and in other parts of the world (see references in Stein 1992a), its formation was never analyzed in detail or explained. On the basis of archaeological and geoarchaeological information, this light/dark stratification could be created by any of three possible processes. Stated as hypotheses, they are:

1. People deposited different things in the lower layer than they deposited in the upper layer, reflecting a change in subsistence or technology. This behavioral change is reflected in the archaeological record as a change in the color of the matrix. If this hypothesis is true, then artifact change should coincide with changes in the color of the matrix.
2. People deposited similar artifacts throughout the time represented in these deposits, and the two-layered stratigraphy was created after deposition by noncultural soil formation that weathered the profile from the surface downward. Such events created a light-colored matrix near the surface that is heavily weathered and a dark-colored matrix near the base that is not weathered. If this hypothesis is true, then agents of weathering should be detected by chemical alterations from the surface downward.
3. People deposited similar artifacts throughout the time represented in these middens. The process responsible for creating the two-toned stratigraphy is infiltration of groundwater at the base of the midden, which darkened

only the lower half of the midden. Infiltration resulted because the areas into which people threw midden materials were either already low coastal areas susceptible to groundwater infiltration or high areas that were later inundated when the base level of the groundwater changed through a mechanism such as sea level change. If this hypothesis is true, then agents of weathering should be detected by chemicals leached from the base.

THE SITE AND DEPOSITS

The English Camp site (45SJ24) on San Juan Island, Washington (Figure 4-2), is a Northwest Coast shell midden located in the northwest portion of the island on Garrison Bay (Figure 4-3). The shell midden, which is part of the San Juan

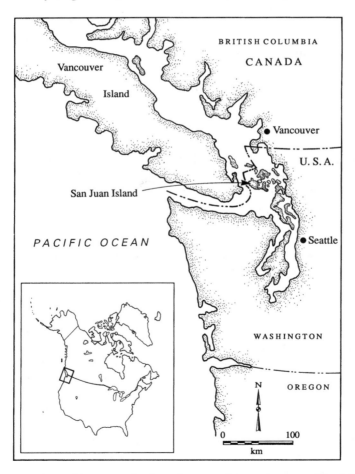

Figure 4-2. Location of San Juan Island, Washington, showing location of English Camp.

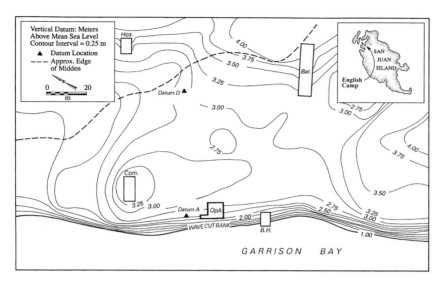

Figure 4-3. Topographic map of the English Camp (formerly British Camp) site, Garrison Bay, San Juan Island. (OpA) Excavation unit Operation A; reconstructed historic buildings: (Hos.) hospital, (Bar.) barracks, (B.H.) block house, (Com.) commissary. Field measurements and map made by Fran H. Whittaker.

Island National Historical Park-English Camp, underlies a later British occupation that occurred between 1859 and 1871 (Thomas and Thomson 1992). The British occupation was an outgrowth of the dispute between the United States and Britain over the location of the international border between Vancouver Island and the mainland. Both countries sent troops to occupy the San Juan Islands until the border issue was resolved (Vouri 1999, 2005). The British encampment happened to be on a site occupied by Northwest Coast peoples for at least the previous 2000 years (Stein et al. 2003). The site was referred to by the soldiers as English Camp, which the National Park Service reinstated when management came under the agency's management. The name, however, has been changed back to British Camp at the request of current inhabitants of the area. I will refer to it here as English Camp, but the previous name of British Camp refers to exactly the same place. The English Camp shell midden was selected for this analysis because previous excavations indicated that a light/dark stratigraphic boundary was present, and this two-layered stratigraphy had been used to group artifacts and define cultural phases.

The strategy for collecting samples from the site was developed in 1984, using an excavation trench of 12 continuous 2 m × 2 m units with no balks separating the units (Figure 4-4).

All layers within each unit were removed layer by layer across the entire group of units. Each of the layers (called facies during excavation) was defined on the basis of the quantity of shells, rock, and matrix color (for excavation

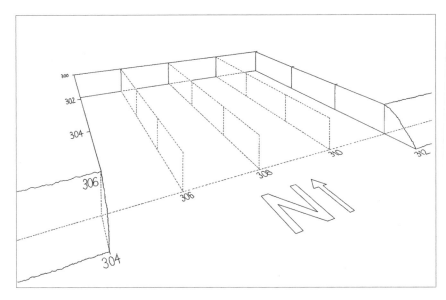

Figure 4-4. Three-dimensional representation of Operation A, showing its orientation to wave-cut bank and the grid coordinates. Vertical lines on the walls of Operation A and dashed lines mark the 12 continuous 2 m × 2 m units. Each of these units is named by the NW coordinate with the East/West coordinate given first. Note unit 310/300 in the upper right corner. Drafted by Timothy D. Hunt.

and sampling strategy, see Stein et al. 1992). The sediment samples were not taken from equal-interval depths, but rather from each layer. The layers are at various times adjacent to, overlying, or underlying each other (Stein 2001). Though layers are given sequential alphabetical labels, they are not all super-imposed one atop the other; some are actually side by side. Data concerning these layers could therefore not be graphically displayed in alphabetical order or in tabular form. Such an arrangement would give the impression that the layers are superpositionally arranged (in alphabetical order), when they really are only in a general sense arranged in that manner.

This sampling strategy requires that all data be displayed using Harris matri-ces (Harris 1989). A Harris matrix is a graphic representation of layers in which a rectangular box is used to represent each layer (no matter what size) and lines are drawn to show the vertical relationships of all the boxes (Figure 4-5).

These rectangles can be arranged in superpositional order in the same sequence as they are found in the archaeological record (Stein et al. 1992). The rectangles can be oriented on a two-dimensional graph in a representation of their three-dimensional order. All data from this research are displayed inside the boxes, which allows the reader to visually assess the stratigraphic changes in data throughout a unit.

To evaluate the hypotheses concerning human behavior and weathering, certain data were selected. To address weathering, the parameters of grain size,

UNIT 310300
8 Liter Buckets (per facies)

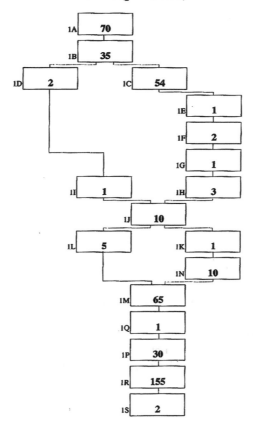

Figure 4-5. Harris matrix of unit 310/300. The Harris Matrix is a graphic representation of the superpositional relationship of all the facies. Each box represents a facies, the label of which abuts the lower left corner of the box. The number in the box is the number of 8-liter buckets excavated from that facies; the volume of that facies in liters would thus be that number times 8. Figure originally drawn by K. Kornbacher; computer-generated version designed by Lance Lundquist.

organic matter percentages, and carbonate percentages were selected. Carbonate is highly susceptible to weathering, but breaks apart only in the presence of an acid. Organic matter is a source for such acid, but is effective only in an aqueous environment, in which solutions can facilitate chemical reactions (Stein 1992d). Grain size determines the retention of both water and organic matter; deposits composed predominantly of sand will not hold either, and deposits with predominantly clay will not let water and organic matter flow to and from the layers (Stein 1987).

To address human behavioral changes, chipped stone debitage was chosen as the appropriate artifact type because historically diagnostic materials (e.g., carved artifacts or projectile points) were found only in low densities and not in the majority of layers included in the Harris matrices. Debitage, on the other hand, was found in every layer in relatively large numbers and therefore provided large samples to compare between layers. Also, for the last four decades, cultural historians have argued that a change in frequency of chipped-stone artifacts occurred and have used this technological shift to separate new phases (Carlson 1960; King 1950; Mitchell 1971). The reason for the change is not entirely understood, but the change was proposed to be related to a change in reliance on other technologies such as bone or wood technologies or to a change in subsistence (Burley 1980; Matson and Coupland 1995; Mitchell 1990). This research, combined with other high-resolution dating studies (Stein et al. 2003), calls into question whether the change occurred at all.

THE DATA

Samples collected from the excavation units in Operation A were separated into gravel fractions (grains larger than 3 mm) and matrix (grains smaller than 3 mm, the sand, silt, and clay fractions). Gravel fractions were sorted by size and composition using nested screens and water washing in the field (Stein et al. 1992). The matrix portion was sorted by using grain-size analysis (Stein 1992e, 2001). The organic matter and carbonate percentages were determined on a small subsample of the matrix using loss-on-ignition (Stein 1984). Data from only 1 of the 12 excavation units are presented here, as an example of the technique. Additional data can be found in Stein (1992b).

Throughout the excavation, the color of the matrix in each layer was noted, the allowable descriptions being restricted to black, brown, tan, and gray. Layers at the surface had predominantly gray matrix; layers at the base had black matrix. The layers in the middle, however, were problematic, in that they often changed color with shifting light or weather conditions. On average, the shift from gray to black matrix was observed in layers located about 80 cm below the surface, a surface that was flat in this unit of the site. That depth was influenced, however, by the amount of shell in the layers and therefore by the amount of matrix. In one corner of the unit, a facies with large shells and little matrix did not display the black matrix, while throughout the rest of the unit (where facies contained less shell and more matrix), the depth of the change from gray to black matrix was at 80 cm below the surface. It was difficult to assign a color to layers with coarse-grained shells and little matrix. In addition, layers with tan matrix (tan because ash was present) displayed no change from light tan to dark tan in either the top or the base of the site and were found throughout the site. One reason for using chemical tests as an independent measure of weathering was that assigning color to the layers in the intermediate zone was so difficult.

The percentages of sand, silt, and clay for the 2 m × 2 m unit 310/300, located in the northeast corner of Operation A (Figure 4-4), are displayed in a Harris diagram (Figure 4-6). The percentage of clay fluctuates widely but displays no systematic increase or decrease from the surface to the base of the midden. The percentage of sand fluctuates around 50% (with the notable exception of facies 1Q). There is no observable systematic change in grain size that might suggest a correlation with a two-toned stratigraphic break of light and dark.

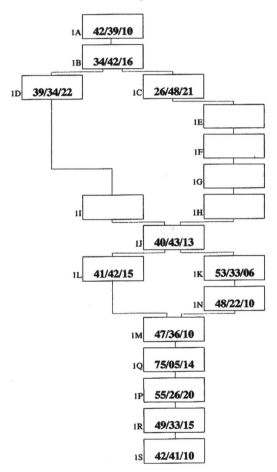

UNIT 310300
sand/silt/clay %

1A 42/39/10

1B 34/42/16

1D 39/34/22   1C 26/48/21

1E

1F

1G

1I   1H

1J 40/43/13

1L 41/42/15   1K 53/33/06

1N 48/22/10

1M 47/36/10

1Q 75/05/14

1P 55/26/20

1R 49/33/15

1S 42/41/10

Figure 4-6. Percentages of sand, silt, and clay of the sediment less than 1/8-inch in size, for samples from excavation unit 310/300 in Operation A. The numbers are the percentages of sand/silt/clay for the sample (or samples) from each layer (facies). The facies are labeled as in Figure 4-4.

UNIT 310300
Organic Matter Percentage (%)

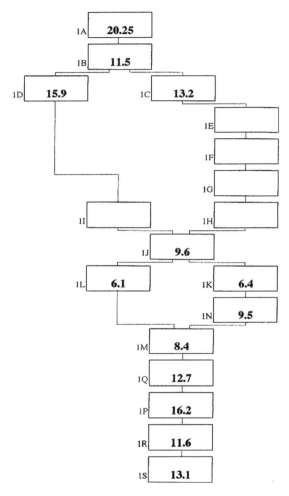

Figure 4-7. Percentage of organic matter in sediment less than 1/8-inch in size, for samples from excavation unit 310/300 in Operation A. The numbers are the percentages of organic matter for the sample (or samples) from each layer (facies). The facies are labeled as in Figure 4-4.

The percentage of organic matter (Figure 4-7) varies within a range of 6.1–16.2% (excluding the surface layer), but displays no systematic increase or decrease from the surface to the base of the midden. Organic matter percentages are highest at the upper surface (24%), caused by the root zone of the grass maintained by the National Park Service. No other systematic differences appear from the upper to the lower deposits.

The percentage of carbonate does show systematic changes (Figure 4-8), with consistently higher values near the surface and lower values near the base

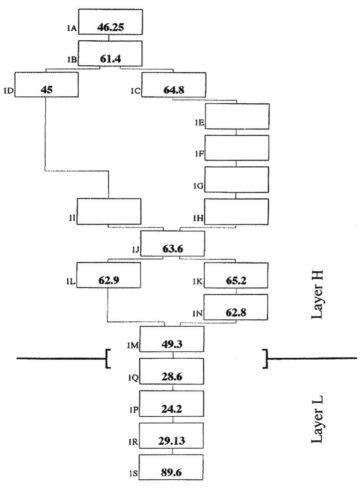

UNIT 310300
Carbonate Percentage (%)

Figure 4-8. Percentage of carbonate in sediment less than 1/8-inch in size, for samples from excavation unit 310/300 in Operation A. The numbers are the percentages of carbonate for the sample (or samples) from each layer (facies). The facies are labeled as in Figure 4-4.

of the excavation. The average percentage of carbonate in the upper layers (1A–1L and 1N) is 59%; that in the lower layers (1P–1R) is 27%. The exceptions to this trend are two layers with tan matrix (1M and 1S). The matrix in these tan layers seems to be composed of a large proportion of crushed shell, little clay, and less organic matter. Besides examining the carbonate percentages in the fine-grained fraction of each layer, the larger gravel-sized shell was examined for evidence of carbonate leaching. The exterior of individual shells in the lower layers is disintegrating, which makes it difficult to assign them to species (Ford 1992) and corresponds with lost carbonate (Figure 4-8).

Figure 4-9 shows the weight of lithics found in each excavated layer in unit 310/300. The debitage is divided into four size categories: 1-inch, 1/2-inch,

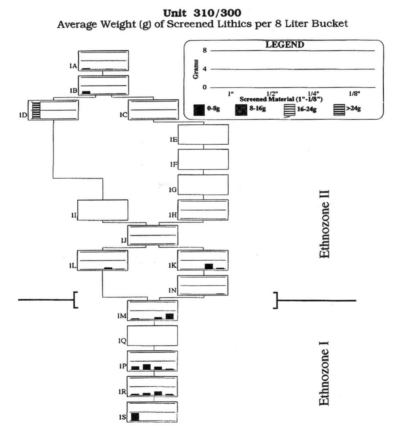

Figure 4-9. Debitage found in layers from excavation unit 310/300 in Operation A is displayed by weight in each of the four sizes screened at the site. The heights of the bars represent, from left to right, the weight in grams of debitage found in the 1-inch, 1/2-inch, 1/4-inch, and 1/8-inch screens. The shading of the bar indicates whether the weight is 0–8 g, 8–16 g, 16–24 g, or .24 g. Figure designed by Lance Lundquist.

1/4-inch, and 1/8-inch. A change in the weight of debitage in the layers is obvious; more debitage of all sizes is found in the lower layers and less in the upper layers.

CONSTRUCTION OF STRATIGRAPHIC UNITS

These data can be used to identify changes in the stratigraphic sequence, but they cannot, by themselves, address the causes of those changes. The data support the first hypothesis: that the layers were created by different cultural events in that differences in chipped-stone artifact frequencies occurred during the deposition of this particular shell midden. The data also support the hypothesis that a difference in weathering occurred, as measured in the amount of carbonate, but suggest that weathering did not occur from the surface and progress downward as predicted in the second hypothesis. The weathering occurred from the base of the midden and progressed upward as predicted in the third hypothesis. Thus, artifacts and carbonate data changed systematically in various portions of the shell midden sampled in unit 310/300; however, the relationship of those changes now has to be examined.

The stratigraphic method is used to correlate changes noted in sequences that were constructed using disparate data (Farrand 1993; Stein 1987, 1990, 1992c, 2000b). The stratigraphic method groups layers that are alike using a variety of data, such as physical appearance, chemistry, or artifactual content (Browman 2002). At English Camp, carbonate data were used to determine one grouping and boundaries. The resulting groups are stratigraphic units called *lithostratigraphic units* (Stein 1992c), or units based on the lithological characteristics of layers. Another type of stratigraphic unit is created using material culture. Such units are called *ethnostratigraphic units* (Stein 1992c), or units based on the artifactual characteristics of layers.

No single type of unit is necessarily "correct." Stratigraphy is the creation of multiple groupings, noting where boundaries of all the different kinds of units relate and articulate. If all the boundaries of all the different types of stratigraphic constructs occur at the same place in the sequence, then certain hypotheses are supported. If they all fall at different depths, other hypotheses are supported. To reiterate, in this example, two kinds of units are constructed: the lithostratigraphic unit (based on carbonate data) and the ethnostratigraphic unit (based on chipped-stone data). The boundaries' coincidences or differences then provide the means to evaluate the hypotheses concerning the causes of the two-toned stratigraphy.

The carbonate percentages indirectly measure the effects of weathering. In Figure 4-8, the carbonate percentages change from values ranging from approximately 45% to 65% in the upper portion of the Harris matrix to values ranging from approximately 30% to 25% in the lower portion of the Harris matrix. This change is defined as the lithostratigraphic boundary and occurs in unit 310/300 between facies 1Q and 1M, about 70 cm below the present ground surface. The two groups thus constructed are called Layer H (for high values), which comprises all layers above the boundary, and Layer L (for low values), which comprises all layers below the boundary.

As shown in Figure 4-9, the weight of chipped stone (predominantly flakes) greater than 1/8-inch in size shows a change between facies 1M and overlying facies 1N and 1L, about 50 cm below the present ground surface. The ethnostratigraphic units defined by these frequencies are named Ethnozone I for the layer below the boundary containing high weights of lithics and Ethnozone II for the layers above the boundary containing low weights of lithics.

Thus, the boundaries of the two stratigraphic sequences constructed here do not coincide. The artifactual changes occur in layers about 20 cm higher in elevation than those in which the carbonate changes occur. Artifactual material does not change in the same vertical position as the carbonate does. These data suggest that cultural events did change at this site, but that the light/dark layers of the midden and the weathering phenomenon are independent of the cultural change (Figure 4-1).

DISCUSSION

The data described above, as well as the data from the other 11 units excavated, suggest that the light/dark stratigraphic division is not a product of cultural events; it is the product of a weathering phenomenon that systematically leached carbonate from the lower layers of the site (Holliday 2004). Carbonate is one of the most mobile chemicals and is usually leached from surface deposits. Its removal starts first at the surface, where rainwater falls and dissolves the carbonate locked in mineral or shell. As the water travels through the substrate, the carbonate is carried downward and removed from the surface layers.

At the English Camp shell midden, carbonate was not removed from the surface, as expected, but rather was removed from the lower layers. There are two possible explanations for this unusual carbonate distribution. One would be that carbonate was originally deposited in smaller amounts in the lower deposits, so lower percentages are detected. The other would be that carbonate was originally deposited in all layers of the shell midden in equal amounts, but was removed from only the lower deposits.

The carbonate data used to test the hypotheses are based on only the fine-grained fraction of the midden and are presumed to be the product of mechanical and chemical breakdown of shell. Carbonate was measured only for material that fell through the 1/8-inch screen. The samples measured contain mineral grains (quartz, metamorphosed chert, ferromagnesium minerals, feldspars, and some limestone) (Latas 1992) of sand, silt, and clay sizes, as well as shell, bone, and charcoal fragments. The carbonate measured in this sample is believed to have as its source the shell within the midden. Some limestone is found in the area, and the glacial drift underlying the area has some carbonate. These mineral sources, however, are very low. The shell, on the other hand, makes up 60–90% of the coarser-size fraction (the gravel-size material) in almost every layer in the site. Shell is the most likely source of the carbonate in the fine-grained fraction.

The question concerning the same amount of carbonate deposited throughout the midden can best be examined if shell larger than 1/8-inch is examined and compared to the fine-grained carbonate. Data collected during screening in the field suggest that shell (and therefore carbonate) was equally distributed throughout the shell midden. Each layer in the site was screened through various-sized mesh, with the weight of shell recorded in the field. These data indicate a highly variable amount of shell in all the layers across the site (Ford 1992); however, specific species of shellfish (mussel [*Mytilus* spp. and *M. edulis*], horse clam [*Tresus* spp. and *T. gapperi*], venus clams [*Protothaca staminea, Saxidomus* spp., *S. giganteus,* and *S. nuttalli*], and barnacles [*Balanus* spp.]) are found in equal frequencies throughout the midden. Only cockles (*Clinocardium* spp. and *C. nuttalli*) and sea urchins (*Strongylocentrotus* spp.) change in their frequencies, with lower frequencies of cockle and higher frequencies of sea urchin found in the lower layers. These data, provided for only two of the twelve 2 m × 2 m units excavated, suggest that most of the shellfish were originally distributed randomly throughout the site. At least there is no overall loss of large-size shellfish in the site.

Because the large-size shellfish were, and still are, distributed in great frequencies throughout the site, the fact that carbonate is not found in the fine-grained portions of some layers suggests that weathering has occurred. Small-size particles have greater surface areas than large-size particles and therefore disappear faster when weathered. As mentioned before, the exterior surfaces of large shell fragments show evidence of weathering. The process has not progressed to the point where the large fragments have been removed from the midden through leaching. Given enough time, however, this process will cause all the shell to dissolve and leach from the deposits. The lower portions of the site will no longer look like a shell midden.

The hydration of an archaeological site from below is not usually the dominant process in humid temperate climates (Holliday 2004), and is surprising considering that no evidence of surface leaching (and therefore soil formation) was found. One would expect water falling as rain to hydrate at least some sediment at the surface, with progressively less hydration taking place as the water moved downward into the subsurface. When the shell midden is observed during torrential rainstorms, however, one sees that water falling on the porous shell-bearing surface flows through the strata too quickly to hydrate the clays and organic matter in its upper portion. Some water is absorbed at the surface and at the root zone, but the remainder travels rapidly through large pores and vesicles, never hydrating the organic matter and clay. My observations suggest that for many shell-bearing sites with large, connected pore spaces, hydration occurs only when coastal lowlands are submerged and the shell-bearing strata are saturated by groundwater.

Once water accumulates in the lower portions of a shell midden, weathering proceeds by saturating the organic matter and clay, which in turn changes the color of the fine-grained fraction from a light to a dark color. Water allows the organic matter to decompose at an increased rate, which in turn produces organic acids as by-products. These acids travel in solution through the large pores into the matrix as well as around small shell fragments and large shell exteriors. There, the organic acids dissolved in solution attack the chemical bonds of particles, in particular the carbonate in the shell. The carbonate dissolves into various chemicals that are removed from the deposit either as solutions in the brackish groundwater or as gases.

The water involved in the leaching cannot come strictly from saline seawater. Calcium is supersaturated in seawater, meaning that no additional calcium could be dissolved from the carbonate and leached from the shell midden if only seawater infiltrated the midden. The fresh water flowing off the surrounding slopes must enter the shell midden from the landward side and mix with seawater to make brackish water. Such a mixture would allow the chemical reactions described above to take place.

Still left to answer are the questions as to when and for how long the lower portion of English Camp shell midden has been inundated by groundwater. Information extracted from cores in nearby tidal marshes (Whittaker and Stein 1992) suggests that the sea level has risen 2 m over the last 2000 years. Cannon (2000) found evidence for constant rise of sea level farther north on the coast of British Columbia. The radiocarbon dates from charcoal in English Camp indicate that the site was occupied at least for the last 2000 years, although perhaps not continuously (Deo et al. 2004; Stein et al. 2003). The original occupants must have lived in a landscape overlooking a sea that was 1–2 m lower than it is today. They deposited midden in areas well above the water table, aggrading the landscape, and perhaps prograding the shoreline. But the water table was lower then than it is now. Eventually that situation changed; when the sea level rose, the lower portions of the shell midden were inundated. With the inundation came weathering and the discoloration that goes along with it.

CONCLUSIONS

In this example, data from one excavation unit at one shell midden in the Northwest Coast demonstrate the need to consider geoarchaeological phenomena when doing all kinds of archaeology. The point of the stratigraphic method is to use a separate sequence for each type of data and compare the locations where changes occur. In this case, the changes caused by cultural depositional events had to be considered independent of those caused by postdepositional events to construct multiple stratigraphic sequences. All types of data collected from the excavation units were displayed separately, and locations where changes occurred in each data type were identified. This study compared grain-size data, organic matter, carbonate, and weight of debitage. Further research could be done, however, in which such categories as the amount of bird, fish, and plant remains as well as the quantity of bone tools and microblades could be considered. Results would indicate which of these artifacts, resources, and weathering values change at the same or different locations. This procedure is the stratigraphic method.

At English Camp, the lithic debitage changed from large amounts in the lower layers to small amounts in the upper layers, but that change did not occur at the same place that the carbonate percentages changed. Therefore, the changes in carbonate percentages and the color of the lithostratigraphic units are not related to the behaviors that lead to the deposition of the artifacts. The significance of this interpretation is that the stratification of these archaeological sites is related only partially to cultural events. Cultural events dictate whether sufficient quantities of organic matter and clay particles are deposited in any one site or layer, but the infiltration of groundwater entering the strata from below is dependent on local topography and sea level, as well as on cultural depositional events.

Thus, grouping artifacts using the light/dark visual criteria will likely result in an inappropriate comparison of artifactual material and descriptions of cultural boundaries. One of the reasons the Marpole and San Juan phases are so poorly defined is that their boundaries are confused as a result of excavation practices. Some older artifacts are lumped into the more recent phase and more recently manufactured artifacts are lumped into the older phase. If matrix color is used to group artifacts, then a change in artifacts will be found, but it will appear as a sudden change, occurring lower in the deposits than it actually exists, and will be assigned an inappropriate older age. The physical stratigraphy (lithology) should not be used to group artifactual data. The stratigraphic method clarifies the situation, allowing cultural changes to be separated from lithological changes.

The reason the light/dark, two-toned stratification is observed so frequently on the Northwest Coast, and at other coasts around the world, is related primarily to the proximity of human occupations to the marine shoreline and to the dynamic nature of these shorelines. In any location where sites contain shells and are porous, where clay is present, where organic matter is a product of human activity, and where the occupation occurred close to the groundwater table, this two-toned stratification could be observed. All that is required is for either the land to subside or the sea level to rise.

When excavating sites, archaeologists excavate layers following natural (physical) stratigraphic breaks. The research at English Camp shell midden suggests that at least one of these stratigraphic boundaries is not related directly to cultural events and should not be used to group artifacts for comparative purposes. This discovery was made by combining stratigraphy, geoarchaeology, and archaeology. If archaeologists do not wish to continue using a single stratigraphic sequence to define cultural and depositional units, and thus force into one sequence all the variability represented in time, technology, and subsistence, then geoarchaeology and stratigraphy need to be considered. In this way, the artifacts can be properly understood.

ACKNOWLEDGMENTS

This research was conducted at San Juan Island National Historical Park–English Camp, supported by the University of Washington and the National Park Service, especially with the assistance of Jim Thomson and Kent Bush. The excavation of Operation A was accomplished as part of a summer archaeological field methods class instructed annually from 1984 to 1989 by G. Thomas Jones, Charlotte Beck, Pam Ford, and Margaret Nelson. The sediment analysis was conducted in the Archaeological Sediments Laboratory at the University of Washington. The Harris matrices were drawn by Kim Kornbacker and linked to the database by Lance Lundquist. The photos were prepared by Christopher Lockwood. The ideas expressed in this chapter benefitted greatly from the insights offered by Mary Parr, Gregg Sullivan, and Fran (Whittaker) Hamilton.

REFERENCES

Browman, D. L., 2002, Origins of Stratigraphic Excavation in North America: The Peabody Museum Method and the Chicago Method, in: *New Perspectives on the Origins of Americanist Archaeology* (D. L. Browman, ed.), University of Alabama Press, Tuscaloosa, pp. 242–264.

Burley, D. V., 1980, *Marpole: Anthropological Reconstruction of a Prehistoric Northwest Coast Culture Type, Department of Archaeology, Simon Fraser University, Publication 8*, Burnaby, British Columbia.

Butzer, K., 1964, *Environment and Archaeology: An Introduction to Pleistocene Geography*, Aldine, Chicago.

Butzer, K., 1982, *Archaeology as Human Ecology*, Cambridge University Press, Cambridge, England.

Cannon, A., 2000, Settlement and Sea-Levels on the Central Coast of British Columbia: Evidence from Shell Midden Cores, *American Antiquity* 65:67–77.

Carlson, R. L., 1960, Chronology and Culture Change in the San Juan Islands, Washington, *American Antiquity* 25:562–586.

Claassen, C., 1998, *Shells*, Cambridge University Press, Cambridge, England.

Deo, J. N., Stone, J. O., and Stein, J. K., 2004, Building Confidence in Shell: Variations in the Marine Radiocarbon Reservoir Correction for the Northwest Coast Over the Past 3,000 Years, *American Antiquity* 69:771–786.

Farrand, W. R., 1993, Discontinuity in the Stratigraphic Record: Snapshots from Franchthi Cave, in: *Formation Processes in Archaeological Context* (P. Goldberg, D. T. Nash, and M. D. Petraglia, eds.), *Monographs in World Archaeology* 17, Prehistory Press, Madison, Wisconsin, pp. 85–96.

Ford, P. J., 1992, Interpreting the Grain Size Distributions of Archaeological Shell, in: *Deciphering a Shell Midden* (J. K. Stein, ed.), Academic Press, San Diego, California, pp. 283–326.

Goldberg, P., Holliday, V. T., and Ferring, C. R., 2001, *Earth Sciences and Archaeology*, Kluwer Academic/Plenum Publishers, New York.

Harris, E. C., 1989, *Principles of Archaeological Stratigraphy*, 2nd ed., Academic Press, New York.

Holliday, V. T., 2004, *Soils in Archaeological Research*, Oxford University Press, Oxford, England.

King, A. R., 1950, Cattle Point: A Stratified Site in the Southern Northwest Coast Region, *American Antiquity Memoir* 15 (Supplement 7):1–94.

Latas, T. W., 1992, An Analysis of Fire-Cracked Rocks: A Sedimentological Approach, in: *Deciphering a Shell Midden* (J. K. Stein, ed.), Academic Press, San Diego, California, pp. 211–238.

Matson, R. G., and Coupland, G., 1995, *The Prehistory of the Northwest Coast*, Academic Press, San Diego, California.

Mitchell, D. H., 1971, Archaeology of the Gulf of Georgia Area, a Natural Region and its Culture Type, *Syesis* 4 (Supplement 1):1–228.

Mitchell, D. H., 1990, Prehistory of the Coasts of Southern British Columbia and Northern Washington, in: *Handbook of North American Indians, Northwest Coast*, Volume 7 (W. C. Sturtevant, general ed.), (W. Suttles, vol. ed.), Smithsonian Institution Press, Washington, DC, pp. 349–358.

Rapp Jr., G. R., and Hill, C. L., 1998, *Geoarchaeology: The Earth-Science Approach to Archaeological Interpretation*, Yale University Press, New Haven, Connecticut.

Schiffer, M. B., 1972, Archaeological Context and Systemic Context, *American Antiquity* 37:156–165.

Schiffer, M. B., 1987, *Formation Processes of the Archaeological Record*, University of New Mexico Press, Albuquerque.

Stein, J. K., 1984, Organic Matter and Carbonates in Archaeological Sites, *Journal of Field Archaeology* 11:239–246.

Stein, J. K., 1987, Deposits for Archaeologists, in: *Advances in Archaeological Method and Theory*, Volume 11 (M. B. Schiffer, ed.), Academic Press, Orlando, Florida, pp. 337–393.

Stein, J. K., 1990, Archaeological Stratigraphy, in: *Archaeological Geology of North America* (N. P. Lasca and J. Donahue, eds.), Geological Society of America, *Centennial Special Volume 4*, Boulder, Colorado, pp. 513–523.

Stein, J. K., 1992a, Analysis of Shell Middens, in: *Deciphering a Shell Midden* (J. K. Stein, ed.), Academic Press, San Diego, California, pp. 1–24.

Stein, J. K. (ed.), 1992b, *Deciphering a Shell Midden*, Academic Press, San Diego, California.

Stein, J. K., 1992c, Interpreting Stratification of a Shell Midden, in: *Deciphering a Shell Midden* (J. K. Stein, ed.), Academic Press, San Diego, California, pp. 71–93.

Stein, J. K., 1992d, Organic Matter in Archaeological Contexts, in: *Soils in Archaeology: Landscape Evolution and Human Occupation* (V. T. Holliday, ed.), Smithsonian Institution Press, Washington, DC, pp. 193–216.

Stein, J. K., 1992e, Sediment Analysis of the English Camp Shell Midden, in: *Deciphering a Shell Midden* (J. K. Stein, ed.), Academic Press, San Diego, California, pp. 135–162.

Stein, J. K., 1993, Scale in Archaeology, Geosciences, and Geoarchaeology, in: *Effects of Scale on Archaeological and Geoscientific Perspectives* (J. K. Stein and A. R. Linse, eds.), Geological Society of America Special Paper 283, Boulder, Colorado, pp. 1–10.

Stein, J. K., 2000a, *Exploring Coast Salish Prehistory: The Archaeology of San Juan Island*, University of Washington Press, Seattle.

Stein, J. K., 2000b, Stratigraphy and Archaeological Dating, in: *Its About Time: A History of Archaeological Dating in North America* (S. Nash, ed.), University of Utah Press, Salt Lake City, pp. 14–40.

Stein, J. K., 2001, Archaeological Sediments in Cultural Environments, in: *Sediments in Archaeological Context* (J. K. Stein and W. R. Farrand, eds.), University of Utah Press, Salt Lake City, Utah, pp. 1–28.

Stein, J. K., Kornbacher, K. D., and Tyler, J. L., 1992, English Camp Shell Midden Stratigraphy, in: *Deciphering a Shell Midden* (J. K. Stein, ed.), Academic Press, San Diego, California, pp. 95–134.

Stein, J. K., Deo, J. N., and Phillips, L. S., 2003, Big Sites – Short Time: Accumulation Rates in Archaeological Sites, *Journal of Archaeological Science* 30:297–316.

Sullivan, G. M., 1993, *Postdepositional Leaching of Shell in Two Northwest Coast Shell Middens*, M.A. thesis, Department of Archaeology, Simon Fraser University, Burnaby, British Columbia.

Thomas, B. H., and Thomson, J. W., 1992, Historic Treatment of a Prehistoric Landscape, in: *Deciphering a Shell Midden* (J. K. Stein, ed.), Academic Press, San Diego, California, pp. 61–70.

Vouri, M., 1999, *The Pig War: Standoff at Griffin Bay*, Griffin Bay Bookstore, Friday Harbor, Washington.

Vouri, M., 2005, Outpost of Empire: The Royal Marines and the Joint Occupation of San Juan Island, Northwest Interpretive Association, National Park Service, Seattle, Washington.

Waters, M. R., 1992, *Principles of Geoarchaeology: A North American Perspective*, University of Arizona Press, Tucson.

Whittaker, F. H., and Stein, J. K., 1992, Shell Midden Boundaries in Relation to Past and Present Shorelines, in: *Deciphering a Shell Midden* (J. K. Stein, ed.), Academic Press, San Diego, California, pp. 25–42.

Chapter 5

Anatomy of a Southwest Florida Sand Burial Mound: Smith Mound at the Pineland Site Complex

Sylvia J. Scudder

One aspect of archaeological research that is particularly important to interpretations of cultural history and social relationships explores the management of labor and materials required for the construction of monuments, especially of burial mounds. The Pineland Site Complex, on the east coast of the Gulf of Mexico (Figure 5-1), features monumental earthworks, hand-dug canals, and rich black-earth and shell middens covering approximately 25 ha (Luer 1991:71; Marquardt 1992a). Generations of archaeologists have focused on cultural interpretations of Pineland's platform mounds, canals, lakes, and sand burial mounds (Cushing 1896; Luer 1989; Widmer 1988). Cultural and environmental evidence demonstrate that this architectural complexity, and associated social complexity, was supported, not by farming, but by the rich, natural mosaic of local, non-domestic resources (Marquardt 1987, 1988, 1992b; Scarry and Newsom 1992; Walker 1992, 2007). Thus, the evidence from this site runs counter to widespread archaeological assumptions that such complexity could only be supported by domesticated food sources. Elaborating upon the construction activities at Pineland demonstrates that the relationships among monumental construction, social complexity, and food acquisition systems are more complex

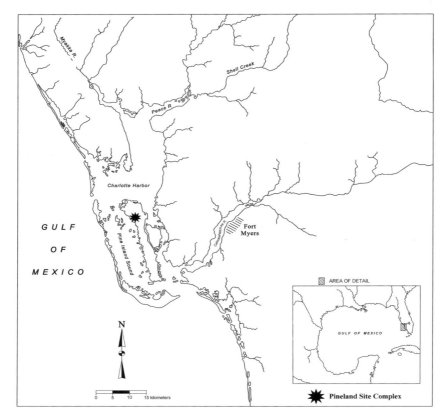

Figure 5-1. Southwest Florida, showing location of Pineland Site Complex. Modified from map by Corbett Torrence, courtesy University of Florida, Institute of Archaeology and Paleoenvironmental Studies.

than presumed in earlier centuries. In this case study, soil morphology as well as chemical and physical soil characteristics testify to the complex and labor-intensive construction history of site. In particular, the study confirms the local origin of the soils used to construct Smith Mound and indicates that the human burial associated with the mound was interred many years before the mound itself was constructed.

THE PINELAND SITE COMPLEX AND SAND BURIAL MOUND

The Pineland Site Complex was once home to Florida's Calusa Indians, a populous, sedentary society that was politically dominant over most of the southern half of the Florida peninsula (Figure 5-1; Luer 1991:71; Marquardt 1992a:19). The earliest dates for the site are approximately A.D. 100, placing its

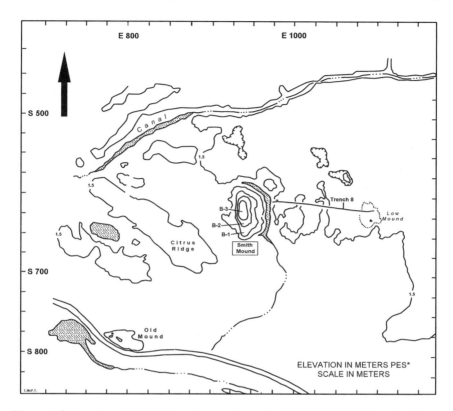

Figure 5-2. Location of Smith Mound excavation units. * PES 5 Pineland Elevation System (Walker 2007). Feature east of Smith Mound is a canal. Modified from map by Corbett Torrence, courtesy University of Florida, Institute of Archaeology and Paleoenvironmental Studies.

genesis in the Caloosahatchee I period, which followed the Terminal Archaic period in Florida (Marquardt 1992b:12). The Late/Terminal Archaic period in Florida saw the earliest manufacture of pottery, increased cultural complexity, and cultural regionalization (Milanich 1994:85–87; 101–104).

One of the central architectural features of Pineland is the sand burial mound known as Smith Mound (8LL36). Smith Mound, which was originally 50 to 60 m in diameter and 6 to 7 m high, was ascended by a spiral causeway, and encircled by an artificial canal (Figure 5-2). The western face of the mound was bulldozed for fill dirt in modern times. In this study, the arrangement and physical and chemical characteristics of the soil horizons found within the mound are compared with the structure and composition of the surrounding soils to evaluate the extent of human manipulation of the local landscape and soils associated with the construction of the mound. Understanding the internal structure of Smith Mound enables a reconstruction of its evolution from original land surface to the monumental architectural feature it became.

GEOLOGY, GEOMORPHOLOGY, SOILS, AND VEGETATION OF PINE ISLAND

The southwest coast of Florida near Pine Island is underlain by the Anastasia Formation, a sandy/shelly limestone of Pleistocene age (Puri and Vernon 1964; White 1970). This limestone bedrock is mantled by Pleistocene and Holocene sands, clays, and marls. Under terrestrial conditions, these sediments weathered into soils that are generally grouped into a series that lies parallel to the coast and reflects local conditions of relief, climate, and parent material. Under marine and nearshore conditions, the sediments have been sorted, worked, drifted into sand bars, washed up as barrier islands and beach ridges, and blown into dunes (Leatherman 1988; Missimer 1973; Widmer 1988).

The barrier islands that presently protect both Pine Island and the southwest Florida mainland, i.e., Sanibel, Captiva, Cayo Costa, and Gasparilla, are the most recent result of the constant rearrangement of nearshore sediments. Missimer (1973:94–95) proposed a Middle Holocene date for their origin. Pine Island itself is thought to be a beach ridge that formed during the Sangamon period (Widmer 1988:181). The presence on Pine Island of well-developed Spodosols (soils with a dark, aluminum-rich subhorizon), unusual in island settings (Henderson 1984), lends credence to the antiquity of the island.

The soils of Pine Island occur in bands parallel to the coast (Henderson 1984:Map Sheet 18). They vary primarily in their depth from ground surface to the upper boundary of the dark spodic (carbon- and aluminum-containing) subhorizon. This is a measure of the depth of the "sand pile" of original sediments in which the soils formed. They ascend from near the shoreline up to the sandy spine of the island, becoming deeper and better-drained, and then descend in opposite order down the landward side of the island. This order is not absolute, but it maintains the coast-parallel orientation that reflects the subtle ridge-and-swale topography created by the coastal processes that have shaped and reshaped the island through time.

Pine Island lies within the Gulf Coastal Lowlands physiographic region of Florida, the Southwest Flatwoods District defined by Brooks (1981). This is a virtually flat landscape underlain primarily by poorly drained "pine flatwoods" Spodosols (Henderson 1984) and somewhat better drained, deeper sandy soils. The native plant community is a tough, scrubby vegetation featuring saw palmetto (*Serenoa repens*), cabbage palm (*Sabal palmetto*), pines (*Pinus* spp.), and a rich variety of native grasses (Scarry and Newsom 1992). The only topographic relief is provided by remnants of ancient dunes and beach ridges.

MATERIALS AND METHODS

Soil morphology and chemical and physical soil characteristics of Smith Mound were analyzed and compared with non-mound soils in order to interpret the internal structure of the mound and its construction phases (Figure 5-2).

Samples were taken in 1992 from profiles of three units (B-1, B-2, B-3) excavated on the east flank of Smith Mound and from below the floor of the excavation units, the deepest of which was 2 m. Sampling did not reach the base of the mound. Off-mound samples were taken near Trench 8, excavated between Smith Mound and another small midden/mound known as Low Mound (8LL1612).

Approximately 350 g of soil were collected from each location. Complete morphologic descriptions were made for each unit profile sampled, based on the U.S. Soil Conservation Service format (Soil Survey Staff 1992). These descriptions include horizon presence, thickness, and order of occurrence; color and mottling; boundary thickness and topography; approximate texture; structure; and the presence and size of roots and other inclusions such as shells, plant material, and potsherds.

The soils were studied at the Florida Museum of Natural History (FLMNH). Two sets of 50 g subsamples were removed from bulk samples. One set was sent to the University of Florida Institute of Food and Agricultural Sciences Analytical Research Laboratory for determination of organic carbon (OC), pH, and extractable phosphorus, calcium, magnesium, zinc, and copper. There they were studied following U.S. Soil Conservation Service protocols (Soil Survey Staff 1992). A second set of samples was submitted to the University of Florida Environmental Pedology and Land Use Laboratory, Soil and Water Science Department, for particle-size distribution analysis (PSDA) using the pipette method of Day (1965:548–567). Aluminum (Al) content of Smith Mound samples was determined using a single-acid Mehlich-1 extraction (Soil Survey Staff 1992). Total phosphorus (P) content of soils from two of the three Smith Mound units (B-2 and B-3) was determined by the alkaline oxidation method of Dick and Tabatabai (1977).

The human burial was a primary burial of an adult (Hutchinson 2007). The skeleton was intact, lying in a flexed position. Cause of death is not known. The individual was purposefully placed in a burial pit but was not accompanied by burial goods.

RESULTS

Based on sub-surface horizon sequences, soil colors, and patterns of accumulation of organic carbon, aluminum, and calcium, three complete soils were identified. One is the original, natural soil and two are of human origin. A human burial intruded into the subhorizons of the deepest soil, which re-formed its disturbed surface horizon before being covered by a mound of sediments. The high aluminum content of the soil that subsequently formed in those sediments indicates they were spoil from local canal- and lake-excavation activities and were drawn from the low-lying, aluminum enriched subsoil. This provides clear evidence that monumental earthworks were constructed by

this non-farming society. It also indicates that the burial was interred before
the mound was constructed and that mound construction over the burial
occurred some years thereafter.

Soil Morphological Descriptions

The following soil morphological terms were used (Soil Survey Staff 1975).
Horizon designations, in normal vertical sequence, include: A (surface horizon), E (leached subhorizon), B (subsurface horizon of accumulation or
increased color), and C (unweathered parent material). The descriptor "w"
attached to some of the B horizons (e.g., Bw), indicates in situ color development with no indication of any colored materials leaching into the horizon
from above. The descriptor "h" refers to B horizons high in organic carbon
(humus) and aluminum. The use of the prime (') indicates identical horizons
in sequence below the original named horizon. For example, an A horizon is
at the soil's surface. An A' horizon is a former surface horizon buried under a
newer soil. An A" would be buried below both the A and A' sequences. Three
specific soil types are described: Spodosol, a poorly-drained soil with a characteristic dark aluminum- and carbon-containing subhorizon; Inceptisol, a soil
with weak or poorly developed horizons; and Spodic Quartzipsamment, a
sandy soil with dark subhorizon but not enough carbon or aluminum to be
considered a Spodosol.

The Smith Mound soil profiles reveal a sequence of buried soils capped in
some areas by an Inceptisol forming in disturbed sediments thrown up by
recent looters' activities (Figure 5-3). Profiles in excavation units B-1 and B-2
are typical of a Spodic Quartzipsamment: a salt and pepper A horizon and a
white E horizon leaching downward in finger-like projections into the light
brown Bw horizon below. Auger samples below the floors of these units
revealed an E' horizon under the Bw and above the reddish Bw' at the water
table. Unit B-3, uppermost on the flank of the mound, was excavated to the
level of a human burial at 2.18 m below the surface (Figure 5-3). This unit
yielded traces of at least two soils buried in sequence below the Inceptisol
which had formed in the upper 24 cm of spoil. The first buried soil, another
Inceptisol, (the "second soil" of Figure 5-3) was characterized by a strong A'
horizon underlain by a Bw' that gradually faded to a thick Bw' C horizon.
Below this, the second buried soil ("third soil" of Figure 5-3) exhibited an
intact A" horizon overlying thin E' and Bw" horizons that were contiguous
with the Spodic Quartzipsamment of units B-1 and B-2. A second sequence of
subhorizons followed: an E" horizon and subadjacent Bw''' which contained
the human burial. The profile shows the burial intruding into the Bw''' and the
obliterated boundaries of the disturbed E' and Bw" horizons of this soil, sealed
over by the pale gray A".

Auger samples taken from a small window cut into the floor of unit B-3
revealed yet a fourth set of horizons that was either another sequence of the

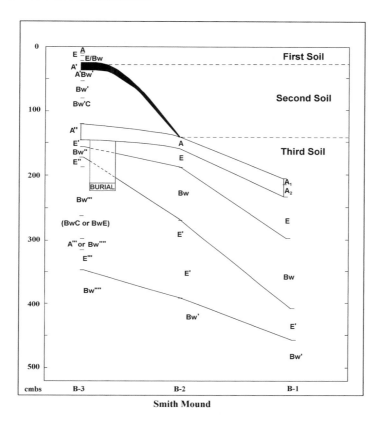

Figure 5-3. Profile of Smith Mound showing stacked soils and soil morphology from each of the three units, B-1, B-2, B-3; cmbs is centimeters below surface. Diagram by Sue Ellen Hunter, courtesy University of Florida, Institute of Archaeology and Paleoenvironmental Studies.

Spodic Quartzipsamment, or a fourth soil beneath it. Due to the near-saturated condition of the soil at this depth (nearly 4 m below surface), the ambiguous chemical nature of the dark horizons in this sequence prevented any further interpretation. However, the topographies of these and all the overlying buried horizon boundaries, except for the intrusive interruption caused by the human burial, were smooth, parallel, and angled toward the top of the mound, as if following a former hill-slope surface.

Physical Analysis

Particle-size distribution in all of the Pineland soil samples is dominated by the fine sand fraction (0.10 to 0.25 mm [Boggs 1987]). Samples from Smith Mound units B-1, B-2, and B-3 all contained 82% or more fine sand (Table 5-1). Very fine sand (0.05 to 0.10 mm) was the second most common fraction, with a

Table 5-1. Smith Mound Particle-size Distribution Analysis, Selected Samples

| Unit/test | Horizon | Depth (cm) | Sand fractions (% by weight) | | | | Silt (wt %) |
			Coarse	Medium	Fine	Very fine	
B-1	A1	0–12	0.2	4.7	82.4	3.8	8.9
	E	22–40	0.3	4.6	90.2	4.4	0.4
	Bw	40–100	0.3	4.2	89.4	5.7	0.4
	Bw (or C)	100	0.1	3.3	90.0	4.6	2.1
	(C or) E'	170	0.2	3.2	88.2	5.3	3.1
	Bw'	220	0.3	2.0	86.8	7.7	3.2
B-2	A	0–23	0.1	2.4	85.6	6.2	5.6
	E	23–41	0.1	2.1	89.5	5.3	2.9
	Bw	41–70	0.2	2.6	88.3	6.3	2.5
	E'1	70–150	2.2	3.2	88.9	5.3	0.4
	E'2	222	1.5	4.5	87.9	5.1	1.2
	Bw'1	272	0.2	3.1	90.2	4.7	1.9
	Bw'2	332	0.2	2.1	90.9	5.6	1.1
B-3	A	0–10	–	–	–	–	–
	E	10–15	0.1	3.8	88.2	5.3	2.6
	A'	24–36	0.7	4.0	87.2	5.3	2.8
	Bw'	53–80	0.2	3.9	88.0	5.9	2.0
	A"	120	1.4	4.1	86.7	5.7	2.2
	Bw"	170	0.4	–	88.8	4.3	–
	Bw'''	220	1.1	4.1	85.2	5.2	4.6
	A''' (or Bw'''')	300	0.3	3.9	88.2	4.4	3.5
	Bw'''''?	347–390	1.4	4.6	84.7	5.6	3.9

range of 3.2% to 7.7%. Medium sand (0.25 to 0.5 mm) ranged from 2% to 6.7%. None of the sand size classes showed any patterns of regular accumulation or depletion with depth. Silt (0.05 to 0.002 mm) was highest in the A horizons of units B-1 and B-2 (8.9% and 5.6%, respectively). Higher silt contents in B-3 generally correlated with A, buried A, and Bw horizons. The Smith Mound soils contained only traces of clay (defined as particles smaller than 0.002 mm).

Chemical Analyses

Smith Mound units B-1 and B-3 each had accumulations of 1.7% organic carbon (OC) in the A horizon (Table 5-2). Square B-2 contained approximately 1% OC in the A horizon. Carbon content of all subsurface horizons was less than 1%, with slight increases in some A', Bw, and Bw' horizons.

The surface horizon samples from Smith Mound contained abundant calcium (Ca), between 450 and 970 mg/kg from the three units. Contents of all other tested elements, with the exception of extractable phosphorus (P), also were high in the A horizon (Table 5-2). The concentration of most of these elements decreased abruptly in the subadjacent E horizon, then increased again in the lowest Bw horizon of each unit. Extractable P content of the surface horizon of Smith Mound was comparable to that of the native soil samples.

Table 5-2. Smith Mound Soil Chemical Analyses, Selected Samples

Unit/test	Horizon	Depth (cm)	pH	% Organic carbon	mg/kg Calcium	Aluminum	Al:OC
B-1	A1	0–12	6.6	1.70	970.0	19.4	11.4
	E	22–40	5.1	0.20	44.5	11.8	59
	Bw	40–100	5.0	0.25	44.5	109.0	436
	(C or) E'	130	4.6	0.20	27.5	73.8	369
	Bw'	220	6.5	0.40	527.0	433.0	1083
B-2	A	0–23	5.5	1.05	450.0	47.2	44
	E	23–41	4.9	0.40	81.9	30.5	76
	Bw	41–70	5.1	0.30	44.7	118.0	393
	E'1	70–150	5.4	0.05	132.0	80.3	1606
	E'2	222	502.0	0.15	45.5	52.0	347
	Bw'1	272	6.3	0.15	296.0	565.0	3766
	Bw'2	332	604.0	0.15	312.0	358.0	2386
B-3	A	0–10	5.1	1.70	508.0	20.2	11.9
	E	10–15	5.5	0.25	137.0	18.2	73
	A'	24–36	5.7	0.30	144.0	111.0	370
	Bw'	53–80	5.4	0.25	98.9	170.0	680
	A"	120	5.9	0.10	95.3	93.1	931
	Bw"	170	–	–	–	–	–
	Bw'''	220	6.0	0.05	57.3	39.9	798
	A''' (or Bw'''')	300	5.8	0.10	99.7	68.8	688
	Bw'''''?	347–390	6.6	0.15	313.0	427.0	2847

Because aluminum content of Smith Mound samples was determined using a single-acid Mehlich-1 extraction, rather than the standard citrate-dithionate method used for other areas in the site, only relative increases and decreases in Al content and Al-to-OC ratios are reported. Surface horizon samples contained more Al than the subadjacent E horizon but less than Bw horizons. In unit B-3, Al content followed those trends: the multiple A horizons had higher Al contents than their subadjacent E horizons but less than the Bw and transitional Bw horizons below them.

The Al:OC ratio was calculated for selected samples. Since virtually all samples had OC contents less than 1%, this ratio is an indication of Al accumulation with depth. In unit B-3, the Al:OC ratio increased with depth through the first two identified soils, as it did in the soils surrounding the mound (Scudder 2007). In the burial-containing third soil, the ratio decreased from the A" horizon through all subadjacent horizons until the deepest Bw at the water table was reached. In that water-saturated horizon, the ratio increased abruptly.

The pH of the Smith Mound soils was similar to that of the surrounding soils and adjacent low mound that were trenched and tested in the i mmediate vicinity (Scudder 2007). The total pH range for the three units was 4.6 to 6.6, with mean values of 5.1, 5.4, and 5.7 for B-1, B-2, and B-3, respectively. In contrast, soils of a sand ridge with midden layers (Citrus Grove Ridge) to the west of Smith Mound were more alkaline, with a pH range of 5.0 to 7.7 (Scudder 2007).

DISCUSSION

Although no rounded remnant sand hills are present near Pineland today, the subhorizon boundaries in the three Smith Mound test units are roughly parallel and angled upward, indicating that they conformed to a preconstruction convex surface as they formed. The mound is comprised of at least three intact soils, one on top of the other. From the surface downward they are: (1) an Inceptisol forming in the disturbed surface spoil; (2) a second Inceptisol with weak Bw horizon development and no leached E horizon, capped by a dark gray A horizon; and (3) a buried Spodic Quartzipsamment with an intact A horizon underlain by an E'-Bw"and E"-Bw'" sequence containing an intrusive human burial. This is a preliminary description leading to a later conclusion.

The clear break in the Bw" boundary created when the human individual was buried is indisputable evidence that the Spodic Quartzipsamment was an intact soil when the burial occurred. If the burial had been covered by fresh sediments on top of the original soil instead of being inserted into the original soil, complete soil horizons would have eventually formed in those sediments, with the body as an in situ inclusion surrounded by natural strata. There would be no broken boundaries or disturbances such as those found during excavation. Thus, the individual was interred below the original surface before mound construction began.

The burial has been ^{14}C dated to A.D. 1020–1170 (calibrated) (uncalibrated 830 6 60; Beta-72995; bone apatite; d^{13}C 5 –17.5), so the thin but continuous gray surface horizon of the Quartzipsamment had a maximum of approximately 800 years to re-form after the burial event. That is an over-estimate of the time that soil was left exposed, because the second Inceptisol that developed in sediments mounded over it would have taken some part of the 800 years to form. It is beyond the scope of this study to determine what portion of the 800 years was devoted to redevelopment of the A" horizon of the Quartzipsamment and what portion to the genesis of the overlying Inceptisol. However, the very thin and weak A" horizon re-sealing the burial formed in a region with abundant organic plant remains on the soil surface. This suggests that minimal time was allowed for the development of that horizon. Intriguing questions of human decision-making or intent in the timing of mound-building over the burial soil remain unanswered by this study: Were the mound and underlying burial simply coincidentally in the same place and un-related or was the mound a belated memorial constructed many decades after the individual died?

Four additional sets of samples corroborate a buried A or darker horizon at approximately the same depth below surface as the former surface (A") of the Spodic Quartzipsamment in unit B-3. Three areas on the upper northwest slope of Smith Mound were augered and examined in 1992. No samples were retained from those tests, but field notes recorded color changes from yellowish or reddish brown to gray at 100 to 140 centimeters below surface (cmbs). In addition, notes for unit A-1, excavated at the crest of Smith Mound in 1990, indicate that sand color changed from tan to "light to dark gray" at 140 cm below the modern surface. The depth of the A" in all of these samples indicates that from 1 to 1.4 m of sediments were piled onto the top of the Spodic Quartzipsamment, forming a cap that covered the evolving mound down to the elevation of unit B-2, where the Quartzipsamment emerged as the surface soil.

The chemical characteristics of the two buried soils encountered in unit B-3 differ considerably (Table 5-2). There is approximately twice as much organic carbon, calcium, magnesium, potassium, phosphorus, and aluminum in the Bw' horizon of the second Inceptisol as there is in A" horizon of the underlying Spodic Quartzipsamment. This is the reverse of the normal chemical weathering sequence of a *single* soil in a humid tropical climate, in which soluble elements (e.g., calcium, magnesium) leach downward in the soil column and accumulate in lower horizons. The relative enrichment of the Inceptisol suggests that the sediments heaped upon the Quartzipsamment were derived from the surrounding low-lying areas. Spoil from the lake or canal construction that characterized the site, particularly as such excavation cut into the deep and chemically-enriched Bw horizon at the water table, was a likely source of these sediments. The dominance of fine sand in all the soils of Smith Mound and its vicinity is additional support for a local source for the mounded sediments.

Additional augering in unit B-3 revealed the presence of yet another set of soil horizons below (or continuing) the burial soil, beginning at about 297 cmbs (Figure 5-3). Difficulties interpreting augered soil samples near the zone of water saturation makes it unclear whether those lowest horizons are another entire soil, beginning with a buried surface horizon (A'''-E'''-Bw'''') or whether they are a third sequence of E and Bw horizons of the overlying Quartzipsamment. Analysis of phytoliths from the lowest dark horizon would help to answer this question directly. There should be an accumulation of plant phytoliths in the dark horizon if it was a former land surface and few phytoliths if it was not (Pearsall 2000:406).

Whatever the origin and genesis of the last set of horizons, the chemical and physical data, along with photographic documentation, clearly indicate that the Quartzipsamment was an intact soil when disturbed by the intrusive human burial. It remained exposed long enough for a thin A horizon to reform and was then covered by sediments that developed into an Inceptisol. The enriched chemical nature of the overlying Inceptisol, its topographic position, and the absence of sediments of like thickness and morphology anywhere else on the site, indicates that the sediments were derived from the immediate vicinity, probably taken from the canal surrounding the mound or from other near-by Calusa excavation projects.

CONCLUSION

Understanding the internal structure of Smith Mound enables a reconstruction of its evolution from original land surface to the monumental architectural feature it became. In what way was the original landscape altered by the construction of Smith Mound? The scope and complexity of human modification of the original Pineland landscape could not have been accomplished without the focused intent and labor of a large segment of the resident population. Similar requirements for labor and resource management were experienced at other archaeological sites elsewhere in the southeastern United States at the same time. But that it was accomplished by a non-farming population relying on local wild resources is important considering the continuing debate over the degree to which fishing communities could develop or sustain culturally-complex societies. This also provides a cautionary tale: the water features of Pineland, superficially similar to such excavations at other sites, were meant for transport and ritual, not irrigation of crops. And the building materials for the monumental burial mound that honored the dead had humble beginnings as spoil from a nearby ditch.

ACKNOWLEDGMENTS

I thank Drs. William Marquardt and Karen Walker, Florida Museum of Natural History, for extending the invitation to work at Pineland, funding the field work and laboratory analyses, and permission to use modified versions of Figures 5-1, 5-2, and 5-3. I appreciate the information provided by Howard Yamataki, of the U.S. Soil Conservation Service. Studied and unstudied materials are available as voucher specimens and for further study by contacting the Florida Museum of Natural History Environmental Archaeology Laboratory.

REFERENCES

Boggs, S., Jr., 1987, *Principles of Sedimentology and Stratigraphy*, Merrill, Columbus, Ohio.
Brooks, H. K., 1981, Map of the Physiographic Divisions of the State of Florida, Map on File, Center for Environmental and Natural Resources, Institute of Food and Agricultural Sciences, University of Florida, Gainesville.
Cushing, F. H., 1896, Exploration of Ancient Key Dweller Remains on the Gulf Coast of Florida, *American Philosophical Society Proceedings* 35:329–448.
Day, P. R., 1965, Particle Fractionation and Particle-size Analysis, in: *Methods of Soil Analysis, Part 1, Agronomy 9* (C. A. Black, ed.), American Society of Agronomy, Madison, U.S. Government Printing Office, Washington, DC, pp. 548–567.
Dick, W., and Tabatabai, M. A., 1977, An Alkaline Oxidation Method for Determination of Total Phosphorus in Soils, *Soil Science Society of America Journal* 41:511–514.
Henderson, W. G., Jr., 1984, *Soil Survey of Lee County, FL*, U.S.D.A. Soil Conservation Service, Tallahassee, Florida.

Hutchinson, D. L., 2007, Precolumbian Human Skeletal Remains from the Pineland Site Complex, in: *The Archaeology of Pineland: A Coastal Southwest Florida Village Complex*, A.D. 100–1600 (W. H. Marquardt, ed.), *University of Florida Institute of Archaeology and Paleoenvironmental Studies Monograph 4*, Gainesville, in preparation.

Leatherman, S. P., 1988, *Barrier Island Handbook*, Coastal Publication Series, Laboratory for Coastal Research, University of Maryland, College Park.

Luer, G. M., 1989, Calusa Canals in Southwestern Florida: Routes of Tribute and Exchange, *Florida Anthropologist* 42(2):89–130.

Luer, G. M., 1991, Historic Resources at the Pineland Site, Lee County, FL, *Florida Anthropologist* 44(1):59–76.

Marquardt, W. H., 1987, The Calusa Social Formation in Protohistoric South Florida, in: *Power Relations and State Formation* (T. C. Patterson and C. W. Gailey, eds.), Archaeology Section, American Anthropological Association, Washington, DC, pp. 98–116.

Marquardt, W. H., 1988, Politics and Production Among the Calusa of South Florida, in: *Hunters and Gatherers,* Volume 1, *History, Evolution, and Social Change* (T. Ingold, D. Riches and J. Woodburn, eds.), Berg Publishers, London, pp. 161–188.

Marquardt, W. H., 1992a, Calusa Culture and Environment: What Have We Learned?, in: *Cultural and Environment in the Domain of the Calusa* (W. H. Marquardt, ed.), *University of Florida Institute of Archaeology and Paleoenvironmental Studies Monograph 1*, Gainesville, pp. 423–436.

Marquardt, W. H., 1992b, Recent Archaeological and Paleoenvironmental Investigations in Southwest Florida, in: *Cultural and Environment in the Domain of the Calusa* (W. H. Marquardt, ed.), *University of Florida Institute of Archaeology and Paleoenvironmental Studies Monograph 1*, Gainesville, pp. 9–57.

Milanich, J. T., 1994, *Archaeology of Precolumbian Florida*, University Press of Florida, Gainesville.

Missimer, T., 1973, *Growth Rates of Beach Ridges on Sanibel Island, Florida*, M.S. thesis, Department of Geology, Florida State University, Tallahassee.

Pearsall, D. M., 2000, *Paleoethnobotany: A Handbook of Procedures*, 2nd ed., Academic Press, San Diego.

Puri, H. S., and Vernon, R. O., 1964, *Summary of the Geology of Florida and Guidebook to the Classic Exposures*, Florida Geologic Survey Special Publication 5, Tallahassee.

Scarry, C. M., and Newsom, L. A., 1992, Archaeobotanical Research in the Calusa Heartland, in: *Cultural and Environment in the Domain of the Calusa* (W. H. Marquardt, ed.), *University of Florida Institute of Archaeology and Paleoenvironmental Studies Monograph 1*, Gainesville, pp. 375–402.

Scudder, S. J., 2007, Soils and Landscapes: Archaeopedology at the Pineland Site, in: *The Archaeology of Pineland: A Coastal Southwest Florida Village Complex*, A.D. 100–1600 (W. H. Marquardt, ed.), *University of Florida Institute of Archaeology and Paleoenvironmental Studies Monograph 4*, Gainesville, in preparation.

Soil Survey Staff, 1975, *Soil Taxonomy*, U.S.D.A. Soil Conservation Service, *Department of Agriculture Handbook 436*, U.S. Government Printing Office, Washington, DC.

Soil Survey Staff, 1992, Methods Manual, *Soil Survey Investigative Report* 42,Version 2.0, Lincoln, Nebraska, pp. 157–159.

Walker, K. J., 1992, The Zooarchaeology of Charlotte Harbor's Prehistoric Maritime Adaptation: Spatial and Temporal Perspectives, in: *Cultural and Environment in the Domain of the Calusa* (W. H. Marquardt, ed.), *University of Florida Institute of Archaeology and Paleoenvironmental Studies Monograph 1*, Gainesville, pp. 265–366.

Walker, K. J., 2007, Excavations at the Pineland Site Complex, in: *The Archaeology of Pineland: A Coastal Southwest Florida Village Complex*, A.D. 100–1600 (W. H. Marquardt, ed.), *University of Florida Institute of Archaeology and Paleoenvironmental Studies Monograph 4*, Gainesville, in preparation.

White, W. A., 1970, *The Geomorphology of the Florida Peninsula. Bureau of Geology, Florida Department of Natural Resources Geological Bulletin 51*, Tallahassee.

Widmer, R. J., 1988, *Evolution of the Calusa – A Non-Agricultural Chiefdom on the Southwest Florida Coast*, University of Alabama Press, Tuscaloosa.

Chapter 6

Archaeozoology, Art, Documents, and the Life Assemblage

RICHARD COOKE, MÁXIMO JIMÉNEZ,
AND ANTHONY J. RANERE

Classic books evaluating human attitudes towards animals (e.g., Berlin 1992; Simoons 1994) make it clear that the ways communities perceive, acquire, use, and distribute animals and their products are unpredictable in space and time. Ethnologists find it difficult to unravel such complexity with informants and tape recorders. It presents even more intractable challenges for archaeozoologists. Does the absence or rarity of animals in an archaeological context mean that they were not readily available, that people did not want to use them, did not bother because other activities were more important, were not allowed to, or did not have the technological capacity to so do? How were animals used by different social groups within larger communities? What are the chances that the species present in an archaeological assemblage were not components of an accessible "life assemblage," a community of live animals in its "natural" proportions (Klein and Cruz-Uribe 1984:1), but were obtained from distant places through trade, exchange, or trekking?

These questions must be addressed to understand how humans, animals, and plants interacted in former environments. A useful methodology is to compare archaeozoological, pictorial, and documentary information about animal use. This is feasible for Parita Bay on the central Pacific coast of Panama (Figures 6-1 and 6-2) where well-preserved archaeozoological remains are found in rubbish dumps, dwellings, workshops, and human burials at several

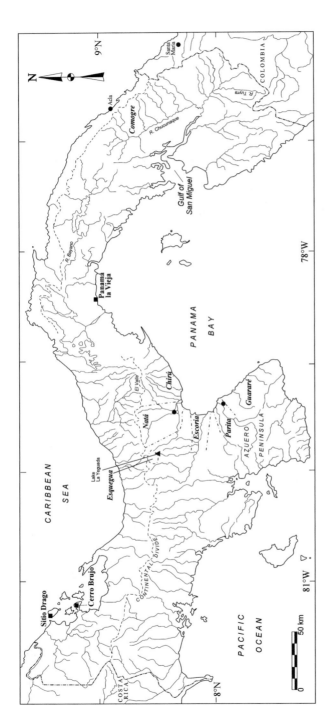

Figure 6-1. Map of Panama showing the approximate location of chiefdoms circa A.D. 1510–1520 (Cooke and Sánchez-Herrera 2004a; Sauer 1966: 221).

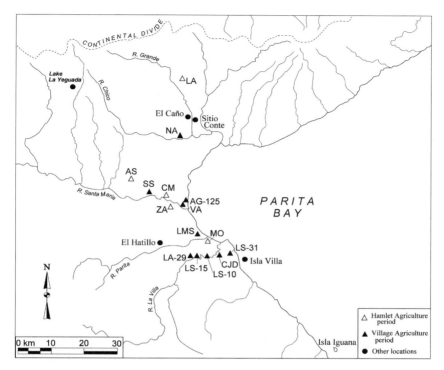

Figure 6-2. Map of central Panama showing the Parita Bay archaeological sites for which vertebrate remains are discussed. Sites belonging to the Hamlet Agriculture period include: AS: Aguadulce Shelter; cm: Cerro Mangote; LA: Cueva de los Ladrones; ZA: Zapotal; and MO: Monagrillo. Sites belonging to the Village Agriculture period include: SS: Sitio Sierra; NA: Natá; LMS: La Mula-Sarigua; CJD: Cerro Juan Díaz; AG-125; VA: Vampiros 1; LA-29; LS-10; LS-15; and LS-31. Open triangles indicate Hamlet Agriculture sites and closed triangles signify Village Agriculture period sites.

sites occupied from about 5000 B.C. to A.D. 1520. These remains provide insights into the identities of the animals used for tools, ornaments and food (Briggs 1989; Cooke 1998, 2004b; Cooke and Ranere 1992a; Lothrop 1937; Mayo 2004). This helps us identify which animals were from the local environment and which were non-local.

After about 200 B.C., the inhabitants of Parita Bay decorated artifacts in the Gran Coclé art style emphasizing animal images. Some images are fantastic or humanized creatures whose meanings defy interpretation. Others are realistic enough that they can be associated with extant animals, sometimes even species. These images provide clues to which animals were cognitively important to people who shared this art style. They also document knowledge of animals whose remains are not found in archaeofaunal collections (Cooke 1998, 2004a, b; Helms 1995, 2000; Linares 1977; Lothrop 1942).

The invading Spaniards described the landscape and animals of Parita Bay (A.D. 1515–1530). Their descriptions were selective, of course, and not necessarily as accurate as modern researchers would wish. Sometimes their accounts exaggerate the numbers, ferocity, or oddness of animals. Despite such biases, the documentary record complements details obtained from archaeological animal remains and art history, thereby improving our knowledge about how the native population did or did not use the local or regional animal "life assemblage."

THE GEOGRAPHIC SETTING

Parita Bay is a small, mangrove-fringed embayment in the northeastern corner of the Azuero Peninsula (Figures 6-1 and 6-2). Precipitation and humidity are strongly seasonal in this tropical setting. During the wet season (May–December) the landscape is green and lush. Fresh water is provided by numerous streams and rivers which rise in the cordilleras and meander towards the bay across narrow flood plains. The frequently-flooding rivers create ephemeral swamps where aquatic animals congregate. The high tidal range (3–5 m) and heavy sediment discharge from rivers, accentuated by several millennia of forest clearance and soil erosion, create a mosaic of coastal habitats (e.g., sandy beaches; mud flats; mangroves; tidal channels; and high tidal flats). During the sunny dry season (January–April) trade winds desiccate the landscape. The water table falls dramatically, streams (but not the largest rivers) dry out, and cut vegetation burns quickly and effectively. In April, the landscape is brown and parched.

BRIEF SUMMARY OF CULTURAL HISTORY
IN AN ENVIRONMENTAL CONTEXT

Early Peoples

Paleoindians, who used fluted projectile points, were active around Parita Bay during the late Glacial period when the climate was less humid and cooler than today. A few stone tools refer to an earlier "Joboid" occupation. Paleoecological models suggest that the littoral vegetation was dry thorn-scrub at this time. The foothills and cordillera were under montane forest. After about 9100 B.C. human activities (especially burning) led to the gradual clearance of forests and proliferation of second-growth vegetation (Cooke et al. 1996; Cooke and Sánchez-Herrera 2004b; Pearson 2002, 2003; Pearson and Cooke 2002; Piperno and Pearsall 1998:78–79, 172–179, 286–293; Ranere and Cooke 2003).

Hamlet Agriculture Period (5000–500 B.C.)

Our earliest archaeofaunas belong to the Hamlet Agriculture period (5000–500 B.C.), which is divided into Late Preceramic (5000 and 2500 B.C.) and Early Ceramic (2500–500 B.C.) phases. Even though the human population was small, habitation sites and camps were scattered across central Panama from Parita Bay to the humid Caribbean slopes (Cooke and Ranere 1992b, c; Griggs 2005). Starch grain and phytolith data from Aguadulce Shelter, Cueva de los Ladrones, and Cerro Mangote indicate that arrowroot (*Maranta arundinacea*), bottle gourd (*Lagenaria siceraria*), maize (*Zea mays*), manioc (*Manihot esculenta*), gourd/squash (*Cucurbita* spp.), and sweet potatoes (*Ipomoea batatas*) were cultivated by the Late Preceramic phase. Another important activity was harvesting nuts from wine palm (*Acrocomia mexicana*) and American oil palm (*Elaeis oleifera*) (Figure 6-2; Dickau 2005; Piperno and Holst 1998; Piperno and Pearsall 1998:286–297; Piperno et al. 2000). Aguadulce and Ladrones are small rock-shelters that were never right on the coast, but were close enough to Parita Bay to grant their occupants access to coastal resources. Cerro Mangote is thought to be a cluster of dwellings occupied between 5000 and 3000 B.C. It was initially about 1.5 km from the coast; but it gradually acquired a more inland position as the coast moved outwards due to sedimentation and delta-formation. The Early Ceramic Monagrillo and Zapotal sites were also first occupied when they were on or near ancient strand lines (Clary et al. 1984; McGimsey 1956; Willey and McGimsey 1954).

Village Agriculture and Social Differentiation Period (200 B.C.–A.D. 1520)

During the subsequent Village Agriculture and Social Differentiation period, the Parita Bay landscape was considerably different from that of the Hamlet Agricultural period due to changes which began as early as 200 B.C.–A.D. 250. Settlements were much larger and more nucleated than during the preceding Hamlet Agriculture period. Archaeofaunal data are available for this period from AG-125, Cerro Juan Díaz, La Mula-Sarigua, LA-29, LS-10, LS-15, LS-31, Natá, Sitio Sierra, and Vampiros (Figure 6-2).

The nucleation process was accompanied by changes consistent with the social and ideological transformations of village life (Cooke 2005). Ceramic vessels were decorated with painted geometric designs and animal images of the Gran Coclé art style (Cooke 2004a; Helms 1995, 2000; Linares 1977). Polished axes, chisels, and adzes, as well as maize-grinding tools were produced in large numbers (Ranere and Cooke 1996). New technologies, such as metallurgy and shell-work for crafting costume and sumptuary ornaments, appeared (Cooke et al. 2003; Cooke and Sánchez-Herrera 1998; Mayo 2004).

These developments promoted social differences exemplified by increasing variability in the sizes, complexity, and functions of sites, as well as by the nature, number, and distribution of grave goods. Evidence for a three-tiered site hierarchy dominated by a single very large site is found in the River Parita valley. By 200 B.C. this site was La Mula-Sarigua and by A.D. 500 it was El Hatillo (Haller 2004; Hansell 1987, 1988; Ladd 1964). In the adjacent La Villa river valley, three very large (>100 ha) sites were evenly spread among smaller villages, hamlets, and activity centers along the alluviated bottomlands (Isaza-Aizprúa 2007). One of these primary villages was Cerro Juan Díaz, where people with special occupations (e.g., shamanism or curing) were buried (Cooke et al. 2000, 2003). In the Santa María river valley, large villages were spaced 2–3 km apart (Weiland 1984). Houses at Sitio Sierra were quite close together (Cooke 1984a; Isaza-Aizprúa 1993).

Differences in wealth are particularly apparent at Sitio Conte where a few adult males were buried about A.D. 700–950 with abundant mortuary accouterments (Briggs 1989; Lothrop 1937). Sitio Conte was a special mortuary zone within a much larger site complex, which includes El Caño and the village of Natá, where the chieftain of the same name lived in A.D. 1516 (Figure 6-1; Breece 1997). The suite of architectural features (i.e., earth mounds, stone-walled terraces, cobble stone pavements, and carved and un-carved stone columns) is unique in the Gran Coclé culture area. Rituals and ceremonies enacted at this site complex may have attracted people from several chiefdoms whose inhabitants recognized a common origin and heritage despite political rivalries (Cooke et al. 2000, 2003; Mayo 2007).

Troops sent westward from Spanish settlements at Santa María and Acla between A.D. 1515 and 1522 reported that savannas stretched from the Darién chiefdom of Comogre to the territory of chief Guararé (Figure 6-1; Andagoya in Jopling 1994:29; Sauer 1966:221). These troops subjugated four chiefdoms. Each chiefdom controlled sections of the Parita Bay coast (Figure 6-1). Archaeofaunal data from the territories of Natá, Escoria, and Parita are the focus of this study. Local vegetation consisted of woodlands on hill tops and along rivers and streams in addition to cultivated fields and grassy tracts. Well-trodden paths to the central cordillera passed through parklands virtually devoid of woody vegetation (Espinosa in Jopling 1994:55, 65, 67). These pre-Hispanic savannas lacked two modern features, cattle (Bos taurus) and introduced grasses such as jaragua (Hyparrhenia rufa). They had never been cleared with iron axes. Thus their floristic composition was quite different from that of present-day savannas.

Spanish soldiers described constant inter-chiefdom raiding (Cooke and Sánchez-Herrera 2004a). Such social unrest probably curtailed the routine procurement of animal products from beyond the boundaries of each chiefdom. Some non-subsistence items, however, were impossible or difficult to obtain without external contacts. For example, manatee bone (Trichechus manatus) used for exquisite carvings buried with high status people must have been acquired from the Caribbean side of the isthmus (Cooke 2004b). This suggests exchange of sumptuary and ritual items among elites.

FISH, DEER, AND IGUANAS

Spanish captains frequently received food offerings from local dignitaries. These invariably consisted of fishes, deer (probably white-tailed deer [*Odocoileus virginianus*]; Sauer [1966:273]), and iguanas (Iguanidae). For example, while Espinosa was encamped at chief Escoria's village, he demanded that crosses be placed on house roofs in obeisance to the new Spanish sovereign. Frightened people from outlying districts appeased him with "very fine victuals . . .deer meat as well as iguanas and fish" (in Jopling 1994:68). This trio (terrestrial, arboreal, aquatic) makes ecological sense in a human-modified littoral environment where the agricultural round influenced scheduling other subsistence activities, affording a high value to animal resources that were easiest or safest to obtain near villages and fields.

Fish and Other Aquatic Resources

Fresh and preserved fish and other aquatic resources contributed substantially to the diet throughout the 6,500 years of our study period (Figure 6-3). Tropical estuaries attract large shoals of small fishes and their attendant predators. Parita Bay's high tidal range and heavy wet-season sediment load create extensive mud-flats and tidal channels which facilitate the use of traps and weirs. Actualistic studies demonstrate that many of the 150 marine fish species identified in our archaeofaunal samples can be caught with such devices. Fishing was an in-shore and river-bank activity. A few species were obtained in marine waters seaward of the turbid middle estuary and also around rocky islets. Small boats probably were used to reach these deeper areas, but there is no evidence for long off-shore voyages. Additionally, some marine fishes travel upstream for considerable distances (Cooke 1992a; Cooke and Jiménez 2004; Cooke and Ranere 1999; Cooke and Tapia-Rodríguez 1994a, b; Jiménez and Cooke 2001). The availability of salt in the high tidal flats aided the preservation and inland transport of fish in an area where rivers inland of the tidal reaches have depauperate fish faunas. Espinosa (in Jopling 1994:57) praised the quality of the salt produced in "handsome and well-tended pans." When he camped at the chief village of Natá in A.D. 1516, people came from the coast with crabs and fishes to exchange for maize.

In addition to fish, Parita Bay estuaries provide abundant mollusks (Carvajal 1998; Hansell 1979), crabs, wading birds, and a few Neotropical mammal species. One of these mammals, the raccoon (*Procyon lotor*), outnumbers white-tailed deer in the collection from Hamlet Agricultural period Cerro Mangote (Cooke and Ranere 1992a:Table 2; Cooke et al. 2007:Table 4).

Two good candidates for specialized shoreline sites in the Santa María drainage are the Vampiros rock shelters and AG-125. They date to the Village

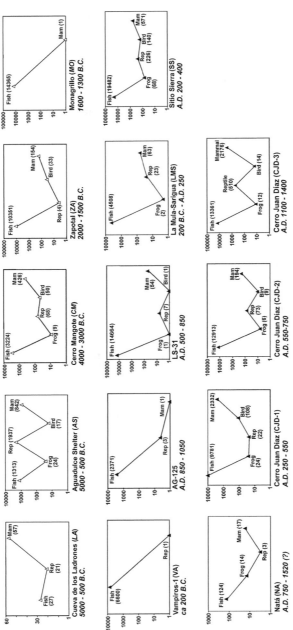

Figure 6-3. Number of Identified Specimens (NISP) from sites in the Parita Bay region. Samples are from single refuse features except those from the Aguadulce Shelter, which combines Late Preceramic and Early Ceramic features (Piperno et al. 2000; Ranere and Hansell 1978). Numbers in parentheses are NISP. Only specimens that could be confidently attributed to a vertebrate class were counted, i.e., fishes (marine and freshwater, including elasmobranchs), frogs (frogs and toads), reptiles (turtles, lizards, and snakes), birds, and mammals. Data are log-transformed except for Cueva de los Ladrones and Natá, which produced very small samples. Zapotal data are from Peres (2001).

Agriculture period and were in Escoria's chiefdom. If we use the most straightforward of proxies (numbers of vertebrate specimens [NISP] recovered by a 1/8-inch mesh), fish predominate in these collections (Figure 6-3). The two Vampiros shelters now are 2 km inland from the coast. Between about 200 B.C. and A.D. 250 they were on or much closer to the coast line (Cooke and Ranere 1984). The most intensive use of these sites was during the period when each was closest to the active marine shore. As the River Santa María delta prograded seawards, the Vampiros shelters were supplanted by another fishing station (AG-125), which occupied the shoreline position between about A.D. 800 and 1000 (Clary et al. 1984; Weiland 1984).

The Vampiros shelters and AG-125 contain remains of all the marine fish taxa recorded at Sitio Sierra located about 13 km upstream from them. Seventy percent of the fish species consumed at Sitio Sierra were of marine origin (Cooke and Jiménez 2004; Cooke and Ranere 1999). Some of these marine species could be taken from nearby freshwater sections of the tidal River Santa María (Cooke and Tapia-Rodríguez 1994a), but others could only be acquired seaward of the turbid water plume in Parita Bay. A straight-line distance of more than 15 km to and from Sitio Sierra and these fishing grounds doubles if one calculates the distance by river. If not salted and dried, fish transported over such distances, especially oily ones, would spoil quickly in the humid, insect-laden tropical environment. We suggest, therefore that much of the marine fish consumed at Sitio Sierra was caught, salted and transported by fisherfolk living at the Vampiros shelters and AG-125. Another possible provisioning site is LS-31, which was probably located briefly on a fossil beach ridge. Isaza-Aizprúa (2007) reported 30 genera and 42 of marine fish species at LS-10 about 8 km up-river. The remains of a few marine taxa occurred in middens at LS-15 and LA-29, villages located 12–14 km inland from the coast. Parita's chiefdom, then, exhibits an inland movement of marine resources similar to that in the neighboring territory of Escoria in the Santa María drainage.

We have not yet recovered the taphonomic details necessary for identifying the methods used for preserving fish; but Spanish documents show that marine salt was readily available in high tidal flats within each of the four Parita Bay chiefdoms, and that salt-fish was one of many items traded from the coast to the interior. Thus, it seems likely that the present-day practice of salting and sun-drying fish extended back to early pre-Hispanic times. Remains of small, shoaling estuarine fish were found in middens dated to the Hamlet Agriculture period (2500–500 B.C.) at Cueva de los Ladrones, located 25 km inland (Cooke and Jiménez 2004; Zohar and Cooke 1997).

Deer

Spaniards were astounded by the large amount of venison consumed in the Parita Bay chiefdoms. Espinosa's "three hundred salted carcasses of deer" stored at Natá might be the exaggeration of a soldier trying to please his general (in Jopling 1994:48). So might his claim that his crossbowmen brought in "10–13 deer a day!" But deer must have been abundant. The fact that deer were found in groups (Espinosa says in "30s and 40s") prompted Sauer (1966:273) to identify them as white-tailed deer rather than smaller, more solitary red brocket deer (*Mazama americana*). Deer images on pottery painted in the Gran Coclé style sport stylized branching antlers, indicating that the artist was identifying male white-tailed deer rather than red brocket deer (Figure 6-6a–e; Cooke 1992b:Figure 6; Helms 2000:33–53). Although it is imprudent to assign the many broken and immature deer specimens to either the white-tailed or brocket taxon, all antler fragments and all complete adult specimens in the Parita Bay collections are referable to white-tailed deer. It assumed for sake of argument that all deer remains are from that species.

Archaeozoology confirms that white-tailed deer were the most frequently taken large- to medium-sized mammal. Only at Hamlet Agricultural period Cerro Mangote was another species (the raccoon) taken more frequently (Table 6-1[Q3]; Cooke 1992b; Cooke and Ranere 1992a). The dominance of deer lessens, however, if Minimum Numbers of Individuals (MNI) is used as the proxy (Figure 6-4 Cooke 1992b). This discrepancy may be due to two characteristics: (1) deer remains have a higher survival potential than the remains of smaller animals in well-trampled settlements where dogs (*Canis familiaris*) were kept and (2) some deposits of animal remains at Cerro Mangote, Sitio Sierra, and Cerro Juan Díaz appear to be the result of short-lived events that led to the rapid burial of nearly complete deer skeletons. Such events may have been feasts. Perhaps the "300 deer" seen by Espinosa in chief Natá's storehouse were stored for this purpose.

Table 6-1. Numbers of Identified Specimens (NISP) from Eight Archaeological Sites around Parita Bay, Panama (includes all vertebrates, except fish)[a]

	1	2	3	4	Total
Frogs and Toads					
Giant toad (*Bufo marinus*)	20	410	606	6	1042
Thin-toed frog (*Leptodactylus insularum*)	24	34	131	5	194
Smokey jungle frog (*Leptodactylus pentadactylus*)	–	2	–	–	2
Veined frog (*Phrynohyas venulosa*)	–	2	7	–	9
Snakes					
Non-poisonous snakes (Colubridae)	12	60	32	9	113
Boas (Boidae)	–	–	5	7	12
Boa constrictor (*Boa constrictor*)	9	11	19	–	39
Poisonous snakes (Viperidae)	–	1	2	–	3
Lizards					
Lizard (*Ameiva* spp.)	7	58	99	6	170
Green iguana (*Iguana iguana*)	95	350	2224	5	2674
Black iguana (*Ctenosaura similis*)	33	221	3628	2	3884
Basilisk lizard (*Basiliscus basiliscus*)	5	5	14	–	24

(Continued)

Table 6-1. (*Continued*)

	1	2	3	4	Total
Crocodiles					
Crocodylidae	–	1	2	–	3
American crocodile (*Crocodylus acutus*)	2	–	2	1	5
Cayman (*Caiman cocodrilus*)	1	1	1	1	4
Sea turtles					
Cheloniidae	2	–	63	3	68
Green turtle (*Chelonia mydas*)	–	–	6	–	6
Hawksbill turtle (*Eretmochelys imbricata*)	3	–	–	–	3
Possible olive ridley turtle (cf. *Lepidochelys olivacea*)	1	–	–	–	1
Freshwater turtles					
Mud turtle (*Kinosternon* spp.)[b]	418	244	590	2	1254
Pond slider (*Trachemys scripta*)	92	97	439	1	629
Birds					
Magnificent frigate-bird (*Fregata magnificens*)	–	–	7	–	7
Brown pelican (*Pelecanus occidentalis*)	–	–	4	–	4
Olivaceous cormorant (*Phalacrocorax olivaceus*)	–	18	20	–	38
Booby (*Sula* spp.)	–	9	57	–	66
Great blue heron (*Ardea herodias*)	–	11	4	–	15
Green heron (*Butorides striatus*)	–	4	3	–	7
Egret (*Egretta* spp.)	3	2	5	–	10
American egret (*Egretta alba*)	–	33	81	–	114
Possible snowy egret (*Egretta* cf. *thula*)	–	1	1	–	2
Yellow-crowned night heron (*Nyctanassa violacea*)	–	1	9	–	10
Black-crowned night heron (*Nycticorax nycticorax*)	–	1	2	–	3
Possible fasciated tiger-heron (cf. *Tigrisoma fasciatum*)	–	1	1	–	2
Black vulture (*Coragyps atratus*)	3	4	14	–	21
Wood stork (*Mycteria americana*)	–	7	4	–	11
White ibis (*Eudocimus albus*)	19	1	18	–	38
Roseate spoonbill (*Ajaia ajaja*)	–	3	–	–	3
Possible pintail (cf. *Anas acuta*)	–	2	1	–	3
Shoveler (*Anas clypeata*)	–	1	–	–	1
Possible blue-winged teal (cf. *Anas discors*)	1	1	1	–	3
Muscovy duck (*Cairina moschata*)	–	19	27	–	46
Whistling-duck (*Dendrocygna* spp.)	–	12	1	–	13
White-faced whistling-duck (*Dendrocygna viduata*)	–	24	1	–	25
Black-bellied whistling-duck (*Dendrocygna autumnalis*)	–	2	2	–	4
Hawk (*Buteo* spp.)	1	1	1	1	4
Roadside hawk (*Buteo magnirostris*)	–	–	5	–	5
Black hawk (*Buteogallus* spp.)	–	1	2	–	3
Mangrove black hawk (*Buteogallus subtilis*)	–	3	2	–	5
Osprey (*Pandion haliaetus*)	–	–	10	–	10
Possible aplomado falcon (*Falco* cf. *femoralis*)	–	5	1	–	6
Yellow-headed caracara (*Milvago chimichima*)	–	1	1	–	2
Crested caracara (*Polyborus plancus*)	–	–	1	–	1
Great curassow (*Crax rubra*)	–	–	3	–	3
Crested guan (*Penelope purpurascens*)	–	16	6	–	22
Grey-headed chachalaca (*Ortalis cinereiceps*)	–	–	1	–	1
Quail (Phasianidae)	–	3	4	1	8
Bobwhite quail (*Colinus* spp.)	–	29	62	–	91

(*Continued*)

Table 6-1. (Continued)

	1	2	3	4	Total
Possible uniform crake (cf. *Amaurolimnas concolor*)	–	1	–	–	1
Sora (*Porzana carolina*)	–	1	–	–	1
Grey-necked wood-rail (*Aramides cajanea*)	–	–	11	–	11
Purple gallinule (*Porphyrula martinica*)	–	–	1	–	1
Possible wattled jacana (*Jacana* cf. *jacana*)	–	–	–	1	1
Possible plover (cf. *Charadrius* spp.)	–	–	1	–	1
Killdeer plover (*Charadrius vociferus*)	–	1	–	–	1
Black tern (*Chlidonias niger*)	–	1	2	–	3
Possible gull (cf. *Larus* sp.)	–	–	1	–	1
Possible tern (cf. *Sterna* spp.)	–	–	2	1	3
Willet (*Catoptrophorus semipalmatus*)	8	–	2	–	10
Dowitcher (*Limnodromus* sp.)	–	–	1	–	1
American curlew (*Numenius americanus*)	–	–	1	–	1
Whimbrel (*Numenius phaeopus*)	–	–	1	–	1
Sandpiper (*Tringa* sp.)	–	1	–	–	1
Possible greater yellowlegs (cf. *Tringa melanoleuca*)	2	–	1	–	3
Solitary sandpiper (*Tringa solitaria*)	–	–	1	1	2
Possible turnstone (cf. *Arenaria interpres*)	1	–	–	–	1
Small sandpiper (*Calidris* spp.)	2	–	–	–	2
Knot (*Calidris canutus*)	1	–	–	–	1
Pectoral sandpiper (*Calidris melanotus*)	–	–	1	–	1
Semipalmated/western sandpiper (*Calidris pusilla* or *mauri*)	2	–	–	–	2
Doves (Columbidae)	–	1	14	–	15
Ruddy quail-dove (*Geotrygon montana*)	4	3	1	–	8
White-tipped dove (*Leptotila* cf. *verreauxi*)	–	7	13	–	20
White-winged or mourning dove (*Zenaida* spp.)	–	12	19	–	31
White-winged dove (*Zenaida asiatica*)	–	–	3	–	3
Mourning dove (*Zenaida macroura*)	–	5	3	–	8
Pigeon (*Columba* spp.)	–	1	1	–	2
Possible pale-vented pigeon (*Columba* cf. *cayennensis*)	–	–	1	–	1
Possible scaled pigeon (*Columba* cf. *speciosa*)	–	–	2	–	2
Ground-dove (*Columbina* spp.)	–	3	12	–	15
Pale-breasted ground-dove (*Columbina minuta*)	–	1	–	–	1
Ruddy ground-dove (*Columbina talpacoti*)	–	2	4	–	6
Amazon parrot (*Amazona* spp.)	–	–	28	–	28
Red-lored amazon (*Amazona autumnalis*)	–	1	–	–	1
Yellow-headed amazon (*Amazona ochrocephala*)	–	1	1	–	2
Macaws (*Ara* spp.)	–	–	105	–	105
Parakeet (*Aratinga* spp.)	–	–	9	–	9
Crimson-fronted parakeet (*Aratinga finschi*)	–	–	2	–	2
Brown throated parakeet (*Aratinga pertinax*)	–	1	7	–	8
Orange-chinned parakeet (*Brotogeris jugularis*)	–	3	3	–	6
Possible mangrove cuckoo (cf. *Coccyzus minor*)	–	2	2	–	4
Ani (*Crotophaga* spp.)	–	10	8	1	19
Striped owl (*Asio clamator*)	–	2	3	–	5
Possible black-and-white owl (cf. *Ciccaba nigrolineata*)	–	–	1	–	1
Pygmy owl (*Glaucidium* spp.)	–	–	4	–	4
Possible ferruginous owl (*Glaucidium* cf. *brasilianum*)	–	–	4	–	4
Screech owl (*Otus* spp.)	–	–	3	–	3

(Continued)

Table 6-1. (*Continued*)

	1	2	3	4	Total
Possible tropical screech-owl (*Otus* cf. *choliba*)	–	6	3	–	9
Barn owl (*Tyto alba*)	–	7	2	–	9
Nightjar or nighthawk (Caprimulgidae)[c]	–	–	1	–	1
Nightjar (*Caprimulgus* sp.)	–	–	1	–	1
White-tailed nightjar (*Caprimulgus cayennensis*)	–	1	–	–	1
Lesser nighthawk (*Chordeiles minor*)	–	1	–	–	1
Ringed kingfisher (*Ceryle torquata*)	–	–	1	–	1
Red-crowned woodpecker (*Melanerpes rubricapillus*)	–	1	1	–	2
Red-legged honeycreeper (*Cyanerpes cyaneus*)	–	–	1	–	1
Grosbeak (*Pheucticus* sp.)	–	–	1	–	1
Rose-breasted grosbeak (*Pheucticus ludovicianus*)	–	–	1	–	1
Streaked saltator (*Saltator albicollis*)	–	–	1	–	1
Possible crested oropendola (*Psaroclius* cf. *decumanus*)	–	–	2	–	2
Possible kingbird type flycatcher (cf. *Tyrannus* spp.)	–	–	2	–	2
Possible fork-tailed flycatcher (cf. *Tyrannus savanna*)	–	1	1	–	2
Great-tailed grackle (*Quiscalus mexicanus*)	–	7	32	–	39
Meadowlark or blackbird (*Sturnella* spp.)	–	1	4	–	5
Possible warbler (cf. *Dendroica* sp.)	–	1	–	–	1
Boat-billed flycatcher (*Megarhynchus pitangua*)	–	4	2	–	6
Greater kiskadee (*Pitangus sulfuratus*)	–	–	1	–	1
Mammals					
Opossums (Didelphidae)[d]	1	–	–	–	1
Possible woolly opossum (cf. *Caluromys derbianus*)	1	–	–	–	1
Opossum (*Didelphis marsupialis*)	8	12	216	–	236
Mexican anteater (*Tamandua mexicana*)	8	–	1	–	9
Possible giant anteater (*Tamandua* cf. *tetradactyla*)	–	–	1	–	1
Nine-banded armadillo (*Dasypus novemcinctus*)	32	30	41	7	110
Cottontail (*Sylvilagus* spp.)[e]	20	30	82	3	135
Squirrel (*Sciurus* spp.)	3	3	17	–	23
Variegated squirrel (*Sciurus variegatoides*)	2	–	6	–	8
Spiny pocket mouse (*Heteromys* spp.)	–	2	*	–	2
Spiny pocket mouse (*Liomys* spp.)	16	50	*	7	73
Rice rat (*Oryzomys* spp.)	10	13	*	1	24
Cane rat (*Zygodontomys brevicauda*)	12	122	*	–	134
Hispid cotton rat (*Sigmodon hispidus*)	6	18	*	–	24
Paca or tepescuintle (*Agouti paca*)	20	33	152	2	207
Agouti (*Dasyprocta punctata*)	5	4	17	1	27
Dogs and foxes (Canidae)	1	–	17	–	18
Dog (*Canis familiaris*)[f]	2	7	45	–	54
Gray fox (*Urocyon cinereoargenteus*)	1	3	16	1	21
Raccoon (*Procyon lotor*)	284	4	21	2	311
Tayra (*Eira barbara*)	–	1	1	–	2
Least weasel (*Mustela frenata*)	–	13	9	–	22
River otter (*Lutra longicaudis*)	–	–	1	–	1
Jaguar (*Panthera onca*)	2	1	1	–	4
Puma (*Puma concolor*)	–	1	6	–	7
Ocelot (*Leopardus pardalis*)	3	2	–	–	5
Jaguarundi (*Herpailurus yaguarondi*)	1	1	5	–	7
Manatee (*Trichechus manatus*)	2	–	–	–	2
Peccary (*Tayassu* spp.)	10	3	8	–	21

(*Continued*)

Table 6-1. (*Continued*)

	1	2	3	4	Total
Possible white-lipped peccary (*Tayassu* cf. *pecari*)	1	–	–	–	1
Collared peccary (*Tayassu tajacu*)	14	7	1	–	22
White-tailed deer (*Odocoileus virginianus*)	1756	2157	*	199	4112
Totals	2992	4320	9232	277	16821

* Small rodents (Muridae and Heteromyidae) and deer are present but not tabulated here because their analysis is not yet complete.

a Materials found in association with burials are not included. Only specimens identified to genus and species are listed except in the case of snakes identified to family and below, dogs and foxes (Canidae), and small doves of the genera *Geotrygon*, *Leptotila*, and *Zenaida* (Columbidae). Tentative identifications to genus or species are designated by "cf." More information about the distribution of some animal taxa can be found in Cooke and Jiménez (2004), Cooke et al. (2007), Cooke and Ranere (1992a), and Jiménez and Cooke (2001). The samples are arranged into four groups as follows:

1: Pooled Hamlet Agriculture period (circa 5000 to 2500 B.C.) data from rockshelters and coastal sites with deep shell middens. These sites are Aguadulce Shelter (AS), Cueva de los Ladrones (LA), Zapotal (ZA), and Monagrillo (MO). The counts are updates of those presented in Cooke and Ranere (1992a:Table 2). Counts from Aguadulce include materials excavated in 1997 by Ranere and Piperno. The counts for Aguadulce and Ladrones sum bones found in Preceramic and Early Ceramic deposits.
2: Pooled Village Agriculture period (circa 200 B.C. and A.D. 1520) data from La Mula-Sarigua (LMS), Sitio Sierra (SS), Natá (NA), and Cerro Juan Díaz (CJD). The counts for Cerro Juan Díaz sum data from the units identified in Figures 6-3 and 6-4 as CJD-1, CJD-2, and CJD-3 (Jiménez 1999). Those for the other sites are updates of data presented by Cooke and Ranere (1992a:Table 3.3).
3. Specimens recovered from other locations at Cerro Juan Díaz. These represent specimens recovered in all excavation units regardless of chronology.
4: Pooled data from 12 Village Agriculture period sites in the lower valley of the La Villa river, including Cerro Juan Díaz (LS-3 [only those units investigated by Isaza]), LS-10, LS-15, LS-31, and LA-29); data from Isaza-Aizprúa (2007).

b Specimens that permit a species-level identification are all referable to the widespread scorpion mud turtle (*Kinosternon scorpioides*) except for one element of the white-lipped mud turtle (*K. leucostomum*) found at Sitio Sierra.

c Apparently a different species from the two nightjars listed below.

d Probably a species of mouse opossum (*Marmosa* spp.) that is not included in the Smithsonian Tropical Research Institute skeletal reference collection.

e All complete elements are referable to the tapeti or Brazilian cottontail (*Sylvilagus brasiliensis*). The Smithsonian Tropical Research Institute skeletal reference collection does not possess a skeleton of Dice's cottontail (*S. dicei*). It is unlikely that this highland endemic was present near the study area under Holocene climatic conditions.

f It is probable but not demonstrable that all of these elements belong to the domestic dog (*Canis familiaris*) although the coyote (*C. latrans*) was first recorded in specialist literature in Panama in 1981 (Vaughan 1983).

Iguanas

Iguanas constitute the third member of the trio of food offerings. Remains of the two locally-available species, the green iguana (*Iguana iguana*) and the black iguana (*Ctenosaura similis*), are frequent in archaeofaunal samples deposited during both the Hamlet Agriculture and Village Agriculture periods (Table 6-1). Lifelike iguanas were sometimes painted on pottery and modeled in gold (Figure 6-6f; Cooke 1992b:Figure 5). Low status individuals made personal ornaments out of iguana long bones (Cooke 2004b:Figure 8f).

Green iguanas prefer gallery forests where they have plenty of leaves to eat and thick foliage in which to hide. Hole-dwelling black iguanas are most abundant around mangrove-edges and rocky shores. The relative abundance of the two species in the archaeofaunal collections reflects these habitat preferences. At coastal Cerro Mangote, the black and green species are equally frequent. At

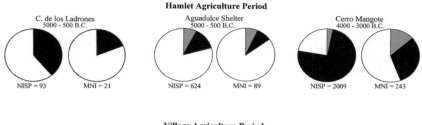

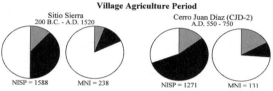

Figure 6-4. Percentages of Number of Identified Specimens (NISP) and Minimum Number of Individuals (MNI) of iguanas (*Iguana iguana* and *Ctenosaura similis*), deer (cf. *Odocoileus virginianus*), and other vertebrates (excluding fishes). Iguanas are in grey, deer in black, and other vertebrates are in white. Data for Cueva de los Ladrones, Cerro Mangote, and Sitio Sierra are from Cooke and Ranere (1992a); data for Aguadulce Shelter include materials excavated in 1997 by Ranere and Piperno; data for Cerro Juan Díaz are from Jiménez (1999). Grayson's (1973) maximum distinction method is used to estimate MNI for all samples except Cerro Juan Díaz, which is a single refuse feature (CJD-2). Dates refer to the total estimated occupation of each site and not to the archaeofaunal samples.

the Aguadulce Shelter, farther inland, the black iguana is rare. More black iguanas were consumed at Cerro Juan Díaz, now 4 km from the coast, than at Sitio Sierra, now 13 km inland (Table 6-1; Cooke et al. 2007; Cooke and Ranere 1992a). In most features at Cerro Juan Díaz, the green iguana outnumbers the black two-to-one. The exception is an excavation unit placed on a prominent hill dominating this site (Operation 31 in Cooke et al. 2000:Figure 8.4), in which the black species is twice as frequent as the green while the estimated average size of both iguana species is much smaller than elsewhere on the site. Green iguanas are now "farmed" by communities in Panama and Costa Rica with considerable success. Perhaps the Cerro Juan Díaz villagers took advantage of the ease with which small iguanas can be captured in and around mangroves and then kept them in cages until they attained an economical size.

BIRDS

Although not listed among the three primary gifts received by Spaniards, birds are an interesting part of the archaeofaunal record and highlight the importance of combining archaeofaunal, artistic, and documentary evidence. Macaw (*Ara* spp.) is the most common bird genus at Cerro Juan Díaz. (Cooke et al.

2007:Figure 5). At Sitio Sierra, a partial skeleton of a scarlet macaw (*Ara macao*) was buried on top of an adult male skeleton; the man may have been a curer or chanter. He also was buried with a flute made from a brown pelican humerus (*Pelecanus occidentalis*; Cooke 1984b:Figure 9, 2004a). No one has seen a macaw in the savannas around Parita Bay within living memory, although isolated populations of scarlet and green macaws (*Ara ambigua*) are found in remnant forests at the southwestern corner of the Azuero Peninsula (Figure 6-1; Ridgely and Gwynne 1993:175–176). The crimson-fronted para-keet (*Aratinga finschi*), also identified at Cerro Juan Díaz (Table 6-1), is found from western Panama to Nicaragua and, thus, may not have been obtained locally. In all likelihood, macaws, parrots, and parakeets were obtained from beyond the subsistence catchment area and were kept at Cerro Juan Díaz as pets, for ritual and ceremonies, or for their brightly-colored feathers (Cooke 1984b, 2004a).

Spaniards observed other penned birds whose remains are found in domes-tic midden and refuse deposits. Accounts of geese (ánsares) probably are refer-ences to muscovy ducks (*Cairina moschata*). These heavy-bodied, pre-Hispanic domesticates of tropical origin are sexually dimorphic; males are considerably larger than females. Domesticated males and wild females do not overlap in size (Stahl 2005). A sample of 46 muscovy specimens from Sitio Sierra and Cerro Juan Diaz includes some elements whose dimensions are referable to domestic males using Stahl's (2005) criteria. One specimen from Sitio Sierra returned an isotopic signal (^{15}N/^{13}C) similar to that of the human population at this site (Cooke et al. 1996:Figure 5). A wild muscovy's diet is unlikely to be so similar to the human diet. Some duck specimens are from very young birds although it is not possible to identify these young birds as muscovy ducks.

Spanish references to "pheasants" (faisanes) and "turkeys" (pavas) were probably to large birds in the Cracidae family such as great curassows (*Crax rubra*), crested guans (*Penelope purpurascens*), and grey-headed chachalacas (*Ortalis cinereiceps*) (Espinosa in Jopling 1994:31, 48; Sauer 1966:272). Northern turkeys (*Meleagris gallopavo*) were kept by the Maya, Aztec, and cul-turally-similar peoples as far south as Costa Rica (e.g., Hamblin 1984:93), but did not reach Panama in pre-Hispanic times. However, three great curassow and 22 crested guan specimens were identified from Cerro Juan Díaz. These large birds are shy denizens of extensive old forests. Both species are very sen-sitive to human hunting and forest fragmentation. Hence they may not have been available near Cerro Juan Díaz. Some of the crested guan remains are from young birds, suggesting they were caged to be fattened before slaughter. Spanish references to "partridges" (perdices) were probably to quail, whose specimens are frequent at Sitio Sierra and Cerro Juan Díaz. Most are from the crested bobwhite (*Colinus cristatus*), a typical denizen of open ground with tall grass and scrubby second-growth. Some are from young birds.

Passerine birds ("passerine" means sparrow-like) are a very large and heterogeneous group that is very informative about habitat. Unfortunately, the hundreds of Neotropical passerine species are difficult to distinguish even with a good comparative skeletal collection. One passerine in the collections that can be identified to species is the great-tailed grackle (*Quiscalus mexicanus*), a gregarious and omnivorous species that congregates around human settlements. As with the muscovy duck, sexual dimorphism creates overlaps with other species in the oriole family, such as crested oropendolas (e.g., *Psaroclius decumanus*). By our reckoning, however, 39 specimens from Sitio Sierra and Cerro Juan Díaz are from the great-tailed grackle. The other passerines that have allowed species-level identification include the boat-billed flycatcher (*Megarhynchus pitangua*) and greater kiskadee (*Pitangus sulfuratus*). Both species have bright yellow plumage on their breasts and frequent vegetation at river edges. Pretty feathers also characterize male rose-breasted grosbeaks (*Pheucticus ludovicanus*), red-legged honeycreepers (*Cyanerpes cyaneus*), and red-breasted blackbirds (*Sturnella militaris*). Brightly colored birds may have been acquired for their plumage, rather than for their meat. The great-tailed grackle was kept in aviaries by the Aztec (Haemig 1973).

Since Cerro Juan Díaz is located close to the marine littoral it is not surprising that the villagers exploited coastal birds. The abundance of boobies (*Sula* spp.) deserves comment because they rarely venture within projectile range of the shore when not nesting. The size of most of the remains suggests they come from the blue-footed booby (*S. nebouxi*). Ornithologists once reported the blue-footed booby nesting on Isla Villa, a rocky islet 12 km southeast of Cerro Juan Díaz (Figure 6-1; Olson 1997). Many booby specimens were cut and polished to make tubes for aprons and gorgets. Pelican and magnificent frigate-bird specimens (*Fregata magnificens*) were also made into ornaments and osprey claws (*Pandion haliaetus*) were deposited in graves (Cooke 2004b:Figure 8c). We can infer, then, that the local people admired the abilities of aquatic birds that are adept fishers.

DIFFERENTIAL ACCESS TO FAUNA

Managing Deer

A curious remark of Spanish soldiers concerns meat-eating and deer-hunting (Cooke 1992b; Sauer 1966:274). Espinosa (in Jopling 1994:65) noticed that in the territory of Parita, "the chieftain and his captains and Indians are like Dominican or Carthusian friars in that they don't eat meat, of any kind or condition, except fish and iguanas, even though there are deer and game all over the place." Another Spanish soldier, Pascual de Andagoya, adds an interesting detail: the warriors were the ones who did not eat meat while the natives "of the field" and "workers" did (in Jopling 1994:34, 65). Their observations are inconsistent, not only with the storehouses full of deer meat seen by Spaniards in this and neighboring chiefdoms, but also with the ubiquity and abundance

of deer remains in midden samples from Cerro Juan Díaz and the sites investigated by Isaza-Aizprúa (2007) in the La Villa valley (Table 6-1). So to what
are the Spanish soldiers referring?

Andagoya's comments allude to a taboo practiced by war bands. Another
comment he made is relevant even if it does not necessarily refer to Parita's
chiefdom. "The lords," he says, "had their hunting reserves where in summer
they would go to chase deer, setting fires to windward, and as the grass is tall,
the fire would quickly spread, and the natives would stand there waiting where
they thought the fire would reach them; and as the deer would run off all
together fleeing the smoke and blinded by it, the very fire would drive them to
where the natives were waiting with their spears tipped with flint points" (in
Jopling 1994:31, translation by Cooke).

The Parita Bay chiefdoms were well populated and had extensive fields during the Village Agriculture period. They also were involved in constant skir-
mishing and raiding. This social unrest probably restricted the range of
hunters. White-tailed deer are browsers with a fairly limited home range. They
prefer second-growth vegetation. In contemporary Panama they are sometimes
kept in captivity, where they will reproduce. The Spanish observations about
taboos and hunting reserves suggest that each chiefdom possessed lands where
deer populations were managed by restricting deer-hunting to certain groups
of people, at certain times of the year, or by dietary rules established for
particular events or activities, such as feasts and other ceremonial gatherings
(see also Pohl and Healy [1980] for an example from Nicaragua).

The "Rich and Famous" and Their Animals

The Sitio Conte and El Hatillo sites were burial grounds for rich and influen-
tial people in the Parita Bay chiefdoms. Briggs (1989) points out that social
rank in this part of Panama was associated not only with finely-crafted mortuary arts but also with the sheer numbers of funeral offerings. Whereas people
of lower station were buried with one or two perforated animal teeth and/or
bone beads (the majority of them from sharks or rays), high-rank individuals
were buried with exquisite carvings made of sperm whale teeth (Physeteridae),
deer vertebrae, and manatee bone, as well as necklaces made of hundreds of
perforated dog and peccary (*Tayassu* spp.) teeth (Cooke 2004b). Some materials used for these artifacts, such as manatee bone, undoubtedly originated well
beyond the Parita Bay chiefdoms.

Spanish captain Espinosa mentions "lions" and "tigers" (in Jopling 1994:65).
Presumably the "lions" are pumas (*Puma concolor*), and the "tigers" jaguars
(*Panthera onca*) and ocelots (*Leopardus pardalis*). Specimens of these big cats and
the smaller jaguarundi (*Herpailurus yaguarondi*) are found in Parita Bay midden
samples, but in very small numbers. The fact that most are modified specimens
such as perforated teeth, metatarsals fashioned into tubes, and polished
phalanges, suggests ritual or decorative use. Mortuary features at Cerro Juan
Díaz, however, contained many perforated teeth including some from jaguars,

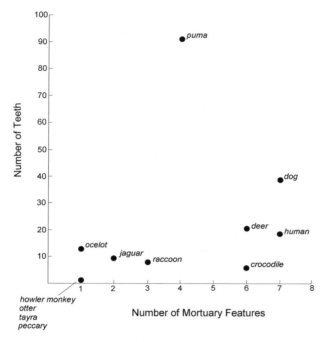

Figure 6-5. The number of mortuary features containing teeth with perforated roots at Cerro Juan Díaz plotted against the number of teeth in each feature. Features were all deposited between circa A.D. 100 and 1100. Scientific names are listed in Table 6-1 for all taxa except howler monkey (*Alouatta* spp.).

pumas, and ocelots (Figure 6-5). One grave contained 57 puma and 12 ocelot teeth. These funerary associations suggest that big cat teeth were a component of ritual paraphernalia involved with shamanic or curing activities (Cooke 1998:100–108; 2004b). Since the Parita Bay region lacked extensive forests at this time, it is tempting to assume that big cat remains represent exchanges with people living in distant forested habitats or long hunting treks. However, pumas are major predators on deer and may have been abundant in the neighboring anthropogenic savannas where deer herds were large. Therefore Espinosa's sightings of "lions" in Parita's chiefdom may not be just tall stories.

CONCLUSION

The pre-Hispanic peoples who deposited the animal remains recovered by this research used about 150 species of fish of marine origin; 14 species of freshwater fish (Cooke and Jiménez 2004); four species of frogs and toads (Cooke 1989); at least three families of snakes; four species of turtles (two freshwater and two marine); four lizards; more than eighty species of birds; and 28 species of mammals. The presence of these animals in several types of features indicates they were used for different purposes. Some were food items for everyday or ceremonial consumption. Some were probably pets with ritual connotations. Many species used regularly for food also provided materials for tools and ornaments. For 6,500 years, Parita Bay hamlets and villages hunted white-tailed deer, not only for fresh and smoked meat, but also for bone, antlers, and, presumably, sinews and hides for making many kinds of utensils and ornaments.

The 28 mammal species recorded in Parita Bay archaeozoological collections represent about half the number which on zoogeographical grounds are likely to have comprised the "life assemblage" under "natural" conditions (Bennett 1968). Spanish chronicler Fernández de Oviedo y Valdés (1853:136) commented that all over Panama fishing was "the principal occupation," being "less work" than hunting wild animals. Hence the optimality of fishing, in addition to the ease with which dried and salted fish could be transported inland, may be one reason why so few terrestrial animals were taken regularly.

Another reason may be hunting pressure in a long-disturbed and seasonally arid environment. It is striking that some medium- to large-sized mammals that are "target" species in other Neotropical regions are absent, rare, or sporadic in assemblages from the Parita Bay region. The presence of the white-lipped peccary (*Tayassu pecari*) is tentatively inferred from only one specimen from Cueva de los Ladrones. Larger numbers of specimens are referable to the collared peccary (*Tayassu tajacu*), which is reported from several sites. Even though this gregarious species was available to local hunters, it was infrequently taken, except at Cueva de los Ladrones, the site situated nearest the cordillera during the Hamlet Agriculture period. Agoutis (*Dasyprocta punctata*) were the favored prey of the inhabitants of Cerro Brujo on the Caribbean coast (Linares 1976; Linares and White 1980). They are very abundant in dry forest remnants around Panama City today; but their remains are spottily distributed in the Parita Bay archaeofaunal collections from both periods. Particularly surprising is the total absence of coati (*Nasua* spp.) specimens. The Maya inhabitants of densely populated Cozumel Island (Mexico) hunted many coatis and also consumed collared peccary meat, possibly from penned-up animals (Hamblin 1984).

Perhaps the depletion of medium- to large mammal "target" species originated during the six thousand years or more prior to the beginning of the Hamlet Agriculture period for which we currently lack archaeofaunal data. Or perhaps human activities from the Paleoindian period onwards curtailed the spread of dry forests when the Holocene period began (Bennett 1968). By the beginning of the Hamlet Agriculture period, forest burning and clearance had become extensive even though human populations were still small and scattered. It is predictable, therefore, that the hypothetical "life assemblage" from which the Parita Bay communities derived their animals after 5000 B.C. should have included species characteristic of wooded anthropogenic savannas with coastal and riverine woods (e.g., iguanas, white-tailed deer, opossums, rabbits, and quail), rather than of extensive dry forest tracts (e.g., red brocket deer, white-lipped peccary, and spider monkeys [*Ateles* spp.]).

The prolonged importance of white-tailed deer and iguanas in the local diet are surprising considering that deer is now locally extirpated and green iguanas are under severe hunting pressure. The white-tailed deer may have been a managed species, as described in Spanish documents regarding the monopoly of venison by elites. It is possible, therefore, that the taboos against meat consumption and the existence of hunting reserves observed by Spanish soldiers around Parita Bay in the early 16th century A.D. were mechanisms ensuring that sufficient venison was available in the storehouses for important social events when large numbers of people had to be fed. Penning palatable birds and perhaps iguanas would serve the same purpose.

In some cases the archaeozoological, artistic, and the documentary record are complementary: the white-tailed deer, for example, is identified as an important animal in all three data sets. But there are also discrepancies. For example, although no tapir (*Tapirus bairdii*) specimens are reported, Andagoya refers to tapirs (vacas dantas) in the chiefdom of Parita (in Jopling 1994:34). The only monkey specimen found in Parita Bay archaeozoological collections is a perforated howler monkey (*Alouatta* spp.) tooth from Cerro Juan Díaz grave (Cooke 2004b). Lifelike monkeys are depicted, however, on locally manufactured pottery, e.g., the night or owl monkey (*Aotus lemurinus*) from Cerro Juan Díaz (Figure 6-6g). This suggests that local artists were well acquainted with these primates. Likewise the rarity of crocodile (*Crocodylus acutus*), cayman (*Caiman cocodrilus*), and marine turtle (Cheloniidae) remains in Parita Bay middens is inconsistent with the many life-like depictions of these animals on pre-Hispanic art objects. Small populations of howler monkeys and white-faced capuchins (*Cebus capucinus*) survive in forest remnants in or near our study area. Sea turtles and crocodilians are currently widespread around Parita Bay (Cooke 1992b, 1998, 2004a; Cooke et al. 2007). Therefore the present-day situation raises doubts as to whether pre-Hispanic hunters, who lacked firearms, would have severely impacted these animals. Even so, although monkeys, sea turtles, and crocodilians seem the best candidates for cognitive rejection by local pre-Hispanic hunters, our arguments will remain circular unless control samples of animal remains *deposited independently of humans* are found. This possibility is remote in an area where no large bone-hoarding animals are known.

The bird remains in the Parita Bay archaeofaunal collections provide good supporting evidence for the longevity of anthropogenic savannas because they allude to an avian community which is remarkably similar to the current avifauna of this area. Where bird genera or species can be identified, they are invariably characteristic of lowland savannas with scant or patchy arboreal vegetation near coastal habitats (Table 6-1). Thus, local hunters exploited a predictably heterogeneous group of marsh, swamp, riverine, and coastal birds. The exceptions to this projection of the contemporary avifauna backwards in

Figure 6-6a–e. Mammal images painted on pottery of the Early Conte style of the Gran Coclé tradition. They highlight branching antlers, which represent the white-tailed deer (*Odocoileus virginianus*). In 'd' the deer protrudes its tongue as though it were being chased. f: twinned image of a lizard. The striped tail and stylized dewlap suggest the green iguana (*Iguana iguana*). g: modeled polychrome vessel painted in the Cubitá polychrome style of the Gran Coclé tradition found at Cerro Juan Díaz. It represents a mammal with large round eyes, probably the night or owl monkey (*Aotus lemurinus*).

time are the white-faced whistling-duck (*Dendrocygna viduata*), great curassow, crested guan, macaws, and crimson-fronted parakeet, none of which are present in the area today (e.g., Cooke and Olson 1984). These birds may have been absent or rare within the normal hunting catchment of the Parita Bay chiefdoms in pre-Hispanic times because of forest fragmentation and/or human predation. They may have been acquired from distant localities and thus do not represent the local "life assemblage."

In Chapter 7 of the first edition of *Environmental Archaeology* we strove to identify harmony and discord among data sets relevant to the environmental history of Panama (palaeobotany, archaeozoology, and stable isotope chemistry) working under the premise that these realms of inquiry are complementary even though they are never isomorphic. Interpreting how a given human population uses animals over long periods of time also benefits from comparing different data sets (osteological, artistic, and documentary). The explanations this methodology invites for human-animal interactions in pre-literate societies are not as complete or convincing as those provided by living human informants. Even so, they enhance our ability to reconstruct the historical ecology of particular regions by allowing us to relate "sampled" and "life" assemblages of animals and plants to several facets of human behavior which exhibit considerable cross-cultural variability in space and time.

REFERENCES

Bennett, C. F., 1968, *Human Influences on the Zoogeography of Panama*. University of California Press, Berkeley, CA.

Berlin, B., 1992, *Ethnobiological Classification. Principles of Categorization of Plants and Animals in Traditional Societies*, Princeton University Press, Princeton, New Jersey.

Breece, L. A. H., 1997, *An Assessment of the Archaeological Potential of Natá, a Precolumbian Town in Central Panama*, Ph.D. dissertation, Department of Anthropology, University of California, Los Angeles.

Briggs, P. S., 1989, *Art, Death and Social Order: The Mortuary Arts of Pre-Conquest Central Panama*, British Archaeological Reports International Series (Oxford) 550.

Carvajal, D. R., 1998, Análisis de Cuatro Componentes en el Rasgo CH Excavado Mediante la Microestratigrafía: El Caso de Cerro Juan Díaz. Tesis de grado, Universidad Nacional de Colombia, Facultad de Ciencias Humanas, Santa Fé de Bogotá. (Available as a PDF file through the Smithsonian Tropical Research Institute library, Panama.)

Clary, J., Hansell, P., Ranere, A. J., and Buggey, T., 1984, The Holocene Geology of the Western Parita Bay Coastline of Central Panama, in: *Recent Developments in Isthmian Archaeology* (F. W. Lange, ed.), *British Archaeological Reports International Series* (Oxford) 212:55–83.

Cooke, R. G., 1984a, Archaeological Research in Central and Eastern Panama: A Review of Some Problems, in: *The Archaeology of Lower Central America* (F. W. Lange and D. Z. Stone, eds.), University of New Mexico Press (School for American Research), Albuquerque, pp. 263–302.

Cooke, R. G., 1984b, Birds and Men in Prehistoric Central Panama, in: *Recent Developments in Isthmian Archaeology* (F. W. Lange, ed.), *British Archaeological Reports International Series* (Oxford) 212:243–281.

Cooke, R. G., 1989, The Use of Anurans for Food by Tropical Amerindians: Archaeological and Ethnographic Evidence, *ArchaeoZoología* 3:133–142.

Cooke, R. G., 1992a, Prehistoric Nearshore and Littoral Fishing in the Eastern Tropical Pacific: An Ichthyological Evaluation, *Journal of World Prehistory* 6:1–49.

Cooke, R. G., 1992b, Preliminary Observations on Vertebrate Food Avoidance by the Precolombian Amerinds of Panama, with Comments on the Relevance of this Behaviour to Archaeozoology and Palaeoenvironmental Reconstruction, in: *Archaeology and Environment in Latin America* (O. Ortiz-Troncoso and T. van der Hammen, eds.), Instituut voor Pre- en Protohistorische Acheologie Albert Egges van Giffen, Universiteit van Amsterdam, Amsterdam, The Netherlands, pp. 59–107.

Cooke, R. G., 1998, The Felidae in Pre-Columbian Panama: A Thematic Approach to Their Imagery and Symbolism, in: *Icons of Power: Felid Symbolism in the Americas* (N. J. Saunders, ed.), Routledge, London, pp. 77–121.

Cooke, R. G., 2004a, Observations on the Religious Content of the Animal Imagery of the 'Gran Coclé' Semiotic Tradition of pre-Columbian Panama, in: *Behaviour Behind Bones. The Zooarchaeology of Ritual, Religion, Status and Identity* (S. O'Day, W. van Neer and A. Ervynck, eds.), Oxbow Press, Oxford, England, pp. 114–127.

Cooke, R. G., 2004b, Rich, Poor, Shaman, Child: Animals, Rank, and Status in the 'Gran Coclé' Culture Area of pre-Columbian Panama, in: *Behaviour Behind Bones. The Zooarchaeology of Ritual, Religion, Status and Identity* (S. O'Day, W. van Neer and A. Ervynck, eds.), Oxbow Press, Oxford, England, pp. 271–284.

Cooke, R. G., 2005, Prehistory of Native Americans on the Central American Land-bridge: Colonization, Dispersal and Divergence, *Journal of Archaeological Research* 13:139–188.

Cooke, R. G., Isaza-Aizprúa, I. I., Griggs, J., Desjardins, B., and Sánchez-Herrera, L. A., 2003, Who Crafted, Exchanged and Displayed Gold in pre-Columbian Panama? in: *Gold and Power in the Intermediate Area* (J. Quilter and J. M. Hoopes, eds.), Dumbarton Oaks, Washington, DC, pp. 91–158.

Cooke, R. G., and Jiménez, M., 2004, Teasing out the Species: Is it worth the Effort?, *Archaeofauna* 13:19–35.

Cooke, R. G., Jiménez, M., and Ranere, A. J., 2007, Influencias Humanas Sobre la Vegetación y Fauna de Vertebrados de Panamá: Actualización de Datos Arqueozoológicos y su Relación con el Paisaje Antrópico Durante la Epoca Precolombina, in: *Evolución en los Trópicos* (E. Leigh, E. A. Herre, J. B. C. Jackson and F. Santos-Granero, eds.), Smithsonian Tropical Research Institute, Panama, in press.

Cooke, R. G., Norr, L., and Piperno, D. R., 1996, Native Americans and the Panamanian Landscape, in: *Case Studies in Environmental Archaeology*, 1st ed. (E. J. Reitz, L. A. Newsom and S. J. Scudder, eds.), Plenum Press, New York, pp. 103–126.

Cooke, R. G., and Olson, S. J., 1984, An Archaeological Record for the White-faced Whistling-Duck (*Dendrocygna viduata*) in Panama, *The Condor* 86:493–494.

Cooke, R. G., and Ranere, A. J., 1984, The "Proyecto Santa Maria": A Multidisciplinary Analysis of Prehistoric Adaptations to a Tropical Watershed in Panama, in: *Recent Developments in Isthmian Archaeology* (F. W. Lange, ed.), British Archaeological Reports International Series (Oxford) 212:3–30.

Cooke, R. G., and Ranere, A. J., 1992a, Human Influences on the Zoogeography of Panama: An Update Based on Archaeological and Ethnohistorical Evidence, in: *Biogeography of Mesoamerica. Proceedings of a Symposium (Mérida, Yucatán, México, October 26–30, 1984)* (S. P. Darwin and A. L. Welden, eds.), Special Publication of the Mesoamerican Ecology Institute, Baton Rouge, Louisiana, pp. 21–58.

Cooke, R. G., and Ranere, A. J., 1992b, Prehistoric Human Adaptations to the Seasonally Dry Forests of Panama, *World Archaeology* 24:114–133.

Cooke, R. G., and Ranere, A. J., 1992c, The Origin of Wealth and Hierarchy in the Central Region of Panama (12,000–2,000 BP), with Observations on its Relevance to the History and Phylogeny of Chibchan-speaking Polities in Panama and Elsewhere, in: *Wealth and Hierarchy in the Intermediate Area* (F. W. Lange, ed.), Dumbarton Oaks, Washington DC, pp. 243–316.

Cooke, R. G., and Ranere, A. J., 1999, Precolumbian Fishing on the Pacific Coast of Panama, in: *Pacific Latin America in Prehistory:The Evolution of Archaic and Formative Cultures* (M. Blake, ed.), Washington State University Press, Pullman, pp. 103–122.

Cooke, R. G., and Sánchez-Herrera, L. A., 1998, Coetaneidad de Metalurgia, Artesanías de Concha y Cerámica Pintada en Cerro Juan Díaz, Panamá, *Boletín del Museo del Oro* (Colombia) 42:57–85.

Cooke, R. G., and Sánchez-Herrera, L. A., 2004a, Panama Indígena (1501–1550), in: *Historia General de Panama*, Volumen 1, Tomo 1 (A. Castillero C., ed.), Comité Nacional de Centenario de la República, Presidencia de la República, Panamá, pp. 47–78, (available in PDF at <www.stri.org/StaffScientists/RichardCooke>).

Cooke, R. G., and Sánchez-Herrera, L. A., 2004b, Panamá Prehispánico, in: *Historia General de Panama*, Volumen 1, Tomo 1 (A. Castillero C., ed.), Comité Nacional de Centenario de la República, Presidencia de la República, Panamá, pp. 3–46. (available in PDF at <www.stri.org/StaffScientists/RichardCooke>).

Cooke, R. G., Sánchez-Herrera, L. A., and Udagawa, K., 2000, Contextualized Goldwork from 'Gran Coclé', Panama: An Update Based on Recent Excavations and New Radiocarbon Dates for Associated Pottery Styles, in: *Precolumbian Gold: Technology, Style and Iconography* (C. McEwan, ed.), British Museum Press, London, pp. 154–176.

Cooke, R. G., and Tapia-Rodríguez, G., 1994a, Marine and Freshwater Fish Amphidromy in a Small Tropical River on the Pacific Coast of Panama: A Preliminary Evaluation Based on Gill-net and Hook-and-line Captures, in: *Fish Exploitation in the Past* (W. Van Neer, ed.), Annales du Musée Royal de l'Afrique Centrale, *Sciences Zoologiques,* Tervuren, Belgium, 274:99–106.

Cooke, R. G., and Tapia-Rodríguez, G., 1994b, Stationary Intertidal Fish Traps in Estuarine Inlets on the Pacific coast of Panama: Descriptions, Evaluations of Early Dry Season Catches and Relevance to the Interpretation of Dietary Archaeofaunas, Proceedings of the 6th Meeting of Fish Working Group of the International Council for Zooarchaeology, *Offa* 51:287–298.

Dickau, R., 2005, *Resource Use, Crop Dispersals, and the Transition to Agriculture in Prehistoric Panama: Evidence from Starch Grains and Macroremains,* Ph.D. dissertation, Department of Anthropology, Temple University, Philadelphia, Pennsylvania.

Fernández de Oviedo y Valdés, G., 1853, *Historia Natural y General de Las Indias, Islas y Tierra Firme del Mar Océano,* 2nd vol. (J. Amador de los Ríos, ed.), Real Academia de Historia, Madrid, Spain.

Grayson, D. K., 1973, On the Methodology of Faunal Analysis, *American Antiquity* 38:432–439.

Griggs, J., 2005, *The Archaeology of Central Caribbean Panama,* Ph.D. dissertation, Department of Anthropology, University of Texas, Austin, Texas.

Haemig, P. D., 1973, Aztec Emperor Auitzotl and the Great-tailed Grackle, *Biotropica* 10:11–17.

Haller, M., 2004, *The Emergence and Development of Chiefly Societies in the Rio Parita Valley,* Ph.D. dissertation, Department of Anthropology, University of Pittsburgh, Pennsylvania.

Hamblin, N. L., 1984, *Animal Use by the Cozumel Maya,* University of Arizona Press, Tucson.

Hansell, P., 1979, *Shell Analysis: A Case Study from Central Pacific Panama,* M.A. thesis, Temple University, Philadelphia, Pennsylvania.

Hansell, P., 1987, The Formative in Pacific Central Panama: La Mula-Sarigua, in: *Chiefdoms of America* (R. D. Drennan and C. Uribe, eds.), University Press of America, Lanham,Maryland, pp. 119–139.

Hansell, P., 1988, *The Rise and Fall of an Early Formative Community: La Mula-Sarigua, Central Pacific Panama,* Ph.D. dissertation, Department of Anthropology, Temple University, Philadelphia, Pennsylvania.

Helms, M. W., 1995, *Creations of the Rainbow Serpent: Polychrome Ceramic Designs from Ancient Panama,* University of New Mexico Press, Albuquerque.

Helms, M. W., 2000, *The Curassow's Crest: Myths and Symbols in the Ceramics of Ancient Panama,* University Press of Florida, Gainesville.

Isaza-Aizprúa, I. I., 1993, *Desarrollo Estilístico de la Cerámica Pintada del Panamá Central con Énfasis en el Período 500 a.C.–500 d.C.,* Tesis de grado, Universidad Autónoma de Guadalajara, México (Available through the library of the Smithsonian Tropical Research Institute, Panama).

Isaza-Aizprúa, I. I., 2007, *The Ancestors of Parita: Settlement Archaeology in the La Villa Valley, Azuero Peninsula*, Ph.D. dissertation, Department of Anthropology, Boston University, Boston, Massachusetts.

Jiménez, M., 1999, *Explotación de Vertebrados Acuáticos y Terrestres por los Indígenas Precolombinos en Cerro Juan Díaz, Los Santos, Durante el Periodo 300–700 d.C.*, Graduation thesis, Escuela de Biología, Universidad de Panamá.

Jiménez, M., and Cooke, R. G., 2001, La Pesca en el Borde de un Estuario Neotropical: El Caso de Cerro Juan Díaz (Bahía de Parita, Costa del Pacífico de Panamá), in: *Noticias de Arqueología y Antropología*, Grupo NaYa, Buenos Aires, Argentina, (available in CD-ROM from <www.stri.org/StaffScientists/RichardCooke>)

Jopling, C. (comp.), 1994, *Indios y Negros en Panamá en los Siglos XVI y XVII: Selecciones de los Documentos del Archivo General de Indias*, C. Lutz, Woodstock, Vermont.

Klein, R. G., and Cruz-Uribe, K., 1984, *The Analysis of Animal Bones from Archaeological Sites*, University of Chicago Press, Chicago, Illinois.

Ladd, J., 1964, *Archaeological Investigations in the Parita and Santa María Zones of Panama*, Smithsonian Institution Bureau of the American Ethnology Bulletin 193, Washington, DC.

Linares, O. F., 1976, Garden Hunting in the American Tropics, *Human Ecology* 4:331–349.

Linares, O. F., 1977, *Ecology and the Arts in Ancient Panama: On the Development of Rank and Symbolism in the Central Provinces, Studies in Precolumbian Art and Archaeology* 17, Dumbarton Oaks, Washington, DC

Linares, O. F., and White, R. S., 1980, Terrestrial Fauna from Cerro Brujo (CA-3) in Bocas del Toro and La Pitahaya (IS-3) in Chiriqui, in: *Adaptive Radiations in Prehistoric Panama* (O. F. Linares and A. J. Ranere, eds.), *Peabody Museum of Archaeology and Ethnology Monographs* 5, Harvard University Press, Cambridge, Massachusetts, pp. 181–193.

Lothrop, S. K., 1937, *Coclé: An Archaeological Study of Central Panama, Part 1, Memoirs of the Peabody Museum of Archaeology and Ethnology* 7, Harvard University Press, Cambridge, Massachusetts.

Lothrop, S. K., 1942, *Coclé: An Archaeological Study of Central Panama, Part 2, Memoirs of the Peabody Museum of Archaeology and Ethnology* 8, Harvard University Press, Cambridge, Massachusetts.

McGimsey, C. R. III, 1956, Cerro Mangote: A Preceramic Site in Panama, *American Antiquity* 22:151–161.

Mayo, J. del C., 2004, *La Industria Prehispánica de Conchas en Gran Coclé, Panamá*, Ph.D. dissertation, Department of Prehistory, Universidad Complutense, Madrid.

Mayo, J. del C. (ed.), 2007, Gran Coclé. Paisaje Cultural del Istmo de Panamá, *Revista Española de Antropología Americana* 37(1): 91–189.

Olson, S. L., 1997, Avian Biogeography in the Islands of the Pacific Coast of Western Panama, in: *The Era of Allan R. Phillips: A Festschrift* (R. W. Dickerman, comp.), Horizon Communications, Albuquerque, New Mexico, pp. 69–82.

Pearson, G. A., 2002, *Pan-Continental Paleoindian Expansions and Interactions as Viewed from the Earliest Lithic Industries of Lower Central America*, Ph.D. dissertation, Department of Anthropology, University of Kansas, Lawrence.

Pearson, G. A., 2003, First Report of a New Paleoindian Quarry Site on the Isthmus of Panama, *Latin American Antiquity* 14: 311–322.

Pearson, G. A., and Cooke, R. G., 2002, The Role of the Panamanian Land-bridge During the Initial Colonization of the Americas, *Antiquity* 76:931–932.

Peres, T. M., 2001, *Coastal Subsistence and Settlement in the early Ceramic: A Zooarchaeological Study from Central Pacific Panama*, Ph.D. dissertation, Department of Anthropology, University of Florida, Gainesville.

Piperno, D. R., and Holst, I., 1998, The Presence of Starch Grains on Prehistoric Stone Tools from the Humid Neotropics: Indications of Early Tuber Use and Agriculture in Panama, *Journal of Archaeological Science* 25:765–776.

Piperno, D. R., and Pearsall, D. M., 1998, *The Origins of Agriculture in the Lowland Tropics*, Academic Press, San Diego, California.

Piperno, D. R., Ranere, A. J., Holst, I., and Hansell, P., 2000, Starch Grains Reveal Early Root Crop Horticulture in the Panamanian Tropical Forest, *Nature* 407:894–897.

Pohl, M., and Healy, P., 1980, "Mohammed's Paradise": The Exploitation of Faunal Resources in the Rivas Region of Nicaragua, in: *Archaeology of the Rivas Region of Nicaragua* (P. F. Healy, ed.), Wilfrid Laurier University Press, Waterloo, Ontario, pp. 287–292.

Ranere, A. J., and Cooke, R. G., 1996, Stone Tools and Cultural Boundaries in Prehistoric Panama: An Initial Assessment, in: *Paths to Central American Prehistory* (F. Lange, ed.), University Press of Colorado, Niwot, Colorado, pp. 49–77.

Ranere, A. J., and Cooke, R. G., 2003, Late Glacial and Early Holocene Occupation of Central American Tropical Forests, in: *Under the Canopy. The Archaeology of Tropical Rain Forests* (J. Mercader, ed.), Rutgers University Press, New Brunswick, New Jersey, pp. 219–248.

Ranere, A. J., and Hansell, P., 1978, Early Subsistence Patterns Along the Pacific Coast of Panama, in: *Prehistoric Coastal Adaptations* (B. L. Stark and B. Voorhies, eds.), Academic Press, New York, pp. 43–59.

Ridgely, R. S., and Gwynne, J. A., Jr., 1993, *Guía de las Aves de Panamá*, Asociación Nacional para la Conservación de la Naturaleza, Panama City, Panamá.

Sauer, C. O., 1966, *The Early Spanish Main*, University of California Press, Berkeley.

Simoons, F. J., 1994, *Eat Not this Flesh. Food Avoidances from Prehistory to the Present*, University of Wisconsin Press, Madison.

Stahl, P., 2005, An Exploratory Osteological Study of the Muscovy Duck (*Cairina moschata*) (Aves: Anatidae) with Implications for Neotropical Archaeology, *Journal of Archaeological Science* 32:915–929.

Vaughan, C., 1983, Coyote Range Expansion in Costa Rica and Panama. *Brenesia*, 21:27–32.

Weiland, D., 1984, Prehistoric Settlement Patterns in the Santa María Drainage of Panama: A Preliminary Analysis, in: *Recent Developments in Isthmian Archaeology* (F. W. Lange, ed.), *British Archaeological Reports International Series* (Oxford) 212:31–53.

Willey, G. R., and McGimsey, C. R. III, 1954, *The Monagrillo Culture of Panama, Papers of the Peabody Museum of Archaeology and Ethnology* 49(2), Harvard University Press, Cambridge, Massachusetts.

Zohar, I., and Cooke, R. G., 1997, The Impact of Salting and Drying on Fish Skeletons: Preliminary Observations from Parita Bay, Panama, *Archaeofauna* 6:59–66.

Chapter 7

Using Land Snails and Freshwater Mussels to Chart Human Transformation of the Landscape: An Example from North Mississippi, U.S.A.

EVAN PEACOCK AND JOCHEN GERBER

Environmental archaeologists face what are at once the challenges and the promises of investigating human/nature relationships at scales unobtainable by other disciplines. The challenges are considerable. Meshing different kinds of paleoenvironmental data, with their own attendant biases, into a coherent explanatory framework requires an enormous amount of intellectual work and methodological rigor. The promises are no less considerable. Not only does environmental archaeology provide robust contexts for understanding past human behavior, but it also provides information useful in applied settings today. Long-term data on the development of plant and animal communities, and on the effects of human actions on those communities, are critical for ecological restoration, where the most difficult question often is "restoration of what?" Environmental archaeology provides answers to that question, or at least provides a corpus of information upon which to base informed management decisions for natural resources under ever-increasing human pressure on the environment (Peacock 1998).

The Black Prairie physiographic province of Mississippi and Alabama provides a useful example in this regard (Figure 7-1). Blackland prairies are distinctive ecosystems found in several parts of the southeastern United States. These prairies are characterized by chalk bedrock, calcareous clay subsoils, rich, black topsoils, gently rolling surface topography, and a host of prairie plant species, many of which are threatened or endangered today. Today, blackland prairies are the focus of considerable conservation and restoration efforts. The importance of historical data in these efforts is acknowledged (see references in Peacock and Schauwecker 2003), but the utility of environmental data derived from archaeological contexts is only now being recognized (Hogue 2003; Peacock 1993; Peacock and Reese 2003).

One environmental characteristic of the Black Prairie physiographic province of Mississippi is a puzzle. Historical records consistently note a lack of surface water in the province during the summer months when the many small streams dry up, leaving impermeable chalk creek beds exposed (see references in Peacock and Schauwecker 2003). Despite this apparent limiting factor, late 17th- and early 18th-century A.D. Native American sites are abundant in the province (e.g., Rafferty 2003), and many are located far away from the larger, perennial streams. Archaeobotanical and bone isotope analyses indicate that these small settlements, usually referred to as "farmsteads," were part of a maize-based (*Zea mays*) farming and settlement pattern (e.g., Hogue 2003; Hogue and Peacock 1995; Rafferty 2001, 2003). Little is currently known about

Figure 7-1. Location of the Black Belt, Mississippi and Alabama, southeastern United States, and archaeological sites 22OK904 and 22OK905. The Black Belt in Mississippi is known as the Black Prairie.

this pre-European economic strategy, how it specifically functioned in the Black Prairie environment and what stresses natural or human-induced environmental change would have caused for the Native American farmstead inhabitants.

To address these questions, we analyzed mollusk remains from archaeological sites in the Black Prairie and modern snail shells collected from different environmental settings. These data are used to evaluate descriptions of historical landscapes, to characterize the site-specific environments inhabited by Native American farmers, and, to some extent, to characterize the larger Black Prairie environment as a whole, providing a context for interpreting archaeological settlement patterns. The implications of the work for ecological restoration efforts also are discussed.

HISTORICAL DESCRIPTIONS OF THE STUDY AREA

The Mississippi Black Prairie is part of a narrow, arc-shaped physiographic province, more generally referred to as the Black Belt, that curves from eastern Mississippi into central Alabama (Figure 7-1). It is an area with a fascinating historical legacy. The non-acidic prairie soils are extraordinarily fertile, a fact which led to intensive farming after the 17th century. Before emancipation in the mid-1800s, slave populations in the Black Belt were among the largest in North America (Fogel 1989; Gray 1933), and many counties in Mississippi and Alabama retain large, descendent African-American populations (Tower 1961). The peculiar environmental characteristics of the Black Belt continue to shape people's use of the landscape (Schmitz et al. 2003).

Early historical accounts (see references in Peacock 1992) provide vivid descriptions of the Black Belt landscape as a "park-like" mosaic of open grassland with interspersed clumps of trees and extensive cane brakes (*Arundinaria* sp.). Despite its charming appearance, the Black Belt presented one major obstacle to historical settlement: a lack of permanent water. This characteristic was noted time and again by explorers, travelers, and missionaries, and is reflected in a lack of water references in place names in the prairie today (Schmitz et al. 2003:196).

One Native American solution was the excavation of large pits. Although these may have served primarily as sources of clay for use in constructing wattle and daub houses (Jennings 1947:54), they were water-storage features replenished by undertaking arduous trips to more permanent water sources (Schmitz et al. 2003). Some natural depressions also were used in this manner:

They [the Chickasaws] live nearly in the center of a very large and somewhat uneven savannah, of a diameter of above three miles; this savannah at all times has but a barren look, the earth is very Nitrous, and the savages get their water out of holes or wells dug near the town . . . they live so far from waters, but they learn their children to swim in clay holes, that are filled in wet seasons by rain (an observation from 1771 by Romans [1999:124–125]).

Similar features were used by early 19th-century settlers:

There are prairies . . . of many miles in extent; the residents on which suffer greatly in dry seasons from the scarcity of water, a want that in a hot climate is peculiarly felt. There are no springs in the prairies, and the inhabitants depend on the rain-water, which, owing to the tenacity of the soil, does not soak into the ground, but accumulates in the hollows until evaporated by the sun. These hollows are sometimes large, and in winter, and during rainy seasons, form permanent ponds of considerable magnitude, but the water is of course very unwholesome.

    I have heard sad accounts of the privations undergone by planters on these "dry and thirsty lands." Not very far from this neighborhood there was a family, whose depend-ence was a large pond of this kind. The weather was excessively hot, and they were panting with thirst all day long, yet dared not use the water but in the most parsimo-nious manner. In any other circumstances it would have been rejected with loathing, for it was green, and stagnant, and lukewarm, and in one part of the pond lay the bloated carcase [sic] of a dead horse, to add to its flavour. This they were reduced to drink until it was absolutely impossible, when the only resource was to send the wag-gons [sic] and team with a large tub almost daily a distance of many miles, at a great expense of labour and time (Gosse 1993:80–81).

Such water-storage pits were eventually replaced by leveed impoundments, followed by cisterns, and, ultimately, bored wells (Schmitz et al. 2003), a pro-gression that allowed settlement to become more extensive over time. The apparent shortage of water described in early documents makes the Native American farming settlement pattern in the Black Belt something of a mystery: the pits described scarcely seem adequate to support such a system. One hypothesis is that the environment was wetter when Native American farming became widespread. The historical accounts may document a last-ditch effort to maintain an earlier system in the face of conditions which had become drier by the 1700s.

    It is important to characterize the tree cover on forested areas in the Black Belt in order to understand how environmental characteristics of the province might have shaped Native American lifeways. Some archaeologists argue, for example, that eastern red cedar (*Juniperus virginiana*), the dominant tree in the Black Prairie today, was also dominant in the past. Using this assumption, a formal settlement pattern model was developed based on expectations of white-tailed deer (*Odocoileus virginianus*) behavior and human predation in such an ecological setting (Johnson 1990; Johnson et al. 1984; Johnson and Sparks 1986). Specifically, Johnson argues that a settlement pattern shift took place shortly before the first appearance of Europeans (the expedition of Hernando de Soto in A.D. 1540). This shift involved populations moving from the major river valleys into dispersed homesteads located in or near cedar glades in the Black Prairie. Although the reasons for the hypothesized move are unclear, Johnson argues for a subsistence reorientation based on the sup-position that cedar glades are prime habitat for white-tailed deer.

    An alternative hypothesis holds that a post oak-hickory association formerly was present on "acid caps," relict alluvial soils surmounting some of the higher land forms in the province. In this hypothesis, the cedar glades so common in the prairie today are interpreted as the product of human-induced erosion and subsequent vegetative community change (see discussion in Peacock 1992; Peacock and Miller 1990). The associated settlement pattern model holds that instead of an abrupt settlement shift, settlement in the Black Prairie was continuous over a period of at least several centuries (Peacock and Rafferty 1996; Rafferty 1996).

The ecological aspects of this latter scenario are more consistent with historical accounts. For example, an early land surveyor described the Black Belt in Alabama as "an extensive body of level rich land, of fine black or chocolate colored soil. The principle growth is hickory; black oak, post oak, dogwood, and poplar are also common, but pine timber is rather scarce" (W. Roberts, quoted in Darby 1818:133). Other 18th- and 19th-century accounts consistently mention oak, hickory, and other hardwoods (see references in Peacock 1992), while reports of cedar are rare.

These anecdotal accounts are bolstered by a more formal set of historical records, the General Land Office (GLO) survey notes. In the early 19th century, instructions were issued by the U.S. government to surveyors establishing land lines still in use today (e.g., Fitz 1832). Townships (six-mile-square parcels of land) were established, each township divided into 36 one-mile-square sections. At the corners of the sections, surveyors were required to mark "witness trees" and to record the species and diameters of trees so marked. The result is a systematic record of forest conditions before extensive modern impacts. In the Black Belt, the GLO notes cover the period from approximately 1832 to 1834. In both Mississippi and Alabama, the notes consistently present the same picture: a mosaic landscape of open prairie patches interspersed with oak/hickory-dominated hardwood stands, with cedar mentioned only rarely (Peacock and Miller 1990).

How is it that, in the face of these historical records, a settlement model could be constructed based upon an entirely different set of environmental conditions, i.e., cedar glades? In constructing his model, Johnson correctly noted a factor that must be considered whenever historical records are used: recorder bias. Historical accounts are never purely objective. People's observations of the world are conditioned by many things: their purpose in making the observations, their knowledge of nature and their observational skills, their cultural background, their familiarity with the area being described, and so on. A comparison of records is often revealing in this regard: where a casual observer might dwell upon the lack of water in an area, an emigrant's guide might extol the agricultural richness of the soil. Even the GLO notes, compiled under formal government directions, are biased by surveyor skill and other factors. For example, surveyors were required to mark witness trees by blazing them with a hatchet. Cedars exude a very tenacious, sticky sap and have a multitude of low, stiff branches that impede access to the trunk. These characteristics make them relatively difficult to mark; hence, surveyors might well bypass a cedar if a suitable hardwood tree were near at hand (Bourdo 1956). Johnson argues that cedars were, in fact, dominant in the Black Belt in the past just as they are today, but are largely absent from the GLO notes because of surveyor bias (Johnson 1990).

This debate highlights the need for using multiple paleoenvironmental data sources: reliance on a single source often leads to arguing from negative evidence, as every single source is inherently biased in one way or another. If multiple data sources present a similar picture of past environmental conditions, confidence in the results is much higher and subsequent archaeological interpretations more defensible. Mollusk remains are a common constituent of archaeological sites worldwide, and are frequently employed in paleoenvironmental analysis. Land snail shells are particularly suitable for two reasons: (1) small land snails were rarely used as food, and thus do not suffer from the cultural biases that shape most other faunal assemblages and (2) land snails do not travel far, and thus reflect local environmental conditions. With these principles in mind, we now turn to our analysis of modern and archaeological mollusks from the Mississippi Black Prairie.

METHODS

Snail (Gastropoda) shells and freshwater mussel (Unionidae) valves were recovered during salvage excavations (Rafferty and Hogue 1998, 1999; Hogue and Peacock 1995) at several sites near Starkville, Mississippi (Figure 7-1). Site 22OK904 was on a small rise surrounded by pasture; at the time of excavation the rise had a few large, scattered hardwoods and numerous cedar saplings with a grass understory. Most of 22OK905 was in pasture, with a dense stand of young cedars on one end. Occupations at these sites range in age from about A.D. 1300 to 1750.

Mollusks were recovered from general excavation levels by water-screening soil through 0.54 cm (1/4-inch) and fine-mesh (1.59 mm) screen. All feature (pit and posthole) fill was subjected to flotation, and shells were hand-picked out of the resultant light and heavy fractions. Snail shells were placed in water in a sonic cleaner for five minutes to clear dirt from the apertures (Figure 7-2).

Figure 7-2. A specimen of wing snaggletooth (*Gastrocopta procera*). Diagnostic, preserved features such as the folds or "teeth" visible within the aperture are used to identify archaeological land snail species.

The results presented in this case study are based upon the final, complete analysis of mollusks from the two sites and should take precedence over earlier reports (e.g., Peacock and Melsheimer 2003; Peacock et al. 2005). Two assemblages from each site are represented, a modern (unbleached) assemblage obtained primarily from the upper zones and an archaeological (bleached) assemblage obtained from excavation levels and feature fill (Table 7-1). Some apparently modern snails were found in the features, probably the result of downward movement caused by bioturbation. Examination under a microscope using a powerful light source allowed the bleached and unbleached shells to be separated with a high degree of certainty and very little overlap between the two

Table 7-1. Archaeological and Modern Gastropod Taxa from 22OK904 and 22OK905[a]

Taxa	22OK904A	22OK904M	22OK905A	22OK905M
Helicinidae				
Globular drop (*Oligyra orbiculata*)	1		2	18
Lymnaeidae				
Fossaria (*Fossaria* sp.)			16	
Ellobiidae				
Ice thorn (*Carychium exile*)				
Succineidae (Ambersnail)[1]			241	538
Pupillidae				
White-lip dagger (*Pupoides albilabris*)	13	871	223	14894
Strobilopsidae				
Bronze pinecone (*Strobilops aeneus*)				
Maze pinecone (*Strobilops labyrinthicus*)				
Southern pinecone (*Strobilops texasianus*)				2
Vertiginidae				
Armed snaggletooth (*Gastrocopta armifera*)	244	39	457	293
Bottleneck snaggletooth (*Gastrocopta contracta*)	11	3	16	63
Comb snaggletooth (*Gastrocopta pentodon*)	44	4	57	17
Wing snaggletooth (*Gastrocopta procera*)		127	123	7718
White snaggletooth (*Gastrocopta tappaniana*)	1			
Snaggletooth (*Gastrocopta* sp.)	2	2	3	47
Black vertigo (*Vertigo milium*)		1		
Palmetto vertigo (*Vertigo oralis*)		1		
Ovate vertigo (*Vertigo ovata*)		1		
Striate vertigo (*Vertigo rugosula*)		5		4
Orthalicidae				
Whitewashed rabdotus (*Rabdotus dealbatus*)		2		89
Punctidae				
Small spot (*Punctum minutissimum*)	2		1	
Helicodiscidae				
Compound coil (*Helicodiscus parallelus*)	2		300	1
Coil (*Helicodiscus* sp.)	10			
Coil (*Lucilla* sp.)[2]	44	11	1140	40
Gastrodontidae				
Medium striate (*Striatura meridionalis*)				
Perforate dome (*Ventridens demissus*)			1	
Pyramid dome (*Ventridens intertextus*)				

(Continued)

Table 7-1. (*Continued*)

Taxa	22OK904A	22OK904M	22OK905A	22OK905M
Globose dome (*Ventridens ligera*)				
Dome (*Ventridens* sp.)				
Quick gloss (*Zonitoides arboreus*)	1		21	4
Euconulidae				
Wild hive (*Euconulus chersinus*)	13	13	114	39
Oxychilidae[3]				
Carved glyph (*Glyphyalinia* cf. *indentata*)	29	12	92	53
Pale glyph (*Glyphyalinia lewisiana*)			66	13
Imperforate glyph (*Glyphyalinia solida*)			31	
Bright glyph (*Glyphyalinia wheatleyi*)				
Globose button (*Mesomphix* cf. *globosus*)				
Unidentified Oxychilidae	2	1	2	1
Pristilomatidae				
Minute gem (*Hawaiia minuscula*)	607	14	337	53
Polygyridae				
Gulf Coast liptooth (*Daedalochila leporina*)				
Lowland pillsnail (*Euchemotrema leai*)	1	1	5	9
Shagreen (*Inflectarius inflectus*)			13	
Texas liptooth (*Linisa texasiana*)				1
Yellow globelet (*Mesodon clausus*)			11	
White-lip globe (*Mesodon thyroidus*)			3	
Hairy slitmouth (*Stenotrema hirsutum*)				
Magnolia threetooth (*Triodopsis hopetonensis*)	1	5	20	
Threetooth (*Triodopsis* sp.)				
Unidentified Polygyridae	4	3	6	1
Unidentified Gastropoda	10	1	14	15
Totals	1042	1117	3315	23913

[a] "A" after the site number indicates archaeological assemblages and "M" indicates modern assemblages from the two sites, 22OK904 and 22OK905, as discussed in the text.

faunas (details in Peacock et al. 2005). As will be seen below, the presumably modern samples from the two sites are very similar to verifiably modern samples taken from non-archaeological contexts in the prairie. A total of 4,357 archaeological snails (1,042 from 22OK904 and 3,315 from 22OK905) and 25,030 modern snails (1,117 from 22OK904, 23,913 from 22OK905) yield a combined total of 29,387 snails analyzed from the two sites. Snails from other sites mentioned in this chapter have not yet been analyzed[4].

A variety of modern micro-habitats in the Black Prairie were sampled for snails to provide data on the habitat requirements of particular species (Table 7-2). Sampling locales were selected subjectively as representative of each habitat. Cedar glades are rises covered primarily with cedar trees, with a grass ground cover where the soil is not eroded. "Prairie" samples are taken from well-preserved, open grasslands on the Tombigbee National Forest in Chickasaw County, Mississippi. The bottomland hardwood sample was taken in a forest of mixed sweetgum (*Liquidambar styraciflua*), boxelder (*Acer negundo*), sugarberry (*Celtis laevigata*), Osage-orange (*Maclura pomifera*), American and winged elm (*Ulmus americana* and *U. alata*), and pecan (*Carya illinoensis*).

Table 7-2. Other Modern Assemblages of Gastropods Obtained from Various Habitats in the Mississippi Black Prairie[a]

Taxa	UH1	UH2	UH3	UH4	UM1	UM2	BH	OO1	OO2	OO3	CG1	CG2	P1	P2	P3	P4	CANE	PINE
Helicinidae																		
Globular drop (*Oligyra orbiculata*)											96							1
Ellobiidae																		
Ice thorn (*Carychium exile*)							2											
Succineidae (Ambersnail)		209					1		11	8	182	6	47	16	210	35		
Pupillidae																		
White-lip dagger (*Pupoides albilabris*)			1		1		3			2	680	110	86	58	195	12	1	
Strobilopsidae																		
Bronze pinecone (*Strobilops aeneus*)				3			2											
Maze pinecone (*S. labyrinthicus*)				1														
Southern pinecone (*S. texasianus*)		1					8	5										
Vertiginidae																		
Armed snaggletooth (*Gastrocopta armifera*)								3	10	17								
Bottleneck snaggletooth (*G. contracta*)		37			9		9			14	33	3	2				1	
Comb snaggletooth (*G. pentodon*)				2	18	9	4				24	36						
Wing snaggletooth (*G. procera*)			1		1	3	1		1	3	293	28	12	7	76		2	
White snaggletooth (*G. tappaniana*)	1		1							9	4							
Snaggletooth (*Gastrocopta* sp.)		3			2	1						90			4	1	1	
Black vertigo (*Vertigo milium*)		2			10													
Striate vertigo (*V. rugosula*)												6						
Punctidae																		
Small spot (*Punctum minutissimum*)		1			4						1	27						
Helicodiscidae																		
Compound coil (*Helicodiscus parallelus*)		4	11				2				6	3	3				2	
Coil (*Lucilla* sp.)							1	2										
Gastrodontidae																		
Medium striate (*Striatura meridionalis*)		21	1	7	30	4					2				1			20
Perforate dome (*Ventridens demissus*)		9	3								44	14					8	
Pyramid dome (*V. intertextus*)	10						1											

(Continued)

Table 7-2. (Continued)

Taxa	UH1	UH2	UH3	UH4	UM1	UM2	BH	OO1	OO2	OO3	CG1	CG2	P1	P2	P3	P4	CANE	PINE
Globose dome (*V. ligera*)							4											
Dome (*Ventridens* sp.)							6											
Quick gloss (*Zonitoides arboreus*)				1				1	2									
Euconulidae																		
Wild hive (*Euconulus chersinus*)		4			16			6	6	16	29	3	4				13	
Oxychilidae																		
Carved glyph (*Glyphyalinia* cf. *indentata*)	9	30	2		24	3	4	1	10	12	78	5	7	3	1		24	
Bright glyph (*G. wheatleyi*)	3																2	2
Globose button (*Mesomphix* cf. *globosus*)		4			1													
Unidentified Oxychilidae																1		
Pristilomatidae																		
Minute gem (*Hawaiia minuscula*)			21		26	23	11	2	1	7	16	12	7					
Polygyridae																	7	
Gulf Coast liptooth (*Daedalochila leporina*)				1		1								4	4			
Lowland pillsnail (*Euchemotrema leai*)					2											1	1	
Yellow globelet (*Mesodon clausus*)										1				1				
Hairy slitmouth (*Stenotrema hirsutum*)						1												
Magnolia threetooth (*Triodopsis hopetonensis*)											41	3					2	
Threetooth (*Triodopsis* sp.)																	7	
Unidentified Polygyridae		3	1					2		13		2	1	7		1		
Unidentified Gastropoda	1	3		14		11				1	47	33	4		1			2
Totals	23	332	21	50	145	59	61	22	39	105	1576	381	173	92	491	54	71	25

[a] BH – Bottomland Hardwoods; CANE – Canebrake; CG – Cedar Glade; OO – Osage-orange; PINE – Pine; P – Prairie; UH – Upland Hardwoods; UM – Upland Mixed Hardwoods.

The upland hardwood sampling locales are more variable (Table 7-2). Sample #1 in that category was taken from a ridge with various hardwood species, but the tree cover within 10 m of the sample consisted almost entirely of Osage-orange. This is a common tree in the Black Prairie and may occur in discrete stands, as was the case for three additional sampling locales. Opinions differ on whether this species is native to the Black Prairie or was introduced by Euro-Americans (Brown 2003; Schambach 2003), an important question given its excellence as a wood for making bows (Schambach 2003). Upland hardwood samples #2 and #3 came from a broad, upland flat with a mix of shagbark and mockernut hickories (*Carya ovata* and *C. tomentosa*), American and winged elms, post oak, and small understory trees such as red mulberry (*Morus rubra*) and huckleberry (*Vaccinium* sp.). Upland hardwood sample #4 came from an isolated clump of hardwoods surrounded by tallgrass prairie: post oak, honey locust (*Gleditsia triacanthos*), and Osage-orange were the main trees, along with blackjack oak (*Quercus marilandica*) and sugarberry.

Several other areas were sampled (Table 7-2). Two samples were taken from "upland mixed" stands of hardwoods and cedars. One sample was taken from a stand of native cane. Finally, one sample was taken from a stand of mature loblolly pines (*Pinus taeda*) planted on an acid cap surrounded by tallgrass prairie.

Approximate 2-liter samples of leaf/grass litter were taken randomly from 1-meter square plots in each area. These samples were washed in a 500-micron screen with the largest plant detritus being washed and then removed. The remaining debris was hand-sorted. The mollusks were identified under low-power magnification; representative samples were identified using the extensive reference collection at the Field Museum of Natural History, Chicago, Illinois.

The identification of mollusks from archaeological contexts is not without problems. Some species are only identifiable by dissection of the soft body, in which case identification can only go to the genus or family level with shell alone. Where identification to species is possible using shell characteristics, some factors still need to be borne in mind: (1) certain shell characteristics such as color and structure of the periostracum (the outermost proteinaceous shell layer) are virtually always lost in archaeological material; (2) poor preservation may further change shells so that some diagnostic features (e.g., shell microsculpture) are lost or damaged; and (3) a large portion of the mollusks recovered are from immature individuals or their shells are fragmentary whereas identification handbooks and field guides describe mature, complete specimens with fully developed features. Thus, considerable experience is needed to make correct identifications of archaeological specimens. Obtaining modern samples from the surrounding landscape is very useful. In most cases it is possible to match at least some of the archaeological and recent forms so that identification of the latter allows for identification of the former.

Another problem with land snail analysis is that mollusk taxonomy is far from stable. Even in relatively well-known areas such as North America, new species are still being described. Conversely, taxonomists revising a systematic group may combine previously separated species. Finally, ongoing research and evolving methods of taxonomy and systematics lead to frequently changing and sometimes controversial hypotheses regarding evolutionary relationships among species. As researchers adapt molluscan nomenclature to new systematic models, names of molluscan groups at all taxonomic levels change as well. For beginners it is highly recommended to have identifications and nomenclature checked by a specialist.

The land snail data are examined using Detrended Correspondence Analysis (DCA) using the program PC-ORD (McCune and Mefford 1997). Correspondence analysis is a way of displaying ecological data, in which each sampling locale is represented as a single point. Each point is measured against all other points in terms of species characteristics and is mathematically positioned in multidimensional space. An ordination diagram represents a two-dimensional slice through the resulting cloud of data points. Axes are drawn through the cloud in such a way as to represent as much variation as possible. Axis 1 lies as far away from all the points as a straight line can; Axis 2 must be perpendicular to Axis 1, but otherwise positioned so as to account for as much of the remaining variation in the data as possible (Gauch 1982; Gower 1987; Ludwig and Reynolds 1988). The scale is not set because distance varies with the axes represented. In general terms, sample points are described as being either positive or negative along individual axes, with the distance between them being a relative measure of similarity or dissimilarity. Patterns in the data are subjectively interpreted by the researcher. This process can be continued for any number of axes, but in most paleoenvironmental applications only the variation represented by the first two or three axes is meaningful. Detrending is a complicated mathematical procedure that corrects for artificial patterns in the data which can occur if samples are taken along an ecological gradient (ter Braak 1985; ter Braak and Prentice 1988). Many of the modern snail samples were taken along transects that ran through different adjacent environmental zones, so detrending is appropriate for these data. Samples with twenty or more identifiable shells were used and rare species were downweighted.

RESULTS

The DCA results can be seen in Figure 7-3. The two archaeological samples are labeled with the site numbers followed by "a" and the modern samples from the two sites are followed by "m." The distribution of the sample points is a direct reflection of different environmental conditions. The modern cedar glade and prairie samples fall together on the left-hand side of the diagram, along with the two modern samples from 22OK904 and 22OK905. The pine stand, not surprisingly, is clearly separated from all other sample points. The remaining points are spread out along both axes, reflecting subtle variations in the soil conditions and vegetation of the sampling locales. The archaeological

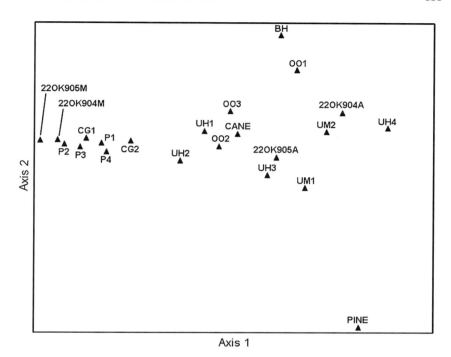

Figure 7-3. An ordination diagram showing the hyperspace distribution of snail sampling locales in the Black Prairie produced by Detrended Correspondence Analysis. The distribution of the points is based upon the characteristics of the snail assemblages from each locale, which in turn reflect environmental differences such as soil type and vegetation cover. Axis length is related to similarities between assemblages, with Axis 1 accounting for 69.8% and Axis 2 accounting for another 24.6% of the inter-assemblage variation. Most of the variation along Axis 1 reflects the difference between prairie and hardwood snail species. The distribution along Axis 2 is primarily structured by the bottomland hardwood and pine stand assemblages, which have markedly different faunas. "A" indicates archaeological assemblages and "M" indicates modern assemblages from the two sites, 22OK904 and 22OK905, discussed in the text. Other modern assemblages were obtained from various habitats in the Black Prairie. BH – Bottomland Hardwoods; CANE – Canebrake; CG – Cedar Glade; OO – Osage-orange; PINE – Pine; P – Prairie; UH – Upland Hardwoods; UM – Upland Mixed Hardwoods.

snail assemblages do not fall into any particular group, but clearly indicate that in the past the site locales were forested with hardwoods.

Major snail species associated with prairie and cedar glade settings include globular drop (*Oligyra orbiculata*), white-lip dagger (*Pupoides albilabris*), wing snaggletooth (*Gastrocopta procera*), and whitewashed rabdotus (*Rabdotus dealbatus*). Minority species include palmetto vertigo (*Vertigo oralis*), ovate vertigo (*V. ovata*), and striate vertigo (*V. rugosula*). The pine stand sample consists almost solely of medium striate (*Striatura meridionalis*), a species apparently able to tolerate highly acidic conditions. The bottomland hardwood sample is distinguished by southern pinecone (*Strobilops texasianus*), globose dome (*Ventridens ligera*), and pyramid dome (*V. intertextus*). Two specimens of ice thorn (*Carychium exile*) also are represented, the only specimens found in either modern or archaeological contexts. The other samples are more difficult to characterize because the species data do not group distinctly, an indication that the ecological preferences of some species (e.g., minute gem [*Hawaiia minuscula*]) are fairly broad (cf., Hinton 1951). Of particular interest is the recovery of 16 archaeological specimens of fossaria (*Fossaria* sp.) from a pit feature at 22OK905. This is a genus of amphibious snails not found in any of the modern samples. Its presence in archaeological contexts could indicate wetter conditions in the past (e.g., it might indicate the former presence of a seep spring at the site). Another explanation is that these mollusks were brought from elsewhere, perhaps when the pit was stocked with water from a creek. However, the small size of the pit (about 1.5 m across and 40 cm deep) argues against its use as a water-storage feature such as those described at Chickasaw settlements in the 18th and early 19th centuries.

The freshwater mussels provide some support for the hypothesis that conditions were wetter in the past. Although not present in large amounts, mussel shell is found in a variety of contexts at most of the archaeological sites tested. These data are presented in Table 7-3 as number of valves recovered.

Table 7-3. Freshwater Mussel Valves from Archaeological Sites in the Black Prairie[a]

	22OK534	22OK595	22OK793	22OK904	22OK905	22OK912
Elephantear (*Elliptio crassidens*)	1	1	-	1	1	-
Southern pigtoe (*Fusconaia cerina*)	-	-	-	-	-	1
Rough fatmucket (*Lampsilis straminea straminea*)	1	-	-	1	2	-
Freshwater mussel (*Lampsilis* sp.)	-	-	-	1	2	-
Freshwater mussel (*Obovaria* sp.)	-	-	-	-	-	1
Washboard (*Megalonaias nervosa*)	-	-	1	-	-	-
Southern clubshell (*Pleurobema decisum*)	-	1	-	-	-	-
Southern mapleleaf (*Quadrula apiculata*)	-	-	-	1	-	-
Freshwater mussel (*Quadrula* sp.)	1	-	-	1	-	-
Pistolgrip (*Tritogonia verrucosa*)	1	-	-	-	-	-
Deertoe (*Truncilla truncata*)	-	-	-	-	1	-
Unidentifiable mussel	4	-	1	7	10	-

[a] The numbers refer to the total number of valves (left and right valves) recovered from the sites. In cases where more than one valve is present, attempts at matching valves showed that they were of different sizes, so different individuals are represented. The numbers therefore also constitute the Minimum Number of Individuals (MNI) represented at each site.

The mussel species represented are commonly found in archaeological sites along the Tombigbee River, the largest waterway in the area, located about 32 km east of Starkville (Peacock 2000), but all are found in tributary streams as well (e.g., Miller 2001). Because mussels cannot survive extended periods without water, their presence in Black Prairie archaeological sites suggests that smaller streams historically recorded as intermittent once flowed year-round. It is possible that the mussels were obtained from far away, although this appears to have been uncommon in the southeastern United States (Peacock 2000). Ultimately, chemical sourcing of mussel shells to specific waterways may make it possible to address this question directly (Peacock et al. 2003).

DISCUSSION

This work contributes to the resolution of an archaeological debate and raises possibilities concerning environmental factors underlying the initiation and spread of farming in the Black Prairie. The archaeological snail faunas from the two sites clearly indicate a hardwood setting. The exact nature of those hard-wood stands is unknown. Given that these sites are "habitat islands" (Butzer 1982), i.e., distinctly human-modified environments, it may be that no modern analogues exist for them. Although it would be useful to examine snail assemblages from additional sites, the data from 22OK904 and 22OK905 indicate that these sites were not located in cedar glades. This seems to falsify Johnson's hypothesis that Native American inhabitants of the Black Prairie settled in cedar glade settings (see Peacock and Miller 1990). Nonetheless, continued analysis of modern snail collections and comparison with other biotic data to refine our knowledge of former plant communities is appropriate. Additionally, further study may resolve whether Osage-orange, a species of considerable economic importance in some parts of the region prior to Euro-American colonization, is native or introduced to the Black Prairie.

The consistent presence of freshwater mussels at Black Prairie sites may be an indication that conditions in the past were wetter than those recorded during the past 500 years. Exactly when farming began in the province is still debated (Hogue and Peacock 1995; Johnson 1996; Peacock and Rafferty 1996; Rafferty 1996), but the presence of an established farming system by circa A.D. 1250–1300 seems reasonable. As more data become available, it will be interesting to compare the Black Prairie to other parts of the region. For example, Stahle and Cleaveland (1994) examined bald cypress (*Taxodium distichum*) tree rings in three southeastern states and found evidence for a circa 60-year drought beginning in the mid-18th century, a time approximately coincident with historical descriptions of water shortages in the Black Belt.

The land snail data from 22OK904 and 22OK905 make it clear that the Black Belt landscape is massively transformed. The historical records, in this case, appear to be quite accurate, meaning that the transformation of a hardwood mosaic into a cedar-dominated mosaic took place primarily as a result of land clearance and erosion over the past five centuries. Given that colonial descriptions of forest conditions are corroborated, historical references to other landscape elements such as the extensive cane brakes (Mohr 1901) may also be accurate, providing additional subjects for ecological restoration.

It is clear, however, that some landscape elements are gone and will be recovered only with great expense and effort. This is the case with the "seep springs" noted historically and perhaps reflected by the presence of an amphibious snail genus at 22OK905. Such springs likely were most common along the edges of the Black Belt where residual alluvial soils were thickest (Schmitz et al. 2003). Those soils and springs, along with their plants and animals, are lost due to human impacts. The accumulation of more paleoenvironmental data from archaeological sites will be the only way to characterize these vanished communities and to provide further ecological baselines for restoration efforts.

CONCLUSIONS

The interpretations offered here would not be possible without modern land snail collections and records of the ecological conditions under which they lived. This is the case with many types of biotic remains found at archaeological sites. Environmental archaeologists must be willing to devote the time and effort needed to establish the ecological requirements of species recovered from the sites where they work. Collaboration with specialists is essential in this regard, and may result in unexpected findings such as expanding the known ranges of species (e.g., Peacock and James 2002). As with historical records, modern analogues for biotic communities identified in the archaeological record cannot be assumed. Such are the challenges of environmental archaeology, but the promises – providing a robust context for archaeological interpretations, expanding our ecological knowledge, and providing information useful in an applied setting – make meeting such challenges more than worthwhile.

ACKNOWLEDGMENTS

This material is based upon work supported by the National Science Foundation under Grant No. 0003833. We thank the following people for their help in the field or in the lab: Jeffrey Alvey, Keith Baca, Randle Beavers, Donovan Broussard, Agnes Burris, Vincent Dongarra, Paul Jacobs, Rebecca Melsheimer, Janet Rafferty, Nicole Rafferty, and Jennifer Seltzer. We would like to thank Dr. Joe Seger, Director of the Cobb Institute, for his continuing support. Finally, we would like to thank the editors for inviting us to contribute to this volume and for their useful comments and suggestions. Materials generated from this work are curated at the Cobb Institute of Archaeology, Mississippi State University.

NOTES

[1] For identification of many species in this family, anatomical data are required. All mollusks belonging to this family are referred to as Succineidae only.

[2] *Lucilla* is more widely known as *Hebetodiscus* or, as a subgenus, *Helicodiscus* (*Hebetodiscus*). In Peacock and Melsheimer (2003) and Peacock et al. (2005), specimens were identified as *Helicodiscus singleyanus* or *Helicodiscus inermis*. However, according to recent publications, the older name *Lucilla* has priority over the other generic names. The distinction between the two species *L. singleyana* and *L. inermis* and, thus, the taxonomic status of the latter, are problematic. Specimens resembling these two nominal species are referred to *Lucilla* sp.

[3] Of the *Glyphyalinia* species in the study area, weathered shells of *G. indentata* and *G. luticola* cannot be separated reliably. All mollusks resembling these two species are referred to *G.* cf. *indentata*.

[4] Other changes from Peacock and Melsheimer (2003) and Peacock et al. (2005) include combining *Gastrocopta armifera* and *G. abbreviata* into *G. armifera*, combining *Helicodiscus parallelus* and *H. notius* into *H. parallelus*, updating *Mesodon inflectus* to *Inflectarius inflectus*, and updating *Polygyra texasiana* to *Linisa texasiana* (Turgeon et al. 1998).

REFERENCES

Bourdo, E. A., Jr., 1956, A Review of the General Land Office Survey and of Its Use in Quantitative Studies of Former Forests, *Ecology* 37:754–768.

Brown, R. L., 2003, Paleoenvironment and Biogeography of the Mississippi Black Belt, in: *Blackland Prairies of the Gulf Coastal Plain: Nature, Culture, and Sustainability* (E. Peacock and T. Schauwecker, eds.), University of Alabama Press, Tuscaloosa, pp. 11–26.

Butzer, K. W., 1982, *Archaeology as Human Ecology*, Cambridge University Press, Cambridge, England.

Darby, W., 1818, *The Emigrant's Guide to the Western and Southwestern States and Territories*, Kirk and Mercein, New York.

Fitz, G., 1832, *Instructions for Surveying in the State of Mississippi*, R. Semple for the Surveyor General, Natchez, Mississippi, reprinted in 1992 by the Mississippi Association of Professional Surveyors, Washington, Mississippi.

Fogel, R. W., 1989, *Without Consent or Contract: The Rise and Fall of American Slavery*, W. W. Norton, New York.

Gauch, H. G., Jr., 1982, *Multivariate Analysis in Community Ecology*, Cambridge University Press, Cambridge, England.

Gosse, P. H., 1993, *Letters from Alabama: Chiefly Relating to Natural History* (H. H. Jackson, ed.), University of Alabama Press, Tuscaloosa, originally published in 1859 by Morgan and Chase, London.

Gower, J. C., 1987, Introduction to Ordination Techniques, in: *Developments in Numerical Ecology* (P. Legendre and L. Legendre, eds.), Springer-Verlag, Berlin, pp. 3–64.

Gray, L. C., 1933, *History of Agriculture in the Southern United States to 1860,* Volume 1, Carnegie Institution, Washington, DC.

Hinton, J. L., 1951, The Terrestrial Shell-bearing Mollusca of the Black Belt of Alabama, M.S. thesis, Department of Biology, University of Alabama, Tuscaloosa.

Hogue, S. H., 2003, Corn Dogs and Hush Puppies: Diet and Domestication at Two Protohistoric Farmsteads in Oktibbeha County, Mississippi, *Southeastern Archaeology* 22(2):185–195.

Hogue, S. H., and Peacock, E., 1995, Environmental and Osteological Analysis at the South Farm Site (22OK534), A Mississippian Farmstead in Oktibbeha County, Mississippi, *Southeastern Archaeology* 14(1):31–45.

Jennings, J. D. (ed.), 1947, Nutt's Trip to the Chickasaw Country, *Journal of Mississippi History* 9:34–61.

Johnson, J. K., 1990, Cedar Glades and Protohistoric Settlement: A Reply to Peacock and Miller, *Mississippi Archaeology* 25(2):58–62.

Johnson, J. K., 1996, The Nature and Timing of the Late Prehistoric Settlement of the Black Prairie in Northeast Mississippi: A Reply to Hogue, Peacock, and Rafferty, *Southeastern Archaeology* 15(2):244–248.

Johnson, J. K., Curry, H. K., Atkinson, J. R., and Sparks, J. T., 1984, Cultural Resources Survey in the Line Creek Watershed, Chickasaw, Clay and Webster Counties, Mississippi, report submitted to the U.S.D.A. Soil Conservation Service, Washington, DC.

Johnson, J. K., and Sparks, J. T., 1986, Protohistoric Settlement Patterns in Northeastern Mississippi, in: *The Protohistoric Period in the Mid-South* (D. H. Dye and R. C. Brister, eds.), *Mississippi Department of Archives and History Archaeological Report* 18, Jackson, pp. 64–87.

Ludwig, J. A., and Reynolds, J. F., 1988, *Statistical Ecology*, John Wiley and Sons, New York.

McCune, B., and Mefford, M. J., 1997, *PC-ORD, Multivariate Analysis of Ecological Data, Version 3.0*, MjM Software Design, Gleneden Beach, Oregon.

Miller, A. C., 2001, *An Analysis of Freshwater Mussels (Unionidae) along Luxapalila Creek, Mississippi, 1999 Studies*, U.S. Army Engineer Research and Development Center, Vicksburg, Environmental Laboratory Report ERDC/EL TR-01–26, Vicksburg, Mississippi.

Mohr, C., 1901, *Plant Life of Alabama*, Brown Printing Company, Montgomery, Alabama.

Peacock, E., 1992, Some Additional Notes on Forest Reconstruction in the Black Belt, *Mississippi Archaeology* 27(1):1–18.

Peacock, E., 1993, Reconstructing the Black Belt Environment Using Leaf Impressions in Daub, *Southeastern Archaeology* 12(2):148–154.

Peacock, E., 1998, Historical and Applied Perspectives on Prehistoric Land Use in Eastern North America, *Environment and History* 4(1):1–29.

Peacock, E., 2000, Assessing Bias in Archaeological Shell Assemblages, *Journal of Field Archaeology* 27(2):183–196.

Peacock, E., and James, T. R., 2002, A Prehistoric Unionid Assemblage from the Big Black River Drainage in Hinds County, Mississippi, *Journal of the Mississippi Academy of Sciences* 47(2):119–123.

Peacock, E., Meaker, T., Neff, H., and Rafferty, J., 2003, Sourcing Shell-tempered Ceramics Using Laser Ablation-Inductively Coupled Plasma-Mass Spectrometry (LA-ICP-MS), poster presented at the 60th Annual Meeting of the Southeastern Archaeological Conference, Charlotte, North Carolina.

Peacock, E., and Melsheimer, R., 2003, Terrestrial Gastropods from Archaeological Contexts in the Black Belt Province of Mississippi, in: *Blackland Prairies of the Gulf Coastal Plain: Nature, Culture, and Sustainability* (E. Peacock and T. Schauwecker, eds.), University of Alabama Press, Tuscaloosa, pp. 27–47.

Peacock, E., and Miller, W. F., 1990, Protohistoric Settlement Patterns in Northeast Mississippi and the Cedar Glade Hypothesis, *Mississippi Archaeology* 25(2):45–57.

Peacock, E., and Rafferty, J., 1996, Settlement Pattern Continuity and Change in the Mississippi Black Prairie: A Response to Johnson, *Southeastern Archaeology* 15(2):249–253.

Peacock, E., Rafferty, J., and Hogue, S. H., 2005, Land Snails, Artifacts and Faunal Remains: Understanding Site Formation Processes at Prehistoric/Protohistoric Sites in the Southeastern United States, in: *Archaeomalacology: Molluscs in Former Environments of Human Behavior* (D. Bar-Yosef, ed.), Oxbow Books, Oxford, England, pp. 6–17.

Peacock, E., and Reese, M. C., 2003, A Comparison of Three Methods of Paleoenvironmental Analysis at an Archaeological Site on the Mississippi Black Prairie, in: *Blackland Prairies of the*

Gulf Coastal Plain: Nature, Culture, and Sustainability (E. Peacock and T. Schauwecker, eds.), University of Alabama Press, Tuscaloosa, pp. 64–79.

Peacock, E., and Schauwecker, T. (eds.), 2003, *Blackland Prairies of the Gulf Coastal Plain: Nature, Culture, and Sustainability*, University of Alabama Press, Tuscaloosa.

Rafferty, J., 1996, Continuity in Woodland and Mississippian Settlement Patterning in the Mississippi Black Prairie, *Southeastern Archaeology* 15:230–243.

Rafferty, J., 2001, Determining Duration at Prehistoric Sites: Short-term Sedentary Settlement at Josey Farm, NE Mississippi, *Journal of Field Archaeology* 28(3 and 4):347–366.

Rafferty, J., 2003, Prehistoric Settlement Patterning on the Mississippi Black Prairie, in: *Blackland Prairies of the Gulf Coastal Plain: Nature, Culture, and Sustainability* (E. Peacock and T. Schauwecker, eds.), University of Alabama Press, Tuscaloosa, pp. 167–193.

Rafferty, J., and Hogue, S. H., 1998, Test Excavations at Six Sites in Oktibbeha County, Mississippi, report submitted to the Mississippi Department of Transportation, Jackson, by the Cobb Institute of Archaeology, Mississippi State University, Starkville, Mississippi.

Rafferty, J., and Hogue, S. H., 1999, Phase II Archaeological Testing, 22OK904, Oktibbeha County, Mississippi, report submitted to the Mississippi Department of Transportation, Jackson, by the Cobb Institute of Archaeology, Mississippi State University, Starkville, Mississippi.

Romans, B., 1999, *A Concise Natural History of East and West Florida* (K. E. H. Braund, ed.), University of Alabama Press, Tuscaloosa.

Schambach, F. F., 2003, Osage Orange Bows, Indian Horses, and the Blackland Prairie of Northeastern Texas, in: *Blackland Prairies of the Gulf Coastal Plain: Nature, Culture, and Sustainability* (E. Peacock and T. Schauwecker, eds.), University of Alabama Press, Tuscaloosa, pp. 212–236.

Schmitz, D. W., Wax, C. L., and Peacock, E., 2003, Water-resource Controls on Human Habitation in the Black Prairie of North-Central Mississippi, in: *Blackland Prairies of the Gulf Coastal Plain: Nature, Culture, and Sustainability* (E. Peacock and T. Schauwecker, eds.), University of Alabama Press, Tuscaloosa, pp. 194–211.

Stahle, D. W., and Cleaveland, M. C., 1994, Tree-ring Reconstructed Rainfall Over the Southeastern U.S.A. during the Medieval Warm Period and Little Ice Age, *Climatic Change* 26:199–212.

ter Braak, C. J. F., 1985, Correspondence Analysis of Incidence and Abundance Data: Properties in Terms of a Unimodal Response Model, *Biometrics* 41:859–873.

ter Braak, C. J. F., and Prentice, I. C., 1988, A Theory of Gradient Analysis, *Advances in Ecological Research* 18:272–317.

Tower, J. A., 1961, The Changing Black Belt – a Geographical Review, *Journal of the Alabama Academy of Science* 32:479–485.

Turgeon, D. D., Quinn, J. F., Bogan, A. E., Coan, E. V., Hochberg, F. G., Lyons, W. G., Mikkelsen, P. M., Neves, R. J., Roper, C. F. E., Rosenberg, G., Roth, B., Schetema, A., Thompson, F. G., Vecchione, M., and Williams, J. D., 1998, *Common and Scientific Names of Aquatic Invertebrates from the United States and Canada: Mollusks*, 2nd ed., *American Fisheries Society Special Publication* 26, Bethesda, Maryland.

*Chapter **8***

Climate Change and Archaeology: The Holocene History of El Niño on the Coast of Peru

C. Fred T. Andrus, Daniel H. Sandweiss, and Elizabeth J. Reitz

Environmental archaeologists locate their research at the dynamic intersection of human behavior and the environment conserved in archaeological sites. On the Peruvian coast, El Niño/Southern Oscillation (ENSO) is an important variable in that interaction and serves as an example of the wide range of environmental data that can be extracted from archaeofaunal remains. The importance of ENSO variation to archaeology on the coast of Peru has been explored for many years. Rollins et al. (1986a) used data from Ostra Base Camp, a Middle Preceramic, Peruvian fishing site, and nearby sites to suggest for the first time that the modern ENSO pattern was not established until after 3800 B.C. Subsequent controversy over the "Birth of El Niño" hypothesis led us to further field and laboratory research (see Sandweiss et al. 1996, 2001; Reitz and Sandweiss 2001; Sandweiss 2003 for reviews of this debate). In this case study, we examine the most recent zooarchaeological data from Ostra Base Camp for evidence of a change in ENSO patterns. The Ostra data suggest that the ENSO-related pattern of interannual climatic variability during the Middle Preceramic period (7000–3800 B.C.) was different from the patterns prevailing during the subsequent Late Preceramic and Initial periods (3800–800 B.C.) and the pattern which exists today. During the Middle Preceramic period, El Niño events appear to have been less frequent or absent and the seasonal range in sea surface temperatures (SST) was wider with warmer summers north of approximately 10°S. In this case study we draw upon evidence from many different sources to characterize ENSO and the relationship of changing patterns of this global atmospheric and oceanic phenomenon to the development of social complexity in Peru.

Detecting evidence for ENSO, documenting changes in ENSO, and delineating the impact of ENSO using archaeological remains only can be accomplished using a spectrum of environmental data from well-dated, large samples representing a chronological sequence that establishes a regional pattern. Such ideal conditions are rare and ENSO studies require compromises to build toward that elusive goal. The relationships among the archaeological record, site formation processes, analytical approaches, Holocene environments, and environmental change are widely reviewed elsewhere (e.g., Albarella 2001; Efremov 1940; Grayson 1984; Monks 1981; Reed 1963; Weigelt 1989). Suffice it to say that the archaeological record is incomplete and biased. Further, all methods introduce additional variables into the record. In particular, readers should be aware of the importance to environmental studies of cultural filters, site formation processes, excavation biases, sample size, exchange systems, geological forces, and the similarities between evidence for seasonal periodicity and evidence for large-scale ENSO variations. As with all archaeological analyses, the accumulation of evidence from many different sources strengthens efforts to characterize ENSO. Multiple lines of cultural, biological, and geological evidence also identify aspects of the archaeological record that are the consequence of human behavior unrelated to ENSO.

PERUVIAN ENVIRONMENTAL CONDITIONS AND CULTURAL CHRONOLOGY

Typically today, the waters off Peru are temperate, with cool, nutrient-rich waters upwelling near shore in response to steady trade winds. This pattern of oceanic and atmospheric conditions, coupled with the Andean rain shadow, results in a desert coast and a narrow seasonal cycle of sea and terrestrial temperatures. The only quasi-permanent fresh water sources in this region are braided streams that drain high-altitude glaciers. Today, the dry, temperate climate is punctuated approximately every 2 to 7 years by El Niño events; periods persisting for several months or more in which tropical surface water displaces the cool, upwelled water upon which the local ecosystems depend. The trade winds slacken or reverse during El Niño events, bringing rains to the coast and the western slope of the Andes. At such times, temperate fisheries collapse or migrate to cooler waters, most mollusk populations are decimated, and flood waters inundate coastal deserts. El Niño conditions contrast with those of La Niña, in which local coastal waters are cooler than average.

Both of these regimes are the result of coupled oceanic and atmospheric variables. The term El Niño refers to the oceanic component and Southern Oscillation to the atmospheric, with El Niño/Southern Oscillation (ENSO) encompassing the combined phenomena with three phases: El Niño warm events, La Niña cool events, and "normal." To fully consider the impact of climate on the history of Peru, we must consider catastrophic El Niños as well as the oceanic conditions within which ENSO operates. For example, nutrient upwelling, a critical part of ENSO, creates the most productive coastal fishery in the world (e.g., Bakun 1996) and is thus central to understanding the local maritime economy. Past upwelling supported an abundant fish fauna which is often characterized archaeologically by anchovies (Engraulidae: *Engraulis ringens*) and other cool water fishes (Reitz and Sandweiss 2001; Sandweiss et al. 2004).

These marine dynamics are fundamental to coastal Peru's cultural history. This history is divided into preceramic and ceramic stages. The Preceramic stage is subdivided into Early (11,000–7000 B.C.), Middle (7000–3800 B.C.), and Late (3800–2200 B.C.) Preceramic periods based upon a variety of factors; the Late Preceramic period is also known as the Cotton Preceramic (Benfer 1984; Keefer et al. 1998; Sandweiss et al. 1989; Sandweiss et al. 1998). Early and Middle Preceramic coastal sites have neither cotton (*Gossypium barbadense*) nor ceramics. Many Late Preceramic sites contain monumental architecture and industrial crops such as cotton (e.g., Béarez and Miranda 2000; Pozorski 1983; Pozorski and Pozorski 1979a; Quilter 1991; Quilter et al. 1991; Shady Solís et al. 2001). The Initial period (2200–800 B.C.) (e.g., Pozorski and Pozorski 1979b, 1987) is characterized by ceramics, agriculture, monumental structures, and many other aspects of a complex cultural life.

EVIDENCE FOR CLIMATE CHANGE AT OSTRA BASE CAMP

Ostra Base Camp is a Middle Preceramic site on the north coast of Peru (Figure 8-1). The site is associated with a series of collecting stations along the top of what is now a raised sea cliff at the southern end of an uplifted bay north of the Santa River Valley (Sandweiss et al. 1983). Radiocarbon dates place the occupation at circa 5200–4200 B.C. When it was occupied, the site sat on the shore of the bay, but the embayment ceased to exist before 3800 B.C. (Rollins et al. 1986a; Sandweiss 1986; Sandweiss et al. 1983). The site was abandoned at this time and not re-occupied.

Following initial reconnaissance work in 1980 (Rollins et al. 1986a; Sandweiss et al. 1983), Sandweiss (1996, 2003; Sandweiss et al. 1996) conducted field work at Ostra in 1991. Faunal materials were recovered

Figure 8-1. Map of study area showing sites.

using 1/4-inch and 1/16-inch meshed screens. The following discussion focuses on interpretations using Minimum Number of Individuals (MNI; *sensu* White), a quantitative technique based on the symmetry of mollusks and vertebrates described elsewhere (Reitz and Sandweiss 2001; Reitz and Wing 1999:194–199).

Well over 90% of the vertebrate individuals at Ostra Base Camp, as well as at other early and mid-Holocene coastal Peruvian sites, are marine organisms (Reitz 2001; Reitz and Sandweiss 2001). Due to the dominance of fishes and invertebrates in human efforts to obtain animal protein, fat, and fat-soluble vitamins it is likely that any change in the marine environment would influence human behavior.

Presence of Indicator Species

Many inferences about Holocene environments are based on modern biogeography of species present in the archaeofaunal collection. The absence of indicator species is highly problematic and could mean that the animal was not present locally, that its use is undocumented in the archaeological collection under study, or that it was not used. People generally focus their subsistence efforts on animals that are flexible in their environmental tolerances so it is seldom possible to assign specific water conditions with well-defined physical and chemical properties to animals in a collection. A complex of animals typically associated with one general environmental setting, however, can be contrasted with a complex of animals typical of another condition either at the study site or at other sites defining a temporal or spatial continuum in the region.

The importance of examining a complex of animals for confirmation of environmental change is clear in the Ostra Base Camp collection, which contains numerous warm-water mollusks and fishes as well as cool-water taxa. Two of the 13 molluscan species in the collection now live 400–500 km north of Ostra, at about 5°S (Sandweiss et al. 1983). These two species, Pacific calico scallop (*Argopecten circularis*) and semi-rough chione (*Chione subrugosa*), constitute 80% of the 713 invertebrate individuals. The two cool-water mussels (*Choromytilus chorus* and *Semimytilus algosus*) contribute only 1% of the individuals. The mussel *Choromytilus* was frequently traded in ancient Peru. Twelve of the 29 vertebrate taxa are typical of warm-water conditions and six are typical of cool waters (Reitz and Sandweiss 2001). All of the estimated 144 vertebrate individuals are marine. Over half of the Ostra taxa for which trophic level (see below) is estimated are typical of the warm-water regimes currently prevailing further north (Table 8-1; Reitz and Sandweiss 2001).

The Ostra collection is not unique in suggesting that early and mid-Holocene fishing focused on a complex of warm-water marine resources (Table 8-1; Reitz 2001; Sandweiss et al. 1996). The Siches site demonstrates the extent to which warm-water vertebrates could dominant the fishing strategy and also anchors the strong latitudinal gradient in place during the Middle Preceramic. The dominance of warm-water species on the north coast declines after circa 3800 B.C. in favor of cool-water taxa. This coast-wise gradient and the temporal shift in emphasis suggests that coastal waters were warmer in the Middle Preceramic than they were when Alto Salaverry (Late Preceramic) and Pampa de las Llamas (Initial period) were occupied (Reitz and Sandweiss 2001; Sandweiss et al. 1996, 2001).

Oxygen Isotope Analysis of Incremental Growth Structures

In organisms that grow throughout their lives, such as mollusks and fishes, growth advances rapidly under favorable conditions and slowly or not at all under unfavorable ones. Periods of rapid and slow growth are visible in some

Table 8-1. Warm, Cool, and Mixed Water Marine Vertebrates as a Percentage of the Marine Minimum Number of Individuals (MNI) Classified in Terms of Trophic Level Preferences[a]

	Warm%	Cool%	Mixed%	S	H'	MNI	Dates cal yr B.C. ([14]C B.P.)
Early and Middle Preceramic							
Siches site, Honda Phase	89.8	8.9	1.2	24	1.717	492	3950–3350 (5150–4550)
Siches site, Siches Phase	50.5	43.7	5.7	30	1.768	2190	6000–4900 (7100–6000)
Siches site, Amotape Phase	52.2	39.1	8.7	15	1.959	46	8750–8250 (9500–9000)
Ostra Base Camp	57.5	32.1	10.4	26	2.634	134	5250–4250 (6250–5450)
Almejas (1/4" fraction)	44.2	38.5	17.3	23	2.681	104	5850 (7000)
Paloma, Probability Samples, Level 100		61.5	38.5	4	1.220	13	~3500 (~4700)
Paloma, Probability Samples, Level 200		88.6	11.4	5	1.067	35	3950–3500 (5100–4700)
Paloma, Probability Samples, Level 300	3.7	87.0	9.3	11	1.531	54	4100–3950 (5300–5100)
Paloma, Probability Samples, Level 400		88.9	11.1	6	1.050	27	6550–4100 (7800–5300)
Ring Site	3.6	82.9	13.5	28	2.444	304	9250–3850 (10,575–5060)
Late Preceramic Alto Salaverry	6.2	85.9	7.8	11	1.311	64	3100–2150 (4450–3750)
Initial Period Pampa de las Llamas-Moxeke (1/4-inch fraction)	6.2	78.1	15.6	14	1.858	64	2150–1400 (3735–3070)

[a] Sites are arranged geographically from north to south within each time period (see Figure 8-1). Mixed includes marine mammals, sharks, rays, and bony fishes with varied habitat preferences (see Reitz 2001). Number of taxa (S) and Diversity (H') are based on marine MNI. Total MNI for each site assemblage may be higher if terrestrial animals were present. Faunal data are from Pozorski and Pozorski (1979a); Reitz (1995, 1999; 2003, 2004b); Reitz and Sandweiss (2001); and Sandweiss et al. (1989). Dates are from these sources as well as Benfer (1984) and Sandweiss et al. (1996). Age ranges are based on calibration of the uncalibrated means of earliest and latest radiocarbon dates for each site or of excavator's estimates when no ^{14}C dates are available, discounting outliers rejected by the excavators. The sources cited provide lists of dates and standard deviations. Human individuals and invertebrates are omitted from these calculations.

hard tissues as alternating bands of thin increments deposited during periods of slow growth or broad bands deposited during fast growth. In some circumstances, pairs of thin and broad increments represent one annual cycle of growth. In other instances, growth patterns reflect irregular and transient events that induce growth diminishment or cessation growth check. Thus, incremental growth patterns observed in mollusk shells and in the skeletons of bony fishes provide evidence of environmental conditions related to periodicity in environmental parameters experienced by animals during their lives (Reitz and Wing 1999:78–83; Wheeler and Jones 1989:154–161).

Although increments are studied for patterns in growth and season of death, they also serve as proxies for El Niño. For example, modern semi-rough chione and cockle (*Trachycardium procerum*) valves grown during strong El Niños on the coast of Peru often contain a growth check and subsequent alteration of shell curvature that is precipitated during the warmest phase of the event (Rollins et al. 1986b, 1987). Similarly, Meekan et al. (1999) note growth checks in otoliths (balance and hearing structures in the crania of teleost fishes) of reef fishes caught off the Pacific coast of Ecuador near the Galapagos Islands, in response to the warm sea surface temperatures of El Niño. Growth checks are caused by many factors (Rhoads and Lutz 1980) and the modern otoliths described in this paper did not regularly contain visual evidence of the 1997–1998 El Niño in their growth bands (Andrus et al. 2002a); therefore, they are not unambiguous indicators of El Niño. However, one would expect that a significant percentage of temperature-sensitive fishes and mollusks would display such growth anomalies in a period of frequent El Niños Although the Ostra site contains abundant otoliths and mollusks, including the cockle, none displays growth anomalies. This negative evidence is consistent with other evidence for diminished El Niño during the occupation of the site. Although visual analysis of skeletal increments is inconclusive, analysis of their isotopic chemistry is less ambiguous.

Mollusk valves and fish otoliths are composed of calcium carbonate ($CaCO_3$), most often in the form of aragonite. Oxygen isotopes within otoliths and most bivalve mollusks are precipitated in equilibrium with seawater, meaning that the ratio of ^{18}O and ^{16}O expressed as $\delta^{18}O$ in parts per mil (‰) relative to the international standard Vienna Pee Dee Belemnite (VPDB) is a function of seawater temperature and its isotopic composition. There is a negative correlation between $\delta^{18}O$ and temperature; as water temperatures rise, $\delta^{18}O$ values fall. A 0.22‰ change in $\delta^{18}O$ is equivalent to 1°C following Grossman and Ku (1986:66, equation b). Assuming stable seawater isotopic chemistry, the $\delta^{18}O$ values in otolith and mollusk increments serve as a paleothermometer characterizing the thermal conditions under which deposition of each band occurred. Thus seasonal and interannual variation in sea surface temperature are documented in the oxygen isotope records of otoliths and most bivalve shells. Otolith and mollusk samples excavated from the Ostra site were analyzed at high temporal resolution for $\delta^{18}O$ through ontogeny (Andrus et al. 2002b, 2003). The otoliths are from the Peruvian sea catfish (*Galeichthys peruvianus*), a species shown to record ENSO-related sea surface temperature variability (Andrus et al. 2002a). The mollusk valves are from the cockle; modern members of this species record El Niño (Rollins et al. 1986b, 1987).

Figure 8-2 summarizes the findings of Andrus et al. (2002b, 2003). Otoliths dating to 4850 B.C. contain evidence of an average seasonal range in sea surface temperature of over 7°C while modern specimens collected in nearby waters contain evidence of a range of approximately 4.8°C. Most of the seasonal increase in sea surface temperature variation occurs in summer months in which temperatures averaged almost 3°C warmer during the Ostra occupation compared to temperatures today.

These otolith data are consistent with those derived from the cockle valves. The mean $\delta^{18}O$ range recorded in valves that grew during the powerful 1982–1983 El Niño was 1.3‰; (Andrus et al. 2003; Perrier et al. 1994; Rollins et al. 1987). The mean $\delta^{18}O$ range in shells collected from fossil and archaeological deposits at Ostra is 1.5‰; (Andrus et al. 2003; Perrier et al. 1994). Thus

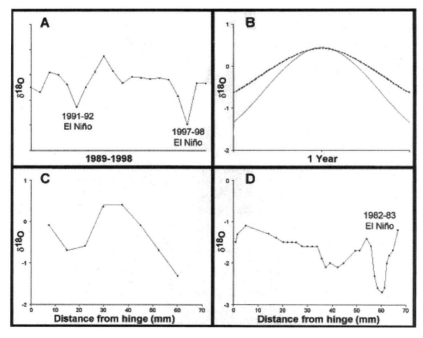

Figure 8-2. Oxygen isotope analysis. Panel A: composite profile from modern sea catfish otoliths (*Galeichthys peruvianus*); Panel B: average seasonal variation in modern (dashed) versus Ostra (solid) otolith profiles; Panel C: profile from Ostra cockle (*Trachycardium procerum*); and Panel D: profile from modern cockle (*T. procerum*) showing 1982–1983 El Niño (negative excursion at circa 60 mm). Plots are derived from data in Andrus et al. (2002a, 2002b, 2003).

ancient shells from Ostra, none of which contain evidence of El Niño in their growth structure, suggest a wider seasonal sea surface temperature range than that experienced by modern cockle during the 1982–1983 El Niño event. As the former bay at Ostra was an evaporative environment with altered oxygen isotope chemistry (Andrus et al 2003; Perrier et al. 1994), absolute temperatures cannot be calculated. Nevertheless, the comparative data corroborate the otolith data.

The isotope data support the assertion that El Niño warm events were less frequent or absent for some millennia prior to circa 3800 B.C. and the seasonal range in sea surface temperature was greater in north-central, coastal Peru. The isotope profiles themselves are too short to infer ENSO frequency, but the seasonal temperature range they suggest helps explain the distribution of warm-water fauna found at Ostra and other sites north of 10°S between circa 7000–3800 B.C.

Trophic Levels Emphasized

A switch from warmer to cooler sea surface temperature conditions after circa 3800 B.C. might explain a possible change in mean trophic level in the marine vertebrates. The mean trophic level in archaeological collections is estimated using modern trophic level data for marine vertebrates published in FishBase (Froese and Pauly 1998; Pauly and Christensen 1995; Pauly et al. 1998, 2000; Reitz 2004a). Broadly speaking, trophic levels range between one and five based on the degree to which organisms feed directly on producers. Primary producers and detritus are at the base of the food chain, a trophic level of one. Zooplankton, benthic herbivores, and detritivores occupy the second trophic level. Carnivores occupy trophic levels three to five. Trophic level estimates are generalizations from the ecological and fisheries literature and must be used with caution. Sometimes it is necessary to use higher taxonomic levels in archaeological applications because the taxonomic identification in the archaeological data, the modern fishery data, and/or FishBase are insufficiently precise or unavailable.

Trophic levels 2.0 through 4.2 are represented in these vertebrate samples. The only species from the lowest trophic level is mullet (*Mugil* sp.). Anchovies and herrings (Clupeidae) are typical of trophic levels 2.2–2.6 and sea catfishes of trophic level 3.2. Common vertebrates from trophic levels 3.3–3.5 are drums (Sciaenidae), jacks (Carangidae), and grunts (Haemulidae). Animals from trophic levels above 3.5 include marine mammals (e.g., sea lions and seals [Pinnipedia, *Otaria byronia*], porpoises [Delphinidae]), sharks (Carcharhinidae), wrasses (Labridae), and mackerels (Scombridae).

Estimates of mean trophic level suggest people living north of 10°S targeted higher-trophic-level fishes than did people living south of 12°S (Figure 8-3; Reitz 2004b). The mean trophic level used during the Middle Preceramic period on the north coast declined by the time people settled the Initial period Pampa

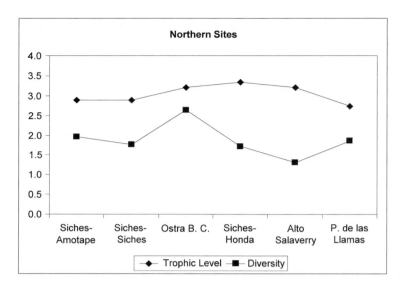

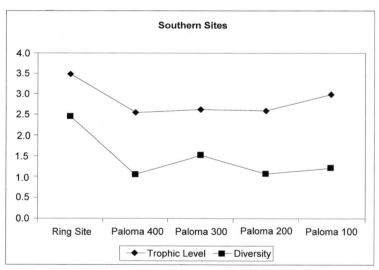

Figure 8-3. Mean trophic levels and diversity based on Minimum Number of Individuals (MNI). Diamonds = trophic level and Squares = diversity. Upper panel contains sites north of 10°S and lower panel contains sites south of 12°S. Sites are arranged chronologically within each panel with the earliest site at the left of each panel.

de las Llamas site. Allowing for variation among the sites, it is clear that people at southern sites deployed a different strategy than did people at contemporary sites on the northern coast and that this strategy changed in the north around 3800 B.C.

The fishing strategy at the southern site of Paloma generally focused on low-trophic-level, cool-water anchovies and herrings while fishing at the northern sites made greater use of higher-trophic-level, warm-water fishes, among which drums figure prominently. The high mean trophic level at the Ring Site is due to abundant sea lion/seal individuals rather than to high-trophic-level fishes. Animals from trophic levels 3.6–4.2 are present in small numbers at all of the northern collections and absent from Paloma. An increase in high-trophic-level drums occurs between roughly 3450–3000 B.C., a phenomenon recorded in Paloma Level 100 and Alto Salaverry collections, followed by an increase in low-trophic-level anchovies at Pampa de las Llamas. This change is not related to the introduction of cotton, which is present at both Alto Salaverry and Pampa de las Llamas.

Richness and Diversity

The richness of exploited species and the degree of subsistence specialization are also sensitive indicators of the environment that should reflect changes in sea surface temperature and ENSO characteristics. Richness (S) is defined here as the number of taxa for which MNI is estimated. Diversity (H') is measured using the Shannon-Weaver Index (Reitz and Wing 1999:233–235) using those marine vertebrates in each collection for which both MNI and trophic level is estimated.

Fishing effort targeting the higher trophic level species is generally characterized by higher diversity while fishing targeting lower trophic level species is more likely to focus on a single, cool-water fish. In the south, anchovies are the targeted low-trophic-level resource, a cool-water fish associated with deep-water upwellings. In the north, low-trophic-level resources are more varied. Low-trophic-level resources in northern Middle Preceramic collections include sea turtles (Cheloniidae), anchovies, herrings, and mullets. The Alto Salaverry focus was on a cool-water drum (*Corvina [Sciaena] deliciosa*; 69% of the marine MNI; trophic level 3.5). By the Initial period, 48% of the marine individuals are anchovies (trophic level 2.2).

DISCUSSION

Based on this case study, we conclude that sea surface temperature at Ostra Base Camp was warmer in the mid-Holocene than at present and that seasonal temperature variation at Ostra was circa 3°C greater than present with most of the increase occurring in the summer months. The warm-water invertebrate

and vertebrate species, higher trophic level fishes, and higher catch diversity
are associated with the warmer sea surface temperature and reduced deep-
water upwelling. After approximately 3800 B.C., the ocean along northern Peru
cooled and deep-water upwelling intensified. This shift resulted in greater use
of cool-water fishes from lower trophic levels and an increasing reliance on
industrial plants such as cotton, associated with the appearance and growth of
monumental architecture and eventually a suite of phenomena characteristic
of later Andean cultures. Evidence for a change in ENSO patterning at Ostra is
supported by signatures of change at other Peruvian archaeological sites;
in non-archaeological data from the Andean region; and in non-archaeological
data from the Pacific Basin and beyond (Sandweiss 2003; Sandweiss et al.
1999, 2001.)

Specifically, the change in data from the north coast appears to reflect a
northward shift of the boundary between the Panamanian province's prevailing warm waters and the Peru-Chilean province's prevailing cool waters.
Before 3800 B.C., the hypothesized boundary between warm and cool waters
lay south of Siches, perhaps near Ostra and Almejas when those sites were
occupied. The boundary moved north to a location closer to Siches after
3850 B.C. during the Honda phase at the Siches site. Alto Salaverry and
Pampa de las Llamas were south of this boundary when they were occupied.

Archaeofaunal data from Ostra Base Camp suggest a different ENSO pattern
was in place when Ostra was occupied from that which prevailed during the
Cotton Preceramic and Initial periods and the one that prevails today. To
explore this possibility further, and to expand it beyond the Peruvian case, we
make the following suggestions for future research.

1. ENSO is part of a global weather pattern and changes in the prevailing
   ENSO pattern found at Peruvian sites likely impacted many parts of the
   planet during the Holocene in different ways at slightly different times.
2. Archaeologists should design their field protocols to enhance the recovery of large botanical, invertebrate, and vertebrate samples from many
   parts of the world that will be suitable for multi-site comparisons. New
   environmental perspectives should be directed toward their study.
3. Interpretations of environmental change should be based on a complex
   of species with specific ecological requirements studied though combinations of botanical, geochemical, and zoological perspectives.

CONCLUSION

Due to the excesses of earlier scholars who carelessly invoked environmental
change as a causal factor in cultural evolution, many environmental archaeologists avoid examining their data for evidence of environmental change.
Evidence from many fields now suggests that the Holocene environment was

constantly changing rather than static. Humans had to accommodate those changes in their daily lives. Cautious use of the modern tools that were not available to our predecessors is appropriate. Only in this way will we be able to explore the balance people achieved between their cultural objectives and a changing environment.

ACKNOWLEDGMENTS

We are grateful to Daniel C. Weinand for his work with the faunal materials from Ostra Base Camp, as well as to Elizabeth S. Wing and Kitty F. Emery for access to the comparative skeletal collection of the Environmental Archaeology Program at the Florida Museum of Natural History. Vertebrate materials were identified using the comparative collections of the Zooarchaeological Laboratory, Georgia Museum of Natural History and the Environmental Archaeology Program, Florida Museum of Natural History. Isotopic analysis was conducted at the Department of Geology Stable Isotope Laboratory, University of Georgia and the University of Alabama Department of Geological Sciences Stable Isotope Laboratory. We thank Uwe Brand for personal communications regarding elemental paleosalinity proxy data for Ostra mollusks. Funds were provided by a grant to Sandweiss from the H. John Heinz III Charitable Trust. Partial funding for Andrus was provided by the United States Department of Energy grant DE-FC09–96SR18546.

REFERENCES

Albarella, U. (ed.), 2001, *Environmental Archaeology: Meaning and Purpose*, Kluwer Academic Publishers, Dordrecht, The Netherlands.

Andrus, C. F. T., Crowe, D. E., and Romanek, C. S., 2002a, Oxygen Isotope Record of the 1997–1998 El Niño in Peruvian Sea Catfish *Galeichthys peruvianus* Otoliths, *Paleoceanography* 17(4):1053–1060.

Andrus, C. F. T., Crowe, D. E., Sandweiss, D. H., Reitz, E. J., and Romanek C. S., 2002b, Otolith δ^{18}O Record of Mid-Holocene Sea Surface Temperatures in Peru, *Science* 295:1508–1511.

Andrus, C. F. T., Crowe, D. E., Sandweiss, D. H., Reitz, E. J., Romanek, C. S., and Maasch, K. A., 2003, Response to Comment on "Otolith δ^{18}O Record of mid-Holocene Sea Surface Temperatures in Peru," *Science* 299:203b.

Bakun, A., 1996, *Patterns in the Ocean: Ocean Processes and Marine Population Dynamics*, California Sea Grant College System Report T-037, La Jolla.

Béarez, P., and Luis Miranda, L., 2000, Análisis arqueo-ictiológico del sector residencial del sitio arqueológico de Caral-Supe, costa central del Perú, *Arqueología y Sociedad* 13:67–77, Museo de Arqueología y Antropología, Universidad Nacional Mayor de San Marcos, Lima, Peru.

Benfer, R. A., 1984, The Challenges and Rewards of Sedentism: The Preceramic Village of Paloma, Peru, in: *Paleopathology and the Origin of Agriculture* (M. N. Cohen and G. J. Armelagos, eds.), Academic Press, New York, pp. 531–558.

Efremov, J. A., 1940, Taphonomy: New Branch of Paleontology, *Pan-American Geologist* 742:81–93.

Froese, R., and Pauly, D. (eds.), 1998, FishBase 98: Concepts, Design and Data Sources, The International Center for Living Resources Management, Kakati City, Philippines, <http://www.fishbase.org/trophic/t.html>.

Grayson, D. K., 1984, Quantitative Zooarchaeology, Academic Press, New York.

Grossman, E. L., and Ku, T. L., 1986, Oxygen and Carbon Isotope Fractionation in Biogenic Aragonite: Temperature Effects, Chemical Geology 59:59–74.

Keefer, D. K., deFrance, S. D., Moseley, M. E., Richardson, J. D. III., Satterlee, D. R., and Day-Lewis, A., 1998, Early Maritime Economy and El Niño Events at Quebrada Tacahuay, Peru, Science 281:1833–1835.

Meekan, M. G., Wellington, G. M., and Axe, L., 1999, El Niño Southern Oscillation Events Produce Checks in the Otoliths of Coral Reef Fishes in the Galapagos Archipelago, Bulletin of Marine Science 64:383–390.

Monks, G. G., 1981, Seasonality Studies, in: Advances in Archaeological Method and Theory, Volume 4 (M. B. Schiffer, ed.), Academic Press, New York, pp. 177–240.

Pauly, D., and Christensen, V., 1995, Primary Production Required to Sustain Global Fisheries, Nature 374:255–257.

Pauly, D., Christensen, V., Dalsgaard, J., Froese, R., and Torres, D.F., Jr., 1998, Fishing Down Marine Food Webs, Science 279:860–863.

Pauly, D., Christensen, V., Froese, R., and Palomares, M. L., 2000, Fishing Down Aquatic Food Webs, American Scientist 881:46–51.

Perrier, C., Hillaire-Marcel, C., and Ortlieb, L., 1994, Paleogeographie littorale et enregistrement isotopique $^{13}C^{18}O$ d'evenements de type El Niño par les mollusques holocenes et recents du nord-ouest peruvien, Geographie Physique et Quaternaire 48:23–38.

Pozorski, S., 1983, Changing Subsistence Priorities and Early Settlement Patterns on the North Coast of Peru, Journal of Ethnobiology 3:15–38.

Pozorski, S., and Pozorski, T., 1979a, Alto Salaverry: A Peruvian Coastal Preceramic Site, Annals of Carnegie Museum 48:337–375.

Pozorski, S., and Pozorski, T., 1979b, An Early Subsistence Exchange System in the Moche Valley, Peru, Journal of Field Archaeology 6:413–432.

Pozorski, S., and Pozorski, T., 1987, Early Settlement and Subsistence in the Casma Valley, Peru, University of Iowa Press, Iowa City.

Quilter, J., 1991, Late Preceramic Peru, Journal of World Prehistory 5:387–438.

Quilter, J., Ojeda, E. B., Pearsall, D. M., Sandweiss, D. H., Jones, J. G., and Wing, E. S., 1991, Subsistence Economy of El Paraíso, An Early Peruvian Site, Science 251:277–283.

Reed, C. A., 1963, Osteoarchaeology, in: Science in Archaeology, 1st ed. (D. Brothwell and S. Higgs, eds.), Basic Books, New York, pp. 204–216.

Reitz, E. J., 1995, Environmental Change at Almejas, Peru, manuscript on file, Georgia Museum of Natural History, University of Georgia, Athens, pp. 96.

Reitz, E. J., 1999, Vertebrate Fauna from Pampa de Las Llamas-Moxeke, manuscript on file, Georgia Museum of Natural History, University of Georgia, Athens, pp. 65.

Reitz, E. J., 2001, Fishing in Peru between 10000 and 3750 BP, International Journal of Osteoarchaeology 11:163–171.

Reitz, E. J., 2003, Resource Use Through Time at Paloma, Peru, Bulletin of the Florida Museum of Natural History 44(1):65–80.

Reitz, E. J., 2004a, "Fishing Down the Food Web": A Case Study from St. Augustine, Florida, USA, American Antiquity 69:63–83.

Reitz, E. J., 2004b, Vertebrate Remains from Honda, Siches, and Amotape Phase Occupations at Sitio Siches PV 7–19, Peru, manuscript on file, Georgia Museum of Natural History, University of Georgia, Athens, pp. 164.

Reitz, E. J., and Sandweiss, D. H., 2001, Environmental Change at Ostra Base Camp, A Peruvian Pre-ceramic Site, Journal of Archaeological Science 28:1085–1100.

Reitz, E. J., and Wing, E. S., 1999, Zooarchaeology, 1st ed., Cambridge University Press, Cambridge, England.

Rhoads, D. C., and Lutz, R. A., (eds.), 1980, *Skeletal Growth of Aquatic Organisms: Biological Records of Environmental Change*, Plenum Press, New York.

Rollins, H. B., Richardson, J. B. III., and Sandweiss, D. H., 1986a, The Birth of El Niño: Geoarchaeological Evidence and Implications, *Geoarchaeology* 1:3–16.

Rollins, H. B., Sandweiss, D. H., and Rollins, J. C., 1986b, Effect of the 1982–1983 El Niño on Bivalve Mollusks, *National Geographic Research* 21:106–112.

Rollins, H. B., Sandweiss, D. H., Brand, U., and Rollins, J. C., 1987, Growth Increment and Stable Isotope Analysis of Marine Bivalves: Implications for the Geoarchaeological Record of El Niño, *Geoarchaeology* 2:181–197.

Sandweiss, D. H., 1986, The Beach Ridges at Santa, Peru: El Niño, Uplift, and Prehistory, *Geoarchaeology* 1:17–28.

Sandweiss, D. H., 1996, Mid-Holocene Cultural Interaction on the North Coast of Peru and Ecuador, *Latin American Antiquity* 7:41–50.

Sandweiss, D. H., 2003, Terminal Pleistocene through Mid-Holocene Archaeological Sites as Paleoclimatic Archives for the Peruvian Coast, *Palaeogeography, Palaeoclimatology, Palaeoecology* 194:23–40.

Sandweiss, D. H., Maasch, K. A., and Anderson, D. G., 1999, Climate and Culture: Transitions in the Mid-Holocene, *Science* 283:499–500.

Sandweiss, D. H., Maasch, K. A., Burger, R. L., Richardson III, J. B., Rollins, H. B., and Clement, A., 2001, Variation in Holocene El Niño Frequencies: Climate Records and Cultural Consequences in Ancient Peru, *Geology* 29:603–606.

Sandweiss, D. H., Maasch, K. A., Chai, F., Andrus, C. F. T., and Reitz, E. J., 2004, Geoarchaeological Evidence for Multidecadal Natural Climatic Variability and Ancient Peruvian Fisheries, *Quaternary Research* 61:330–334.

Sandweiss, D. H., McInnis, H., Burger, R. L., Cano, A., Ojeda, B., Paredes, R., Sandweiss, M., and Glascock, M., 1998, Quebrada Jaguay: Early Maritime Adaptations in South America, *Science* 281:1830–1832.

Sandweiss, D. H., Richardson III, J. B., Reitz, E. J., Hsu, J. T., and Feldman, R. A., 1989, Early Maritime Adaptations in the Andes: Preliminary Studies at the Ring Site, Peru, in: *Ecology, Settlement and History in the Osmore Drainage, Peru* (D. Rice, C. Stanish and P. R. Scarr, eds.), *British Archaeological Reports International Series* (Oxford) 545i, pp. 35–84.

Sandweiss, D. H., Richardson, III, J. B., Reitz, E. J., Rollins, H. B., and Maasch, K. A., 1996, Geoarchaeological Evidence from Peru from a 5000 Years BP Onset of El Niño, *Science* 273:1531–1533.

Sandweiss, D. H., Rollins, H. B., and Richardson, III, J. B., 1983, Landscape Alteration and Prehistoric Human Occupation on the North Coast of Peru, *Annals of Carnegie Museum* 52:277–298.

Shady Solís, R., Haas, J., and Creamer, W., 2001, Dating Caral, a Preceramic Site in the Supe Valley on the Central Coast of Peru, *Science* 292:723–726.

Weigelt, J., 1989, *Recent Vertebrate Carcasses and Their Paleobiological Implications* (J. Schaefer, trans.), University of Chicago Press, Chicago, Illinois.

Wheeler, A., and Jones, A. K. G., 1989, *Fishes*, Cambridge University Press, Cambridge, England.

Part III

Human Ecology

Chapter **9**

Living on the Margins: Biobehavioral Adaptations in the Western Great Basin

CLARK SPENCER LARSEN, ROBERT L. KELLY, CHRISTOPHER
B. RUFF, MARGARET J. SCHOENINGER,
DALE L. HUTCHINSON, AND BRIAN E. HEMPHILL

Since the founding of anthropology, a great deal of attention has been paid to characterizing the adaptive "efficiency" or "affluence" of human populations dependent upon hunting and gathering as a primary means of acquiring sustenance. For many years, the Hobbesian portrayal of hunter–gatherer lifeways as "nasty, brutish, and short" dominated much of the scientific and popular perceptions of foraging human groups. In his widely used archaeology textbook, Braidwood (1967:113) depicted the hunter-gatherer lifestyle as "a savage's existence, and a very tough one . . .following animals just to kill them to eat, or moving from one berry patch to another [and] living just like an animal." However, within a short time following the epic "Man the Hunter" conference in 1966 (Lee and DeVore 1968), this image of hunter-gatherers took an about-face in response to data suggesting that far from being nutritionally deprived and subject to excessive demands of work devoted principally to the food quest, hunter-gatherers apparently have adequate nutrition. Nor are they subject to overbearing amounts of work; in fact, their lives are leisurely, and food is plentiful. According to Sahlins (1972:1–39), they are the "original affluent society."

Stimulated by the 1966 conference and especially by the provocative observations of Lee (1979) on the Ju/'hoansi of southern Africa, researchers initiated ecologically-oriented projects among extant foragers in diverse settings. These studies examined affluence and quality of life among the world's foraging societies (see Hill and Hurtado 1989; Kelly 1995a). Contrary to earlier assessments, these studies pointed to a high degree of variability in hunter-gatherer lifestyles. Far from demonstrating uniformity, they showed that by virtually any measure–food selection, food preference, nutritional quality, mobility pattern, workload, disease stress, and demography–foraging populations do not fit into a single pattern. The only common theme appeared to be a reliance on nondomesticated plants and animals as the primary food source.

The archaeological record contributes an important perspective on hunter-gatherer lifeways. This record gives a crucial diachronic perspective. Archaeologists working in the North American Great Basin have revealed key aspects of human foraging behavior because of the availability of a rich archaeological record (e.g., Bettinger 1989; Heizer and Krieger 1956; Jennings 1957, 1978; Kelly 2001; Simms 1999; Thomas 1985, 1988; see reviews by Grayson 1993; Kelly 1997) supplemented by accounts of native groups written by 19th-century pioneers and explorers (e.g., Simpson 1876) and by 20th-century ethnographers (e.g., Fowler 1992; Lowie 1924; Steward 1938; Wheat 1967).

Throughout many regions of the world, biological anthropologists have made important contributions to studies of adaptive processes in human societies (see Cohen and Armelagos 1984; Katzenberg and Saunders 2000; Larsen 1997; Larsen and Milner 1994; Steckel and Rose 2002; Wing and Brown 1979). Skeletal and dental tissues are remarkably sensitive indicators of human interaction with the environment, providing a cumulative biological history or "memory" of an individual's lifetime, including dietary stress, nutritional quality, disease history, mobility pattern, and physical activity. Many regional studies across the globe rely upon human remains for biocultural inferences about extinct lifeways (e.g., Buikstra 1988; Cybulski 1992; Domett 2001; Hodges 1989; Hutchinson 2002; Lambert 2000; Larsen 2001; Martin et al. 2001; Papathanasiou 2001; Peterson 2002; Pietrusewsky and Douglas 2001; Steckel and Rose 2002; Ubelaker 1994; Whittington and Reed 1997; Williamson and Pfeiffer 2003).

Compared to other regions of North America, few human remains are found in the Great Basin (see Larsen 2006). Consequently, few researchers have attempted to use this source in the investigation of Great Basin hunter-gatherer lifeways. The meager record for mortuary behavior changed dramatically in the mid-1980s. Between 1982 and 1986, record high winter precipitation caused massive floods in several Great Basin wetlands (Hemphill and Larsen 1999), including the Stillwater Marsh, located in the Carson Desert of western Nevada (Figure 9-1). The retreating floodwaters exposed dozens of previously unknown archaeological sites and hundreds of human skeletal remains. From 1985 to 1987, two separate teams collected human remains in danger of destruction in the Stillwater Marsh (see Larsen and Kelly 1995; Tuohy et al. 1987).

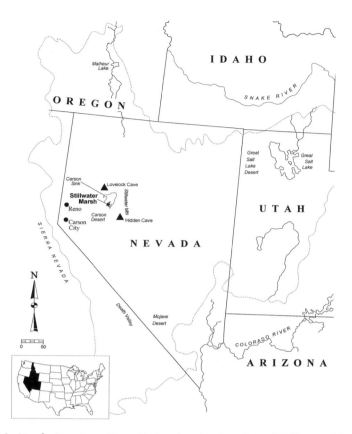

Figure 9-1. North American Great Basin, showing location of Stillwater Marsh and other Great Basin settings discussed in this chapter (adapted from Hemphill and Larsen 1999:Figure 1.1; reproduced courtesy of The University of Utah Press). Dotted line shows the Great Basin.

The study of this unprecedented collection of human remains allows us to characterize hunter-gatherer lifeways in a geographic setting that was relatively unknown from a bioarchaeological perspective. In this case study, we use skeletal and dental data to address competing reconstructions of the pre-Columbian use of wetland resources in the Stillwater Marsh and to document some general aspects of the human condition in this region.

Previously, researchers argued over two reconstructions of ancient lifeways in the Stillwater Marsh. Proponents of one reconstruction, called the *limnosedentary* model, argue that wetland resources could provide sufficient food and other resources to support a sedentary hunter-gatherer population (Heizer and Napton 1970). Alternatively, proponents of the *limnomobile* model assert that although wetlands may have anchored a settlement system, fluctuations in the availability of food in the wetlands in comparison with resources in other valleys, the nearby uplands (e.g., Stillwater Mountains), or in both would have resulted in a more mobile lifeway regardless of the relative productivity of the marsh habitats (Kelly 1995b, 2001; Thomas 1985). Thus, supporters of the first model maintain that sedentary populations obtained most dietary resources from lacustrine contexts of the marsh wetlands. Supporters of the second model contend that populations, although relying partly on marsh resources, spent significant amounts of time and effort in the collection and transport of uplands resources, sometimes over great distances (for alternative viewpoints, see Janetski and Madsen 1990; Kelly 1992, 1995b; Raven 1990). New research suggests that the wetlands, when present, provided the best foraging opportunities for women in most seasons and most years; and that men may have profited by traveling farther afield to seek game, such as bighorn sheep (*Ovis canadensis*), in the mountains (Kelly 2001; Zeannah 1996, 2004; Zeannah et al. 1995). The dialog among Great Basin archaeologists is couched within the larger problem of the role of environment in hunter-gatherer adaptations (see Mandryk 1993) as well as the archaeological documentation of mobility (see Kelly 1992, 1995a; Kent 1992; Rafferty 1985). The question still remains, however, as to how this settlement pattern and division of labor affected the quality of life of those hunter-gatherers who inhabited the Stillwater Marsh.

We will show that land-use patterns and resource procurement as well as the overall quality of life in this region can be investigated by the study of human skeletal remains. In so doing, we underscore the value of osteological samples and their morphological and pathological correlates of behavior, diet, and health to critically inform debates in this region in particular and to analyses of behavioral patterns and adaptive efficiency in hunter-gatherers in general.

BIOARCHAEOLOGICAL APPROACHES TO BEHAVIOR, DIET, AND HEALTH

The primary objective of this study is to test hypotheses about patterns of physical behavior associated with a highly challenging lifeway. Written accounts indicate that populations of the Toedökadö ("Cattail eaters") band of the Northern Paiute occupying the Stillwater Marsh, frequently engaged in long-distance travel to gather resources and led a physically demanding lifeway overall (Fowler 1992; Kelly 1995b). We use two complementary approaches to the study of biomechanical adaptation in order to characterize the physical activity patterns experienced by members of these populations. These approaches include study of osteoarthritis, a degenerative articular joint disorder caused by excessive and repetitive mechanical loading of articular joints (e.g., hip, knee, elbow), and structural analysis of long bone morphology.

The first approach uses pathological changes on joint articular margins and surfaces that are commonly associated with the disorder. These changes include bony lipping along joint margins, joint surface porosity, and polishing (eburnation) of joint surfaces due to direct bone-bone contact following destruction of articular cartilage (Hough 2001). Osteoarthritis develops over the course of an individual's lifetime from daily wear and tear on the joints arising from physically demanding, repetitive activities (Radin et al. 1972; Sharma 2001).

The second approach involves the application of beam theory as developed by civil and mechanical engineers for structural analysis of building materials (Lanyon and Rubin 1985; Ruff 2000). One can model human long bones (e.g., femur, humerus) as hollow beams and measure the structure of these beams via a series of properties called *cross-sectional geometric properties*. These properties measure the strength or resistance of bone to bending and torsion (twisting), the two primary forces to which human long bones are subject in life. Hence, physically active individuals will display higher values of cross-sectional geometric properties in their long bones than those individuals who are not.

The second objective of this study is to reconstruct diet. With this information in hand, inferences about nutritional quality can be made and place of origin of particular foods consumed by native groups can be identified. Some foods in the Carson Desert are available primarily in the marsh and lowlands. These include Indian rice grass (*Oryzopsis hymenoides*), bulrush (*Scripus* sp.), cattail (*Typha latifolia*), needle and thread (*Stipa comata*), water birds, fishes, and aquatic mammals. Other foods are available in the uplands, such as pinyon nuts (*Pinus monophylla*) and mammals, especially bighorn sheep. A few grasses, such as wild rye (*Elymus* sp.), are available in both marsh and nonmarsh areas. Therefore, the central question for this discussion is: What foods were consumed in the Great Basin and were those foods extracted primarily from marsh settings, uplands settings, or some combination of both? This question is closely linked to with our inquiry about the degree of mobility of hunter-gatherers in the region. If it can be demonstrated that a significant component of the dietary regime was obtained from both marsh and uplands habitats, then these populations, or at least a particular component (e.g., adult males), were necessarily mobile.

There is a long history of interest in Great Basin foodways, but aside from documenting the kinds of foods eaten through archaeobotany, zooarchaeology, and ethnographic observation, knowledge of the actual menu of foods consumed and the relative proportions of these foods to overall diet composition is elusive. Food remains often are not preserved, or not preserved in the proportion consumed, thus limiting understanding of diet and nutrition in archaeological settings. The vagaries of qualitative approaches to dietary reconstruction via the identification of plant and animal remains are partly alleviated by stable carbon and nitrogen isotope analysis (Katzenberg 2000; Schoeninger 1995).

Stable isotope analysis is based on the assumption that the isotopic composition of human bone reflects the isotopic composition of the foods consumed. In other regions of North America, isotopic studies for plant and animal communities have produced diagnostic signatures of nitrogen and carbon stable isotope ratios (^{15}N/^{14}N and ^{13}C/^{12}C, respectively). The bimodal distribution of stable carbon isotope ratios (expressed in parts per mil ‰ relative to a standard) in terrestrial plants is the most familiar and is instrumental in tracking the introduction and intensification of maize (*Zea mays*) in the Americas. The distribution reflects the manner in which plants fix carbon in photosynthesis: Maize fixes carbon through the C_4 photosynthetic pathway, although most other economically important plants fix carbon through the C_3 and, to a lesser extent, CAM pathways. Worldwide, marine and terrestrial organisms express a bimodal pattern of stable nitrogen isotope ratios. This pattern facilitates the identification of human consumption patterns relating to sea versus land-based foods as well as other settings (Schoeninger 1995; Schoeninger et al. 1983; Schoeninger and Moore 1992). Data for freshwater marshes are not available, however. Nevertheless, it is possible that carbon and nitrogen cycling in the Carson Desert marshes could be different from that in the surrounding desert and nearby uplands and that these differences could be detected in human skeletal remains.

As our third and final objective, we hope to provide new information about health status inferred from evidence for physiological stress in Stillwater hunter-gatherers. The study of stress, physiological disruption resulting from impoverished circumstances, has taken on increasing importance in interpreting adaptive efficiency in human populations, both living and extinct (see overview in Larsen 1997). We utilize a stress indicator that reflects growth disruption and is known as *linear enamel hypoplasia*. Hypoplasias are deficiencies or defects of enamel that are manifested as circumferential pitting, linear furrowing, or, in extreme cases, lack of enamel (Larsen 1997). The result of a cessation of activity in the cells that lay down enamel during the development of the tooth crown (called ameloblasts), these defects are linked to a plethora of stressors in living and archaeological settings, including, but not limited to, starvation, malnutrition, and various infectious diseases (Larsen 1997).

For this discussion, we present data on the width of these enamel defects. Because hypoplasias arise during the process of the growth and development of the enamel, from the tip of the crown to the junction between the crown and root, widths of defects represent an approximation of stress-episode duration or degree of severity or some combination of both (Blakey and Armelagos 1985; Guita 1984; Larsen and Hutchinson 1992; Suckling et al. 1986). Hypoplastic activity appears to be largely restricted to incisors and canines (Goodman and Armelagos 1985).

The number of Stillwater skeletons is about 35, depending on the type of study. For a fuller accounting of methods of analysis of osteoarthritis, cross-sectional geometric properties, stable isotopes, and dental defects, see Larsen and Kelly (1995) and various papers in Hemphill and Larsen (1999).

THE REGIONAL AND MORTUARY CONTEXT

The Carson Desert is an extensive area of sand dunes, alkali flats, and slightly alkaline marshes covering an area of about 2800 km^2 in west central Nevada. Forming one of the lowest points in this area of the Great Basin, it represents the sink or drainage terminus of the Carson River. The Stillwater Marsh, an ecologically rich area, is the most significant feature of the region (Figure 9-1; Kelly 2001).

Archaeological survey and testing revealed that most mortuary sites were located within a 16 km^2 area of the marsh (Larsen and Kelly 1995; Tuohy et al. 1987). Field reconnaissance identified at least 38 sites with human remains. Human remains were excavated from mostly isolated graves scattered throughout the marsh area rather than cemeteries containing a group or groups of skeletons. This noncemetery burial pattern indicates that these populations did not bury deceased individuals in formal areas. On the other hand, some of the sites containing human remains yielded a large number of skeletons. Brooks et al. (1988), for example, list some 58 individuals in varying states of completeness from a single site (26CH1043). The Stillwater mortuary pattern is likely characteristic of many ethnographically known foraging groups world-wide, although this pattern has not been observed universally (see Charles and Buikstra 1983; Charles et al. 1986), especially when archaeological cases are considered.

The precise length of time that the Stillwater Marsh was used for burial is incompletely understood. Only six skeletons are radiocarbon-dated and they encompass a period of time that spans nearly two millennia, from circa 270 B.C. to A.D. 1644 (corrected dates) (see Brooks and Brooks 1990; Larsen and Kelly 1995). Moreover, few temporally diagnostic artifacts (i.e., projectile points) are associated with human remains. Chronological evidence based on other classes of archaeological sites (e.g., middens), however, suggests that the region was most heavily utilized by populations from two sequential phases, the latter half of the Reveille phase (1300 B.C.–A.D. 700) and the succeeding Underdown phase (A.D. 700–1300) (Raymond and Parks 1990).

This discussion highlights the problems besetting the analysis of human remains found in the Stillwater Marsh, especially regarding the lack of precise temporal control and cemeteries. Because the Stillwater human remains are drawn from many sites that span hundreds (if not thousands) of years of native use of the region, it is possible that they are not representative of the populations from which they were drawn. Brooks et al. (1988) and we (Larsen and Kelly 1995), however, have noted the homogeneity of skeletal measurements as well as cranial morphology in the Stillwater series. These findings suggest that the series can be regarded as a biologically or culturally cohesive unit of study amenable to analysis and comparison.

OSTEOLOGICAL ANALYSIS OF ACTIVITY, DIET, AND STRESS

Activity

Analysis of osteoarthritis and cross-sectional geometric properties indicates a lifeway that was physically demanding. Over three quarters (76%) of individuals in the series are affected by osteoarthritis; all individuals older than 30 years are arthritic. Consideration by specific joint region shows that the lowest prevalences occur in the hands (13%) and the highest in the lumbar vertebrae (67%). The cervical vertebrae and the elbow have frequencies of 50% or more. In males, arthritis frequencies range from the lowest values for the hand (18%) to the highest values for the cervical, shoulder, and elbow joints (all are 61%). In females, values range from none in the ankle to 75% in the lumbar vertebrae. The overall high prevalence of osteoarthritis in the Stillwater series is borne out by comparison with other populations. For example, the prevalence is higher than either hunter-gatherer or agricultural populations in the southeastern United States, where most values are under 20% (cf., Bridges 1992).

Analysis of cross-sectional geometric properties of Stillwater femora and humeri reveals a number of important tendencies. Cortical area (CA) and percent cortical area (PCCA) measure the *amount* or mass of bone in a cross section. Compared with other populations, Stillwater bone mass is at the low end of the range (Figure 9-2). Total subperiosteal area (TA), however, a measure of the outer dimension of a long bone in cross section, is on the upper end of the range of comparative samples. The latter indicates that the bones of Stillwater individuals are very large and robust despite low values of CA and PCCA.

The meaning of large values of TA is clarified by comparing cross-sectional geometric properties called *second moments of area*. These values measure the cross-sectional *distribution* of bone around an imaginary central (or "neutral") axis corresponding to the longitudinal axis of the bone shaft. Low values of second moments of area reflect a distribution of bone close to this central axis; conversely, high values reflect a distribution of bone located far from the central axis. According to beam theory, the higher the value of second moments of area, the greater the strength of the bone, and hence the greater ability of the bone to resist high levels of mechanical loading, particularly bending (I) and torsion (J) that occur during walking, running, and other physical activities. Simply stated, greater values of second moments of area represent stronger bones that develop in response to a lifetime of physically demanding activities. In general, the Stillwater J values, representing an overall average of bending strength in a cross section, are quite high for Holocene human populations, especially for males (Figure 9-3) (e.g., see Bridges et al. 2000; Holt 2003; Ledger et al. 2000; Ruff 1991, 1994, 1999; Ruff and Larsen 2001; Ruff et al. 1993; Stock and Pfeiffer 2001, 2004).

The biomechanical shape of the femoral midshaft in cross-section as denoted by a ratio or index of I_x/I_y (the ratio of bending strength in anteroposterior to

Cross-Sectional Areas

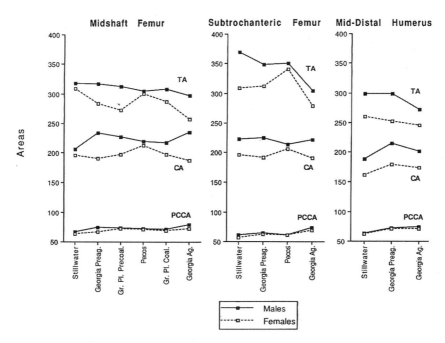

Figure 9-2. Comparison of cross-sectional areas of Stillwater femora and humeri with other skeletal series. Key: (CA) cortical area; (PCCA) percent cortical area; (TA) total subperiosteal area. Cultural units: Stillwater (1300 B.C.–A.D. 1300); Georgia Preagricultural (500 B.C.–A.D. 1150); Great Plains Precoalescent (A.D. 400–1600); Pecos Pueblo (A.D. 1300–1650); Great Plains Coalescent (A.D. 1600–A.D. 1850); Georgia Agricultural (A.D. 1150–1550) (Data from Ruff 1994, 1999; Ruff and Larsen 2001).

mediolateral planes) represents an approximation of degree of mobility in human populations (Ruff 1987). Ruff (1987) has found a decline in the index from hunter-gatherers to agriculturists to Western industrial populations. Agriculturists have values closer to 1 (reflecting a perfectly round cross section, or relatively little mechanical stress) than hunter-gatherers, thus indicating greater anteroposterior orientation of bending strength in femora of hunter-gatherers. Not surprisingly, 20th-century Americans have very low ratios, reflecting their highly sedentary nature. The Stillwater I_x/I_y ratio shows that males are on the high end of the range in comparison with other populations, while females are less so (Larsen et al. 1995; Ruff 1999). Consistent with the great difference in the ratio and the great difference in degree of mobility between males and females, Stillwater and other Great Basin populations express a very high level of

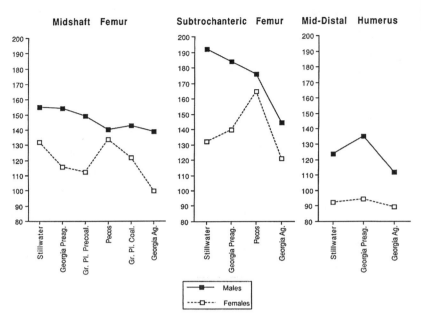

Figure 9-3. Comparison of polar second moment of area in torsional analysis (*J*) in Stillwater femora and humeri with other skeletal series. Cultural units: Stillwater (1300 B.C.–A.D. 1300); Georgia Preagricultural (500 B.C.–A.D. 1150); Great Plains Precoalescent (A.D. 400–1600); Pecos Pueblo (A.D. 1300–1650); Great Plains Coalescent (A.D. 1600–1850); and Georgia Agricultural (A.D. 1150–1550) (Data from Ruff 1994, 1999; Ruff and Larsen 2001).

sexual dimorphism (Figure 9-4). Indeed, these populations are the most sexually dimorphic from any archaeological setting studied by physical anthropologists.

Diet

Stable carbon and nitrogen isotopic analysis of human skeletal samples reveals that a variety of foods were likely consumed by aboriginal populations in this region. To summarize these data, $\delta^{13}C$ values range from –18.9‰ to –14.4‰ (average: –17.1‰), and $\delta^{15}N$ values range from 8.7‰ to 16.7‰ (average: 11.4‰). Several key findings emerge. The juvenile samples analyzed fall completely within the range of adults, suggesting that there are no detectable age-related differences in diet (juvenile average $\delta^{13}C$: –17.6‰ juvenile average $\delta^{15}N$: 12.7‰).

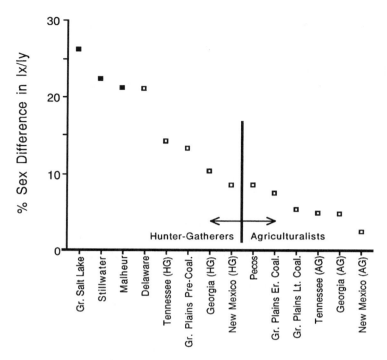

Figure 9-4. Comparison of sexual dimorphism in "shape" (I_x/I_y ratio) of Great Basin (Great Salt Lake, Stillwater Marsh, Malheur Lake in black boxes) midshaft section of femora compared with other skeletal series arranged in descending order (HG: hunter-gatherers; AG: agriculturalists). Cultural units: Great Salt Lake (A.D. 600–1380); Stillwater (1300 B.C.–A.D. 1300); Malheur (A.D. 1000–1700); Delaware (A.D. 750–950); Tennessee (6000–1000 B.C.); Great Plains Precoalescent (A.D. 400–1600); Georgia (500 B.C.–A.D. 1150); New Mexico (A.D. 500–1150); Pecos (A.D. 1300–1650; Great Plains Early Coalescent (A.D. 1600–1740); Great Plains Late Coalescent (A.D. 1740–1850); Tennessee (A.D. 1200–1500); Georgia (A.D. 1150–1550), New Mexico (A.D. 1150–1540) (data from Bridges 1989; Brock and Ruff 1988; Robbins et al. 1989; Ruff 1994, 1999; Ruff and Larsen 2001) (adapted from Ruff 1999:Figure 14.6; reproduced courtesy of The University of Utah Press).

For adults of known sex, the female and male means for both carbon and nitrogen are within 0.1‰ of each other (female average $\delta^{13}C$: −17.2‰ male average $\delta^{13}C$: −17.1‰ female average $\delta^{15}N$: 11.0‰ male average $\delta^{15}N$: 11.0‰). There are no differences by site (Schoeninger 1995, 1999). Thus, there are no apparent differences in food consumption by age, sex, or burial location, at least as they are represented by stable isotope values.

On the other hand, the wide dispersion of isotopic values of the series as a whole indicates that a variety of plants and animals were consumed by people in the Stillwater Marsh. Experimental laboratory animals fed monotonous diets show a range of less than 2‰ in carbon and nitrogen (DeNiro and Schoeninger 1983), far less than the samples analyzed here. On the basis of comparisons with known data of C_3 and C_4 plants and their animal consumers, some Stillwater individuals could have eaten only C_3 plants and their consumers, while others ate a significant amount of foods with a C_4 signature. The nitrogen isotope data suggest a similar variety foods eaten.

Physiological Stress

Two-thirds (64%) of the individuals examined possessed at least one hypoplastic defect. Measurement of widths of hypoplastic events for maxillary and mandibular incisors and canines reveals that they ranged from an average of 0.23 mm (mandibular first incisor) to 0.60 mm (mandibular second incisor). In order to evaluate the relative degree of stress, it is necessary to compare the sample with other dental series. We have measured pre-Columbian dentitions from the Georgia coast and northern Florida, including the hunter-gatherers and agriculturists (Hutchinson and Larsen 2001; Larsen and Hutchinson 1992). Generally, our findings show that the Stillwater hypoplasias are narrower than Georgia coast hypoplasias (e.g., maxillary canine: Stillwater = 0.40 mm; Georgia preagriculturists = 0.56 mm; Georgia agriculturists = 0.70 mm). All other teeth show a similar pattern (see Hutchinson and Larsen 1995, 2001).

BIOCULTURAL IMPLICATIONS OF BEHAVIOR AND HEALTH

Activity

Osteoarthritis. Analysis of osteoarthritis in the Stillwater series indicates that these foragers engaged in physically demanding activities. This level of demand is reflected in the high frequency of osteoarthritis, particularly in the spine. The spine is subjected to high levels of mechanical loading. A number of Stillwater individuals exhibit, in addition to extensive lipping, other pathological changes reflecting mechanical demand. For example, several individuals have vertebral compression fractures, most of which are restricted to the lower thoracic and lumbar vertebrae. In addition, several individuals have depressions on the vertebrae arising from herniated intervertebral disks (called Shmorl's nodes). These pathological modifications are consistent with a population experiencing pronounced levels of mechanical stress, probably generated by the frequent transportation of heavy loads on foot.

The Stillwater remains exhibit a pattern of sexual dimorphism in osteoarthritis prevalence; males are generally more affected than females. In archaeological skeletal series, there is a surprising amount of variation in sexual dimorphism (reviewed in Bridges 1992). Only rarely, however, do females in archaeological populations show more osteoarthritis than males, except with regard to specific joints. For example, one might find osteoarthritis in the temporomandibular joint in communities where females were responsible for hide processing (e.g., Eskimos [see Merbs 1983]). For the most part, however, Stillwater males show more osteoarthritis than females. The statistically significant or near-significant levels of sex differences were revealed for the shoulder (females: 23%; males: 61%), hip (females: 6%; males: 40%), and ankle (females: 0%; males: 45%) joints (chi-square, p 0.05). Considering the kinds of foods eaten in the region as documented by ethnographic and other observations (e.g., Fowler 1992), this pattern probably reflects sex differences in mechanical demands related to resource procurement in general and hunting in particular. In this setting, excessive use of the shoulder in males may have arisen from thrusting activity, perhaps in spear throwing.

Osteoarthritic involvement of the hip and ankle likely reflects the mechanical demands associated with running and walking. The presence of significantly greater osteoarthritis prevalence in males than in females suggests that it was acquired through a lifetime of activity, perhaps in difficult terrain, such as in the uplands surrounding the Stillwater Marsh. It is exceedingly difficult to traverse the muddy areas of the marsh, and this, too, would have placed demands on the lower limbs in ambulatory activities. It is probably not the sole factor, however, because it would have affected males and females equally.

One of the outstanding characteristics regarding the presence of osteoarthritis is the unusually high prevalence in comparison with that in other skeletal samples (cf., Bridges 1992 and studies in Cohen and Armelagos 1984; Steckel and Rose 2002). Some insight into the behavioral context of the high prevalence is provided in the ethnographic literature. For example, Brooks et al. (1988) summarize the various activities recounted by Steward (1938) and Wheat (1967) in their ethnographic studies of Great Basin native populations. Among other behaviors that could lead to articular joint deterioration, activities they reported include carrying heavy loads, cutting animal hides, seed grinding and pounding with mortars and pestles, stone tool production, and the use of the bow and arrow. Given the ubiquity of osteoarthritis in the Stillwater remains, we are not able to say what specific behaviors were responsible, but clearly the Stillwater peoples were highly active. In sum, these findings suggest that the Carson Desert groups may have traversed many kinds of terrain, probably including the nearby mountains. On the other hand, it is our impression that slogging through the marsh on a daily basis over the course of a lifetime could increase the prevalence of the disorder. We must therefore turn to structural analysis in seeking answers to the question surrounding the degree of mobility.

Cross-Sectional Geometric Properties. Comparisons of cross-sectional geometric properties in various archaeological samples show a consistent pattern of elevated bone strength in the Stillwater series relative to comparative samples, which we interpret to reflect high bending and torsional loading modes. Low values of CA and PCCA are somewhat more difficult to interpret. Humans begin to lose bone mass after about age 40, resulting in osteoporosis in older adulthood (Garn et al. 1964). The Stillwater series contains a number of older adults (>40 years), which suggests the presence of age-related bone loss in the sample and, hence, low values of CA and PCCA. However, the average age at death for a number of the North American comparative samples used here is actually greater than that for the Stillwater group (Larsen et al. 1995). Thus, age at death is likely not a significant factor in interpreting these results.

Aside from age, a very important factor in determination of adult bone mass is nutritional quality. In human populations experiencing protein-calorie malnutrition, the growth spurt associated with adolescence is delayed, resulting in reduced amounts of bone, which is reflected by low CA and PCCA values in adults (Garn 1966; Garn et al. 1964). Therefore, in the absence of age factors for interpreting lower bone mass relative to other archaeological samples, we suggest that suboptimal nutrition may explain these values. We speculate that periods of reduced availability of some plants and animals during drought years or during winter months in the Carson Desert contributed to recurrent bouts of episodic undernutrition. More important periods of stress may have been in the early Spring, when foods collected during the previous year began to run out.

A functional/mechanical interpretation for the distribution of bone tissue in Stillwater long bone cross-sections is supported by the presence of consistently greater differences in femur midshaft robusticity compared to humerus middistal robusticity. Especially important in these observations is the presence of low robusticity (low values of TA and second moments of area) and low values of CA and PCCA in the Stillwater humeri. In other words, limitation of high robusticity to the femur indicates a localized effect, that is, a functional interpretation (increased mechanical loading on the lower limb); whereas, the low CA and PCCA values for both the femur and the humerus are more likely related to the effects of nutritional quality.

These findings, by themselves, do not provide additional insight into degree of mobility in the Stillwater populations. They basically show that the bones from Stillwater reflect a physically robust and highly active population. Comparison of the index I_x/I_y, however, reveals some key differences that shed light on the problem. Shape of the femoral shaft is expressed as the ratio of I_x to I_y, which measures the relative anteroposterior-to-mediolateral bending strength. Greater femoral midshaft I_x/I_y values match the conclusion of Ruff (1987) that hunter-gatherers, especially males, have the highest ratio values, reflecting greater bone strength in the anteroposterior direction in foragers due to mobile behaviors, which includes long distance travel. Moreover, comparison of Stillwater males and females shows that the Stillwater populations are highly dimorphic (Figure 9-4), which is consistent with his comparison of hunter-gatherers, agriculturists, and Western industrial populations (see also Ruff 2000). That is, males have higher I_x/I_y ratios than females, indicating that males were more mobile than females. These findings reflect a pronounced degree of sexual division of labor, with males engaged in long-distance activities such as hunting excursions and other types of forays. Females, on the other hand, were likely less mobile and were perhaps relatively more "tethered" to the marsh setting. This particular sexual division of labor is quite common among living hunter-gatherers (Kelly 1995a).

As such, we contend that the limnosedentary and limnomobile hypotheses are oversimplifications because they assume that the mobility patterns of females and males were similar, if not identical. Our findings indicate that this is not the case, and that males were more mobile than females.

Dietary Reconstruction: Stable Carbon and Nitrogen Isotopes

Similarly, stable isotope analysis reveals that diet was probably quite variable. The more general question, however, is whether the resources in the Stillwater Marsh could have fully supported the native groups or whether supplements from outside the marsh were necessary. Examination of stable isotope values suggests that nonmarsh foods were used by some portion of the population. The $\delta^{13}C$ values for the human samples have a range of -18.9 to -14.4. Individuals with the less negative values must have eaten a significant amount of C_4 foods throughout their lifetimes, yet a nonsystematic survey of the plants available in the marsh (Schoeninger 1995, 1999 [based on Raven and Elston 1989]) revealed that the majority are C_3. Of the 10 plants ranked highest by Raven and Elston (1989), the first-ranked plant, cattail, is C_3; of the remaining nine plants, three are C_4 and six are C_3. Although it is not possible to estimate the relative contributions of these plants to overall diet, the presence of such marsh-based C_4 plants does not contradict the use of upland C_4 plants as an explanation for the C_4 signature of these individuals. In addition, two animal samples used as a comparison with the human samples have $\delta^{13}C$ values of -12.7 (hare [*Lepus* sp.]) and -11.2 (diving duck [*Aythya* spp.]), indicating that up to 70% of the diets of these animals was one or more C_4 plants or one or more CAM plants with a C_4 signature. This estimate is based on the assumptions that a value in bone of -21 represents a 100% C_3 diet, that a value in bone of -7 represents a 100% C_4 diet, and that there is a linear relationship between diet composition and the $\delta^{13}C$ values (Schwarcz and Schoeninger 1991; Tieszen 1991). This result suggests that these animals were selectively eating C_4 plants, that they migrated into the marsh from the uplands, or that they were hunted in areas away from the marsh. The more negative $\delta^{13}C$ values in the humans can be accounted for by resources found in the marsh.

The variability in plant $\delta^{15}N$ stable nitrogen values provides important insight into reconstruction of foodways. The $\delta^{15}N$ values for the bulk of the human samples have a range of 8.7–14.4 (with an outlier value of 16.7). These values are more positive than those reported for people eating non-marine-based diets. Interestingly, analysis of cattail and desert-blite (*Suaeda* sp.) reveals very positive values (12.1 and 11.4, respectively) even though other marsh plants (e.g., bulrush) have values of around 6.0 (Schoeninger 1995). Large ranges of variation also are found in environments such as South Africa (Vogel et al. 1990) and Isle Royale in Lake Superior, Michigan (Bada, Schimmelmann, Peterson, and Schoeninger 1990, unpublished data; see Figure 9-1, Chapter 2, this volume). These data suggest that there is variation in the isotopic ratio of the nitrogen that is available to different plants. Kuhlbusch et al. (1991) report, for example, that up to 50% of organically-bound nitrogen is lost as nitrogen gas (N_2) during biomass burning. The remaining nitrogen would presumably be enriched in the heavier isotope (^{15}N) because the lighter isotope (^{14}N) would be lost first. The mechanism of fire followed by plant uptake of soil with enriched organically bound nitrogen could account for the very positive nitrogen isotope values observed in some of the plants discussed here, but it is simply not known whether a fire had occurred in the area where these plants were collected.

In any case, the presence of relatively positive $\delta^{15}N$ values in the Stillwater samples (9 and higher) argues against the use of pinyon as a major component of diet because the $\delta^{15}N$ value for pinyon itself is close to zero. That is to say, if pinyon contributed a major portion of diet, then the $\delta^{15}N$ values for the human skeletal samples analyzed should be much lower than those observed in the series analyzed. In contrast, cattail and desert-blite had very positive $\delta^{15}N$ values (12.1, and 11.4, respectively). Both plants are considered highly valued foods (see Raven and Elston 1989), suggesting that they were the source of the high $\delta^{15}N$ values in the Stillwater Marsh human samples.

In general, then, comparison of data derived from human skeletal remains with data from living and archaeological plants and animals from the region indicates that the diet of the majority of the Stillwater individuals depended on plants and animals similar to those analyzed for this study. These data do not refute a model suggesting that people remained year-round in the vicinity of the marsh. On the other hand, they certainly do not argue against forays into surrounding regions, perhaps for lengthy periods of time on a frequent basis. Indeed, the variation we observed in the isotope values argues for a highly varied diet derived from a combination of both marsh and nonmarsh habitats. Specifically, the more negative human bone collagen $\delta^{13}C$ values (−19 to −17) can be accounted for by a combination of plants and animals from the immediate marsh region. These data support archaeological reports that the majority of identifiable quids found at Hidden Cave are bulrush and cattail (see Thomas 1985), both of which are C_3 plants. This finding is consistent with ethnographic observations indicating that these plants were chewed like celery, especially during the spring (see Fowler 1990).

The small group of individuals with the least negative $\delta^{13}C$ values is not easily interpreted in light of the food items available to them. Perhaps foods consumed that resulted in these isotope values were from exclusively non-marsh (e.g., uplands) areas. Certainly the skeletal structural analysis indicates that the Great Basin populations in general were very mobile, particularly males. Thus, it would be expected that at least some of the individuals would have values reflecting non-marsh food consumption.

Physiological Stress: Enamel Defects

Observations of physiological stress in the Stillwater population indicate that environmental perturbation was not, on one hand, rampant or, on the other hand, infrequent. This finding is supported by comparisons of hypoplasia frequencies in other populations (e.g., Guatelli-Steinberg et al. 2004; Mack and Coppa 1992; Ubelaker 1992). Moreover, compared with the Georgia coastal samples representing subsistence strategies ranging from exclusively hunting and gathering to partial agricultural dependence, the Stillwater series shows relatively low levels of hypoplasia prevalence. Taken as a whole, however, the Stillwater prevalence (64%) is moderately high and even exceeds that in some modern populations from the eastern United States with elevated stress (cf., Steckel and Rose 2002). Thus, although physiological stress was not severe in the Stillwater Marsh populations, it was present at appreciable levels.

A variety of causes are linked to hypoplasia, including nutritional disorders, disease, or other factors predisposing individuals to physiological stress, such as low birthweight and socioeconomic status (see reviews in Ten Cate 1994; Larsen 1997). Laboratory studies show that hypoplasia can be induced by alloxan diabetes (Kreshover and Clough 1953a), fever (Kreshover and Clough 1953b), and parasitic infection (Suckling et al. 1986; Suckling and Thurley 1984), among other factors. Both malnutrition and disease likely lead to hypoplasia owing to their synergistic relationship.

As with living populations, most attempts to link specific diseases or dietary insufficiencies to hypoplasias in archaeological human populations are only marginally successful (see Goodman and Rose 1990, 1991; Larsen 1997). Consequently, researchers generally follow the conservative approach taken by Kreshover (1960), interpreting hypoplasias as nonspecific or general indicators of metabolic stress. Despite the nonspecificity of hypoplasia, however, careful consideration of potential stressors can provide insights into specific causes. In archaeological settings like the Stillwater Marsh, it is unlikely that genetic disorders (which are linked to hypoplastic activity [Witkop 1970]) would be important factors because the individual would likely not have survived much beyond birth. Localized trauma to the oral cavity is associated with hypoplasia (e.g., Skinner and Hung 1989), but facial trauma possibly resulting in hypoplastic activity has not been observed in the Stillwater remains. A number of instances of fractured nasal bones are found (see Brooks et al. 1988), but these fractures are likely not linked with hypoplasia.

A wide range of infectious diseases are linked to hypoplasias (Pindborg 1982). In the Stillwater series, however, skeletal pathologies reflecting infectious conditions are infrequent. Only 16% of individuals (10 of 61) have some form of periosteal inflammation, mostly associated with tibiae. This prevalence is either lower than or equal to that in most other North American hunter–gatherer series (e.g., Rose et al. 1984). The inflammations are largely localized periosteal reactions on single bones caused by pathogenic organisms (e.g., the bacterium *Staphylococcus aureus*) introduced into the bloodstream via skin penetration or by localized trauma (see Steinbock 1976). It is highly unlikely that these infections contributed to the kinds of systemic insults that could result in enamel defects.

It is possible that parasitic infections may be a causal factor in hypoplasia. Parasites are present in coprolites excavated from the nearby Lovelock Cave as well as from other areas of the Great Basin (Reinhard 1990). It remains unclear, however, what the link between hypoplasias and parasites, if any, might be for this region.

A number of studies report a causal link between enamel hypoplasias and protein–calorie malnutrition in living populations. In this regard, Goodman et al. (1991) have compared two groups of individuals from the State of Puebla, Mexico; one group received supplemental diets and the other group received nonsupplemental diets. The latter group had generally more hypoplasias than the former. This finding strongly implicates the linkage between nutritional quality and likelihood of having enamel defects.

Given that genetic factors, trauma, and infectious disease were likely not important factors in explaining the presence of hypoplastic teeth in the Stillwater series, it seems that nutritional deficiencies were potentially the more common sources of metabolic insult. In the Carson Desert setting, the availability of edible plants and animals is closely tied to precipitation levels, which average less than 10 cm/yr (see Raven and Elston 1989). Thus, most water supplying the Stillwater Marsh is from snowpack melt from the Sierra Nevada Mountains to the west. The availability of water fluctuates widely. In instances of excessive water, there can be loss of valuable plants and animals in the marsh due to inundation (as documented in the 1980s floods). In the other extreme, precipitation shortages can result in drying of the lakes resulting in a similar reduction of edible plants, waterbirds, and other fauna. Therefore, we suggest that hypoplasia in the Stillwater series is linked to occasional food shortages perhaps caused by water fluctuations and availability of plants and animals as well as seasonal shortages.

Stillwater hypoplasias are relatively frequent and narrow, especially compared to the Georgia coast. Some investigators suggest that width of hypoplasias provides a reasonable quantification of duration of stress events (Blakey and Armelagos 1985; Hutchinson and Larsen 1988, 1990; Sarnat and Schour 1941, 1942). Guita (1984) links the width of hypoplasia to the length of the period of stress. On the basis of research with laboratory animals, Suckling and coworkers (Suckling 1989; Suckling et al. 1986) argue, however, that severity of stress plays a vital role in determination of hypoplasia width.

Growth disruption duration has been studied via the analysis of growth increments called perikymata (Guatelli-Steinberg et al. 2004). Counting these increments provides a more detailed picture of growth stress duration than does measurement of the hypoplasia. Nonetheless, the overall size of the hypoplasia provides an indication of the stress magnitude in the affected individual. Our findings suggest that the magnitude of stress was relatively low in these populations.

A nutritional interpretation for the presence of hypoplasia is consistent with the measurement of bone mass (CA and PCCA) in the Stillwater remains. Bone mass values provide an important index of nutritional status in living human populations (see Garn et al. 1964). The relatively low values in the Stillwater series suggest that these populations likely experienced some nutritional deprivation. Although numerous factors enter into the etiology of hypoplasias, our assessment argues for nutrition or nutrition-related causes. The hypoplasias are narrow, however, indicating that if nutritional problems were involved, they were of short duration or severity or both.

COMPARISONS WITH OTHER GREAT BASIN SETTINGS: MALHEUR LAKE, OREGON, AND GREAT SALT LAKE, UTAH

The flooding in the mid-1980s exposed archaeological sites in at least two other Great Basin settings, namely Malheur Lake in the Harney Basin in Oregon (Hemphill 1992; Nelson 1999; Oetting 1999) and the Great Salt Lake in Utah (Bright and Loveland 1999; Simms 1999). Comparison of health and lifestyle based on the study of human remains for the three regions shows a number of important similarities and differences. The similarities suggest a common adaptive system for the vast region of the Great Basin, yet with some local response and adaptation unique to the particular settings of Stillwater Marsh, Malheur Lake, and Great Salt Lake (see various chapters in Hemphill and Larsen 1999). In particular, physical behavior and activity appears broadly similar for the three regions of the Great Basin. Osteoarthritis prevalence is high for all settings (Hemphill 1999, 2005), indicating a demanding lifestyle for all three regions. Similarly, analysis of cross-sectional geometry of long bones for all three regions reveals high and homogeneous robusticity (Ruff 1999). On the other hand, relatively low I_x/I_y ratios for Malheur indicates less long distance travel in this setting compared to Stillwater and Great Salt Lake. Regardless, the three groups are all highly sexually dimorphic, indicating very different patterns and degree of mobility in comparing males and females.

Stable isotope data are available for Stillwater and Great Salt Lake only. The differences between the two regions are striking (cf., Coltrain and Leavitt 2002; Coltrain and Stafford 1999; Schoeninger 1999). In this regard, the Great Salt Lake populations representing the Fremont Culture were not exclusive foragers. That is, after A.D. 400, stable carbon isotope ratios are less negative, but become more negative after A.D. 1150. The less negative values between A.D. 400 and A.D. 1150 reflect the period of time when the native populations occupying the region were engaged in maize agriculture. Following the twelfth century, however, the maize-dependent populations reverted to a hunter-gatherer lifeway (or abandoned the region and were replaced by a foraging population). Although the reasons for the disappearance of agriculture are not well understood, a widespread drought in the region likely contributed to the adaptive shift.

Evidence for health status variation shows a general pattern of low prevalence of indicators of poor health and physiological stress. In this regard, all three settings show low prevalence of periosteal reactions. Only in the late prehistoric (pre-A.D. 1800) component from Malheur does there appear to be elevated frequency of periosteal reactions. In this setting, it appears that an increase in infection coincided with increased population size and concentration (Nelson 1999). Across all three settings, the incidence of dental caries is remarkably low, reflecting the low carbohydrate diet (except for the maize agriculturalists in Great Salt Lake; see Bright and Loveland 1999). Stress indicators reveal a pattern of little evidence of iron deficiency anemia (low prevalence of cranial porosities), but relatively high prevalence of systemic stress (high prevalence of hypoplasias) for all three regions. The greater systemic stress is likely linked to seasonal variation in food availability, a common theme of life in marginal settings like the Great Basin (Hutchinson and Larsen 1995; Larsen 2006).

CONCLUSION

The fortuitous exposure and subsequent recovery of human skeletal remains from the west central and other settings in the Great Basin has allowed bioarchaeological research to address some long-standing questions about the relationship between hunter–gatherers and their environment. Several principal findings emerge from this study. First, the human populations inhabiting this region were physically robust, reflecting a demanding lifestyle. Analysis of cross-sectional geometric properties of bone reveals that these populations were mobile, particularly males. It is difficult to envision this mobility as being entirely restricted to the marsh setting. These populations, then, were not engaged in a primarily sedentary lifestyle in the marsh. This view implies that although the marsh contained a variety of economically-important foods and could probably support a population year-round, the Stillwater population was mobile. In fact, research at rockshelters in the Stillwater Mountains finds that when the Stillwater Marsh sites were occupied, foragers used the shelters as short-term camps, with evidence of bighorn sheep hunting (Kelly 2005). Comparison with other Great Basin settings suggests that mobility was commonplace throughout the region, at least as can be determined from these three settings.

Second, diet was diverse. Based just on the analysis of carbon and nitrogen stable isotope ratios, it is not possible to say conclusively whether or not foods were extracted from primarily marsh or nonmarsh settings. Rather, subsistence may have involved various combinations of both. In contrast to other regions of the Great Basin, there can be little doubt that pinyon contributed only marginally to the diets of hunter–gatherers at Stillwater Marsh. This may be because pinyon was not present in abundance during the time period when the Stillwater Marsh sites were occupied as the paleoecological research suggests (reviewed in Kelly 2005). In general, both C_3 and C_4 plants and their animal consumers were eaten, and there was little difference with respect to age, sex, or burial locality.

Finally, the health of these populations was likely relatively good in that physiological stress was present at moderate levels and of short duration or severity or both. The presence of relatively low levels of bone mass in adults suggests some nutritional deprivation, which may reflect seasonal or some other form of variable resource availability.

Study of the Stillwater and other Great Basin skeletal series presents an opportunity to address issues regarding hunter–gatherer adaptation using archaeological data. Winterhalder (1993) remarks that the lack of consensus among anthropologists regarding labor in hunter-gatherers results from the lack of an encompassing framework for work analysis. In a broader sense, our discussion points to recent developments in bioarchaeological analysis that can clarify work behaviors in human societies.

The study of human remains takes us closer to an understanding of human ecology in the Great Basin and consequently toward understanding variation in the hunting and gathering lifestyle. It is not important to say whether the Hobbesian or the original affluent society model better portrays foraging societies in the Great Basin in particular or among hunter–gatherers in general, but rather that there is variability among hunter–gatherers in terms of workload, diet, and health status. Some hunter-gatherer populations also will incorporate agriculture where it is seen as advantageous, at least for a time (e.g., Coltrain and Stafford 1999). The exercise becomes much more valuable when attention is focused on the sources of this variation, rather than on a debate over which model better characterizes *all* hunter-gatherers.

ACKNOWLEDGMENTS

Many individuals deserve special thanks for their role in the study of the Stillwater human remains. Thanks are extended to the staff of the Stillwater Wildlife Management Area (U.S. Department of the Interior, Fish and Wildlife Service), and especially to Anan Raymond, for his help and advice throughout the fieldwork and the following analysis. The fieldwork and research could not have been completed without the trust and understanding of the Fallon

Paiute-Shoshone Tribe and former Tribal Chairman Richard Hicks. The Stillwater data collection and analysis were facilitated by various people. Christine Larsen provided invaluable assistance in data recording. Thanks are extended to Thomas Barcia at the Veterans Administration Hospital, Reno, Nevada, for approval and arrangements for our use of the computed tomography scanner for long bone structural analysis. John Blitz, Al Hengge, Urszula Iwaniec, Matthew Murray, Renee Robinson, Mark Schurr, and Isabel Treichel prepared samples for stable isotope analysis. Modern plants for stable isotope analysis in the Carson Desert were collected by Elizabeth Budy and Anan Raymond. Investigation of the human remains in the Nevada State Museum was made easier due to the work of Michelle Haldeman, Sheilagh Brooks, Amy Dansie, and Don Tuohy in identifying skeletons and in cataloging and arranging skeletal remains prior to our study of them. Katherine Russell's participation in the project as field director, osteologist, and consultant extraordinaire is greatly appreciated. Amelia Hubbard provided editorial assistance in the preparation of this chapter. The fieldwork and laboratory analysis were funded by the National Science Foundation (BNS-8704094). Additional funding came from Northern Illinois University, the University of Louisville, and the University of Wisconsin Alumni Research Foundation.

REFERENCES

Bettinger, R. L., 1989, *The Archaeology of Pinyon House, Two Eagles, and Crater Middens: Three Residential Sites in Owens Valley, Eastern California, Anthropological Papers* 67, American Museum of Natural History, New York.

Blakey, M. L., and Armelagos, G. J., 1985, Deciduous Enamel Defects in Prehistoric Americans from Dickson Mounds, Prenatal and Postnatal Stress, *American Journal of Physical Anthropology* 66:371–380.

Braidwood, R. J., 1967, *Prehistoric Men*, 7th ed., Scott, Foresman, Glenview, Illinois.

Bridges, P. S., 1989, Changes in Activities with the Shift to Agriculture in the Southeastern United States, *Current Anthropology* 30:385–394.

Bridges, P. S., 1992, Prehistoric Arthritis in the Americas, *Annual Review of Anthropology* 21:67–91.

Bridges, P. S., Blitz, J. H., and Solano, M. C., 2000, Changes in Long Bone Diaphyseal Strength with Horticultural Intensification in West-Central Illinois, *American Journal of Physical Anthropology* 112:217–238.

Bright, J. R., and Loveland, C. J., 1999, A Biological Perspective on Prehistoric Human Adaptation in the Great Salt Lake Wetlands, in: *Prehistoric Lifeways in the Great Basin Wetlands: Bioarchaeological Reconstruction and Interpretation* (B. E. Hemphill and C. S. Larsen, eds.), University of Utah Press, Salt Lake City, pp. 103–116.

Brock, S. L., and Ruff, C. B., 1988, Diachronic Patterns of Change in Structural Properties of the Femur in the Prehistoric American Southwest, *American Journal of Physical Anthropology* 75:113–127.

Brooks, S., and Brooks, R. H., 1990, Who Were the Stillwater Marsh People? *Halcyon: A Journal of the Humanities* 12:63–74.

Brooks, S. T., Haldeman, M. B., and Brooks, R. H., 1988, *Osteological Analyses of the Stillwater Skeletal Series, Cultural Resource Series* 2, U.S. Department of the Interior, Fish and Wildlife Service, Washington, DC.

Buikstra, J. E., 1988, *The Mound-Builders of Eastern North America: A Regional Perspective*, Elfde Kroon-Voordracht, Amsterdam.

Charles, D. K., and Buikstra, J. E., 1983, Archaic Mortuary Sites in the Central Mississippi Drainage: Distribution, Structure, and Behavioral Implications, in: *Archaic Hunters and Gatherers in the American Midwest* (J. L. Phillips and J. A. Brown, eds.), Academic Press, New York, pp. 117–145.

Charles, D. K., Buikstra, J. E., and Konigsberg, L. W., 1986, Behavioral Implications of Terminal Archaic and Early Woodland Mortuary Practices in the Lower Illinois Valley, in: *Early Woodland Archeology* (K. B. Farnsworth and T. E. Emerson, eds.), *Kampsville Seminars in Archeology* 2, Center for American Archeology, Kampsville, Illinois, pp. 458–474.

Cohen, M. N., and Armelagos, G. J., 1984, *Paleopathology at the Origins of Agriculture*, Academic Press, Orlando, Florida.

Coltrain, J. B., and Leavitt, S. W., 2002, Climate and Diet in Fremont Prehistory: Economic Variability and Abandonment of Maize Agriculture in the Great Salt Lake Basin, *American Antiquity* 67:453–485.

Coltrain, J. B., and Stafford, T. W., Jr., 1999, Stable Carbon Isotopes and Great Salt Lake Wetlands Diet: Towards an Understanding of the Great Basin Formative, in: *Prehistoric Lifeways in the Great Basin Wetlands: Bioarchaeological Reconstruction and Interpretation* (B. E. Hemphill and C. S. Larsen, eds.), University of Utah Press, Salt Lake City, pp. 55–83.

Cybulski, J. S., 1992, *A Greenville Burial Ground: Human Remains and Mortuary Elements in British Columbia Coast Prehistory, Archaeological Survey of Canada Mercury Series Paper* 146, Canadian Museum of Civilization, Ottawa.

DeNiro, M. J., and Schoeninger, M. J., 1983, Stable Carbon and Nitrogen Isotope Ratios of Bone Collagen: Variations Within Individuals, Between Sexes, and Within Populations Raised on Monotonous Diets, *Journal of Archaeological Science* 10:199–203.

Domett, K. M., 2001, *Health in Late Prehistoric Thailand, British Archaeological Reports International Series* (Oxford) S946, Archaeopress, Oxford, England.

Fowler, C. S., 1990, *Tule Technology: Northern Paiute Uses of Marsh Resources in Western Nevada, Smithsonian Folklore Studies* 6, Smithsonian Institution Press, Washington, DC.

Fowler, C. S., 1992, *In the Shadow of Fox Peak: An Ethnography of the Cattail-Eater Northern Paiute People of Stillwater Marsh, Cultural Resource Series* 5, U.S. Department of the Interior, Fish and Wildlife Service, Washington, DC.

Garn, S. M., 1966, Malnutrition and Skeletal Development in the Pre-School Child, in: *Pre-School Child Malnutrition*, National Academy of Science, Washington, DC, pp. 43–62.

Garn, S. M., Rohrmann, C. G., Behar, M., Viteri, F., and Guzman, M. A., 1964, Compact Bone Deficiency in Protein-Calorie Malnutrition, *Science* 145:1444–1445.

Goodman, A. H., and Armelagos, G. J., 1985, The Chronological Distribution of Enamel Hypoplasia in Human Permanent Incisor and Canine Teeth, *Archives of Oral Biology* 30:503–507.

Goodman, A. H., and Rose, J. C., 1990, Assessment of Systemic Physiological Perturbations from Dental Enamel Hypoplasias and Associated Histological Structures, *Yearbook of Physical Anthropology* 33:59–110.

Goodman, A. H., and Rose, J. C., 1991, Dental Enamel Hypoplasias as Indicators of Nutritional Status, in: *Advances in Dental Anthropology* (M. A. Kelley and C. S. Larsen, eds.), Wiley-Liss, New York, pp. 279–292.

Goodman, A. H., Martinez, C., and Chavez, A., 1991, Nutritional Supplementation and the Development of Linear Enamel Hypoplasias in Children from Tezonteopan, Mexico, *American Journal of Clinical Nutrition* 53:773–781.

Grayson, D. K., 1993, *The Desert's Past: A Natural Prehistory of the Great Basin*, Smithsonian Institution Press, Washington, DC.

Guatelli-Steinberg, D., Larsen, C. S., and Hutchinson, D. L., 2004, Prevalence and the Duration of Linear Enamel Hypoplasia: A Comprehensive Study of Neadertals and Inuit Foragers, *Journal of Human Evolution* 47:65–84.

Guita, J. L., 1984, *Oral Pathology*, 2nd ed., Williams and Wilkins, Baltimore, Maryland.

Heizer, R. F., and Krieger, A. D., 1956, *The Archaeology of Humboldt Cave, Churchill County, Nevada, Publications in American Archaeology and Ethnology* 47(1), University of California, Berkeley.

Heizer, R. F., and Napton, L. K., 1970, *Archaeology and the Prehistoric Great Basin Lacustrine Subsistence Regime as Seen from Lovelock Cave, Nevada, Contributions of the University of California Archaeological Research Facility* 10, Berkeley.

Hemphill, B. E., 1992, *An Osteological Analysis of the Human Remains from Malheur Lake, Oregon* (3 volumes), *U.S. Department of the Interior, Fish and Wildlife Service, Cultural Resource Series* 6, Washington, DC.

Hemphill, B. E., 1999, Wear and Tear: Osteoarthritis as an Indicator of Mobility among Great Basin Hunter-Gatherers, in: *Prehistoric Lifeways in the Great Basin Wetlands: Bioarchaeological Reconstruction and Interpretation* (B. E. Hemphill and C. S. Larsen, eds.), University of Utah Press, Salt Lake City, pp. 241–289.

Hemphill, B. E., 2005, At What Cost a Full Belly? An Investigation of the Seductive Allure of Sedentary Horticulture in the Great Basin, *American Journal of Physical Anthropology*, Supplement 40:113.

Hemphill, B. E., and Larsen, C. S. (eds.), 1999, *Prehistoric Lifeways in the Great Basin Wetlands: Bioarchaeological Reconstruction and Interpretation*, University of Utah Press, Salt Lake City.

Hill, K., and Hurtado, A. M., 1989, Hunter-Gatherers of the New World, *American Scientist* 77:436–443.

Hodges, D. C., 1989, *Agricultural Intensification and Prehistoric Health in the Valley of Oaxaca, Mexico, Museum of Anthropology Memoirs* 22, University of Michigan, Ann Arbor.

Holt, B. M., 2003, Mobility in Upper Paleolithic and Mesolithic Europe: Evidence from the Lower Limb, *American Journal of Physical Anthropology* 122:200–215.

Hough, A. J., Jr., 2001, Pathology of Osteoarthritis, in: *Osteoarthritis: Diagnosis and Medical/Surgical Management*, 3rd ed. (R. W. Moskowitz, D. S. Howell, R. D. Altman, J. A. Buckwalter and V. M. Goldberg, eds.), W. B. Saunders Company, Philadelphia, pp. 69–100.

Hutchinson, D. L., 2002, *Foraging, Farming, and Coastal Biocultural Adaptation in Late Prehistoric North Carolina*, University Press of Florida, Gainesville.

Hutchinson, D. L., and Larsen, C. S., 1988, Determination of Stress Episode Duration from Linear Enamel Hypoplasias: A Case Study from St. Catherines Island, Georgia, *Human Biology* 60:93–110.

Hutchinson, D. L., and Larsen, C. S., 1990, Stress and Lifeway Change: The Evidence from Enamel Hypoplasia, in: *The Archaeology of Mission Santa Catalina de Guale: 2. Biocultural Interpretations of a Population in Transition* (C. S. Larsen, ed.), *Anthropological Papers* 68, American Museum of Natural History, New York, pp. 50–65.

Hutchinson, D. L., and Larsen, C. S., 1995, Physiological Stress in the Prehistoric Stillwater Marsh: Evidence of Enamel Defects, in: *Bioarchaeology of the Stillwater Marsh: Prehistoric Human Adaptation in the Western Great Basin* (C. S. Larsen and R. L. Kelly, eds.), *Anthropological Papers* 77, American Museum of Natural History, New York, pp. 81–95.

Hutchinson, D. L., and Larsen, C. S., 2001, Enamel Hypoplasia and Stress in La Florida, in: *Bioarchaeology of Spanish Florida: The Impact of Colonialism* (C. S. Larsen, ed.), University Press of Florida, Gainesville, pp. 181–206.

Janetski, J. C., and Madsen, D. B. (eds.), 1990, *Wetland Adaptations in the Great Basin: Papers from the Twenty-First Great Basin Anthropological Conference, Museum of Peoples and Cultures Occasional Paper* 1, Brigham Young University, Provo, Utah.

Jennings, J. D., 1957, *Danger Cave, University of Utah Anthropological Papers* 27, Salt Lake City, Utah.

Jenning, J. D., 1978, *Prehistory of Utah and Eastern Great Basin, University of Utah Anthropological Papers* 98, Salt Lake City, Utah.

Katzenberg, M. A., 2000, Stable Isotope Analysis: A Tool for Studying Past Diet, Demography, and Life History, in: *Biological Anthropology of the Human Skeleton* (M. A. Katzenberg and S. R. Saunders, eds.), Wiley-Liss, New York, pp. 305–327.

Katzenberg, M. A., and Saunders, S. R. (eds.), 2000, *Biological Anthropology of the Human Skeleton*, Wiley-Liss, New York.

Kelly, R. L., 1992, Mobility/Sedentism: Concepts, Archaeological Measures, and Effects, *Annual Review of Anthropology* 21:43–66.

Kelly, R. L., 1995a, *The Foraging Spectrum: Diversity in Hunter-Gatherer Lifeways*, Smithsonian Institution Press, Washington, DC.

Kelly, R. L., 1995b, Hunter-Gatherer Lifeways in the Carson Desert: A Context for Bioarchaeology, in: *Bioarchaeology of the Stillwater Marsh: Prehistoric Human Adaptation in the Western Great Basin* (C. S. Larsen and R. L. Kelly, eds.), *Anthropological Papers* 77, American Museum of Natural History, New York, pp. 12–32.

Kelly, R. L., 1997, The Late Holocene Prehistory of the Great Basin, *Journal of World Prehistory* 11:1–49.

Kelly, R. L., 2001, *Prehistory of the Carson Desert and Stillwater Mountains: Environment, Mobility, and Subsistence in a Great Basin Wetland*, University of Utah Anthropological Papers 123, Salt Lake City, Utah.

Kelly, R. L., 2005, Mustang Shelter: Test Excavation of a Rockshelter in the Stillwater Mountains, Western Nevada, manuscript on file, Department of Anthropology, University of Wyoming, Laramie.

Kent, S., 1992, Studying Variability in the Archaeological Record: An Ethnoarchaeological Model for Distinguishing Mobility Patterns, *American Antiquity* 57:635–660.

Kreshover, S. J., 1960, Metabolic Disturbances in Tooth Formation, *Annals of the New York Academy of Sciences* 85:161–167.

Kreshover, S., and Clough, O. W., 1953a, Prenatal Influences on Tooth Development. I. Alloxan Diabetes in Rats, *Journal of Dental Research* 32:246–261.

Kreshover, S., and Clough, O. W., 1953b, Prenatal Influences on Tooth Development. II. Artificially Induced Fever in Rats, *Journal of Dental Research* 32:565–572.

Kuhlbusch, T. A., Lobert, J. M., Crutzen, P. J., and Warneck, P., 1991, Molecular Nitrogen Emissions from Denitrification during Biomass Burning, *Nature* 351:135–137.

Lambert, P. M. (ed.), 2000, *Bioarchaeological Studies of Life in the Age of Agriculture: A View from the Southeast*, University of Alabama Press, Tuscaloosa.

Lanyon, L. E., and Rubin, C. T., 1985, Functional Adaptation in Skeletal Structures, in: *Functional Vertebrate Morphology* (M. Hildebrand, D. M. Bramble, K. F. Liem and D. B. Wake, eds.), Belknap Press and Harvard University Press, Cambridge, Massachusetts, pp. 1–25.

Larsen, C. S., 1997, *Bioarchaeology: Interpreting Behavior from the Human Skeleton*, Cambridge University Press, Cambridge.

Larsen, C. S. (ed.), 2001, *Bioarchaeology of La Florida: The Impact of Colonization*, University Press of Florida, Gainesville.

Larsen, C. S., 2006, Great Basin: Skeletal Biology, in: *Handbook of North American Indians: Environment, Origins, and Population*, Volume 3 (W. C. Sturtevant, general ed.), (D. H. Ubelaker, ed.), Smithsonian Institution Press, Washington, DC, pp. 581–589.

Larsen, C. S., and Hutchinson, D. L., 1992, Dental Evidence for Physiological Disruption: Biocultural Interpretations from the Eastern Spanish Borderlands, in: *Recent Contributions to the Study of Enamel Developmental Defects* (A. H. Goodman and L. L. Capasso, eds.), *Journal of Paleopathology Monographic Publications* 2, Associazione Anthropologica Abruzzese, Chieti, Italy, pp. 151–169.

Larsen, C. S., and Kelly, R. L. (eds.), 1995, *Bioarchaeology of the Stillwater Marsh: Prehistoric Human Adaptation in the Western Great Basin*, Anthropological Papers 77, American Museum of Natural History, New York.

Larsen, C. S., and Milner, G. R. (eds.), 1994, *In the Wake of Contact: Biological Responses to Conquest*, Wiley-Liss, New York.

Larsen, C. S., Ruff, C. B., and Kelly, R. L., 1995, Structural Analysis of the Stillwater Postcranial Human Remains: Behavioral Implications of Articular Joint Pathology and Long Bone Diaphyseal Morphology, in: *Bioarchaeology in the Stillwater Marsh: Prehistoric Human Adaptation in the Western Great Basin* (C. S. Larsen and R. L. Kelly, eds.), *Anthropological Papers* 77, American Museum of Natural History, New York, pp. 107–133.

Ledger, M., Holtzhausen, L.-M., Constant, D., and Morris, A. G., 2000, Biomechanical Beam Analysis of Long Bones from a Late 18th Century Slave Cemetery in Cape Town, South Africa, *American Journal of Physical Anthropology* 112:207–216.

Lee, R. B., 1979, Kalahari Hunter-Gatherers, Harvard University Press, Cambridge, Massachusetts.

Lee, R. B., and DeVore, I. (eds.), 1968, Man the Hunter, Aldine, Chicago, Illinois.

Lowie, R. F., 1924, Notes on Shoshonean Ethnography, Anthropological Papers 20(3), American Museum of Natural History, New York.

Mack, M. E., and Coppa, A., 1992, Frequency and Chronological Distribution of Enamel Hypoplasias from the Ra's al-Hamra-5 (RH5) Skeletal Collection (Oman), in: Recent Contributions to the Study of Enamel Development Defects (A. H. Goodman and L. L. Capasso, eds.), Journal of Paleopathology Monographic Publications 2, Associazione Anthropologica Abruzzese, Chieti, Italy, pp. 131–141.

Mandryk, C. A. S., 1993, Hunter-Gatherer Social Costs and the Nonviability of Submarginal Environments, Journal of Anthropological Research 49:39–71.

Martin, D. L., Akins, N. J., Goodman, A. H., Toll, H. W., and Swedlund, A. C., 2001, Harmony and Discord: Bioarchaeology, Archaeology Notes 242, Office of Archaeological Studies, Museum of New Mexico, Santa Fe.

Merbs, C. F., 1983, Patterns of Activity-Induced Pathology in a Canadian Inuit Population, Archaeological Survey of Canada Mercury Series Paper 119, Canadian Museum of Civilization, Ottawa.

Nelson, G. C., 1999, Physiologic Stress and Environmental Fluctuation in the Northern Great Basin, in: Prehistoric Lifeways in the Great Basin Wetlands: Bioarchaeological Reconstruction and Interpretation (B. E. Hemphill and C. S. Larsen, eds.), University of Utah Press, Salt Lake City, pp. 219–240.

Oetting, A. C., 1999, An Examination of Wetland Adaptive Strategies in Harney Basin: Comparing Ethnographic Paradigms and the Archaeological Record, in: Prehistoric Lifeways in the Great Basin Wetlands: Bioarchaeological Reconstruction and Interpretation (B. E. Hemphill and C. S. Larsen, eds.), University of Utah Press, Salt Lake City, pp. 203–218.

Papathanasiou, A., 2001, A Bioarchaeological Analysis of Neolithic Alepotrypa Cave, Greece, British Archaeological Reports International Series (Oxford) S961, Archaeopress, Oxford, England.

Peterson, J., 2002, Sexual Revolutions: Gender and Labor at the Dawn of Agriculture, Altamira Press, Walnut Creek, California.

Pietrusewsky, M., and Douglas, M. T., 2001, Ban Chiang, A Prehistoric Village Site in Northeast Thailand I: The Human Skeletal Remains, University of Pennsylvania Press, Philadelphia.

Pindborg, J. J., 1982, Aetiology of Developmental Defects Not Related to Fluorosis, International Dental Journal 32:123–134.

Radin, E. L., Paul, I. L., and Rose, R. M., 1972, Role of Mechanical Factors in Pathogenesis of Primary Osteoarthritis, Lancet 1:519–522.

Rafferty, J. E., 1985, The Archaeological Record on Sedentariness: Recognition, Development and Implications, in: Advances in Archaeological Method and Theory, Volume 8 (M. B. Schiffer, ed.), Academic Press, Orlando, Florida, pp. 113–156.

Raven, C., 1990, Prehistoric Human Geography in the Carson Desert. Part II. Archaeological Field Tests of Model Predictions, Cultural Resource Series 4, U.S. Department of the Interior, Fish and Wildlife Service, Washington, DC.

Raven, C., and Elston, R. G. (eds.), 1989, Prehistoric Human Geography in the Carson Desert. Part I. A Predictive Model of Land-Use in the Stillwater Wildlife Management Area, Cultural Resource Series 3, U.S. Department of the Interior, Fish and Wildlife Service, Washington, DC.

Raymond, A. W., and Parks, V. M., 1990, Archaeological Sites Exposed by Recent Flooding of Stillwater Marsh, Carson Desert, Churchill County, Nevada, in: Wetland Adaptations in the Great Basin (J. C. Janetski and D. B. Madsen, eds.), Museum of Peoples and Cultures Occasional Paper 1, Brigham Young University, Provo, Utah, pp. 33–61.

Reinhard, K. J., 1990, Archaeoparasitology in North America, American Journal of Physical Anthropology 82:145–163.

Robbins, D. M., Rosenberg, K. R., and Ruff, C. B., 1989, Activity Patterns in Late Middle Woodland, Delaware, American Journal of Physical Anthropology 78:290–291.

Rose, J. C., Burnett, B. A., Nassaney, M. S., and Blaeuer, M. W., 1984, Paleopathology and the Origins of Maize Agriculture in the Lower Mississippi Valley and Caddoan Culture Areas, in: *Paleopathology at the Origins of Agriculture* (M. N. Cohen and G. J. Armelagos, eds.), Academic Press, Orlando, Florida, pp. 393–424.

Ruff, C. B., 1987, Sexual Dimorphism in the Human Lower Limb Bone Structure: Relationship to Subsistence Strategy and Sexual Division of Labor, *Journal of Human Evolution* 16:391–416.

Ruff, C. B., 1991, *Aging and Osteoporosis in Native Americans from Pecos Pueblo, New Mexico*, Garland, New York.

Ruff, C. B., 1994, Biomechanical Analysis of Northern and Southern Plains Femora: Behavioral Implications, in: *Skeletal Biology in the Great Plains: A Multidisciplinary View* (D. W. Owsley and R. L. Jantz, eds.), Smithsonian Institution Press, Washington, DC, pp. 235–245.

Ruff, C. B., 1999, Skeletal Structure and Behavioral Patterns of Prehistoric Great Basin Populations, in: *Understanding Prehistoric Lifeways in the Great Basin Wetlands: Bioarchaeological Reconstruction and Interpretation* (B. E. Hemphill and C. S. Larsen, eds.), University of Utah Press, Salt Lake City, pp. 290–320.

Ruff, C. B., 2000, Biomechanical Analyses of Archaeological Human Skeletons, in: *Biological Anthropology of the Human Skeleton* (M. A. Katzenberg and S. R. Saunders, eds.), Wiley-Liss, New York, pp. 71–102.

Ruff, C. B., and Larsen, C. S., 2001, Reconstructing Behavior in Spanish Florida: The Biomechanical Evidence, in: *Bioarchaeology of Spanish Florida: The Impact of Colonialism* (C. S. Larsen, ed.), University Press of Florida, Gainesville, pp. 113–145.

Ruff, C. B., Trinkaus, E., Walker, A., and Larsen, C. S., 1993, Postcranial Robusticity in *Homo*. I. Temporal Trends and Mechanical Interpretation, *American Journal of Physical Anthropology* 91:21–53.

Sahlins, M., 1972, *Stone Age Economics*, Aldine, Chicago, Illinois.

Sarnat, B. G., and Schour, I., 1941, Enamel Hypoplasia (Chronologic Enamel Aplasia) in Relation to Systemic Disease: A Chronologic, Morphologic and Etiologic Classification, *Journal of the American Dental Association* 28:1989–2000.

Sarnat, B. G., and Schour, I., 1942, Enamel Hypoplasia (Chronologic Enamel Aplasia) in Relation to Systemic Disease: A Chronologic, Morphologic and Etiologic Classification, *Journal of the American Dental Association* 29:67–77.

Schoeninger, M. J., 1995, Dietary Reconstruction in the Prehistoric Carson Desert: Stable Carbon and Nitrogen Isotopic Analysis, in: *Bioarchaeology in the Stillwater Marsh: Prehistoric Human Adaptation in the Western Great Basin* (C. S. Larsen and R. L. Kelly, eds.), *Anthropological Papers* 77, American Museum of Natural History, New York, pp. 96–106.

Schoeninger, M. J., 1999, Prehistoric Subsistence Strategies in the Stillwater Marsh Region of the Carson Desert, in: *Prehistoric Lifeways in the Great Basin Wetlands: Bioarchaeological Reconstruction and Interpretation* (B. E. Hemphill and C. S. Larsen, eds.), University of Utah Press, Salt Lake City, pp. 151–166.

Schoeninger, M. J., and Moore, K., 1992, Bone Stable Isotope Studies in Archaeology, *Journal of World Prehistory* 6:247–296.

Schoeninger, M. J., DeNiro, M. J., and Tauber, H., 1983, Stable Nitrogen Isotope Ratios of Bone Collagen Reflect Marine and Terrestrial Components of Prehistoric Human Diet, *Science* 220:1381–1383.

Schwarcz, H. P., and Schoeninger, M. J., 1991, Stable Isotope Analyses in Human Nutritional Ecology, *Yearbook of Physical Anthropology* 34:283–321.

Sharma, L., 2001, Epidemiology of Osteoarthritis, in: *Osteoarthritis: Diagnosis and Medical/Surgical Management*, 3rd ed. (R. W. Moskowitz, D. S. Howell, R. D. Altman, J. A. Buckwalter and V. M. Goldberg, eds.), W. B. Saunders Company, Philadelphia, pp. 3–27.

Simms, S. R., 1999, Farmers, Foragers, and Adaptive Diversity, in: *Prehistoric Lifeways in the Great Basin Wetlands: Bioarchaeological Reconstruction and Interpretation* (B. E. Hemphill and C. S. Larsen, eds.), University of Utah Press, Salt Lake City, pp. 21–54.

Simpson, J. H., 1876, *Report of Explorations Across the Great Basin of the Territory of Utah for a Direct Wagon-Route from Camp Floyd to Genoa, in Carson Valley, in 1859*, U.S. Government Printing Office, Washington, DC.

Skinner, M. F., and Hung, J. T. W., 1989, Social and Biological Correlates of Localized Enamel Hypoplasia of the Human Deciduous Canine Tooth, *American Journal of Physical Anthropology* 79:159–175.

Steckel, R. H., and Rose, J. C. (eds.), 2002, *The Backbone of History: Health and Nutrition in the Western Hemisphere*, Cambridge University Press, Cambridge, England.

Steinbock, R. T., 1976, *Paleopathological Diagnosis and Interpretation: Bone Diseases in Ancient Human Populations*, Charles C. Thomas, Springfield, Illinois.

Steward, J. H., 1938, *Basin-Plateau Aboriginal Sociopolitical Groups*, Bulletin 120, Bureau of American Ethnology, Washington, DC.

Stock, J. T., and Pfeiffer, S. K., 2001, Linking Structural Variability in Long Bone Diaphyses to Habitual Behaviors: Foragers from the Southern African Later Stone Age and the Andaman Islands, *American Journal of Physical Anthropology* 115:337–348.

Stock, J. T., and Pfeiffer, S. K., 2004, Long Bone Robusticity and Subsistence Behaviour among Later Stone Age Foragers of the Forest and Fynbos Biomes of South Africa, *Journal of Archaeological Science* 31:999–1013.

Suckling, G., 1989, Developmental Defects of Enamel-Historical and Present-Day Perspectives of Their Pathogenesis, *Advances in Dental Research* 3:87–94.

Suckling, G., Elliot, D. C., and Thurley, D. C., 1986, The Macroscopic Appearance and Associated Histological Changes in the Enamel Organ of Hypoplastic Lesions of Sheep Incisor Teeth Resulting from Induced Parasitism, *Archives of Oral Biology* 31:427–439.

Suckling, G., and Thurley, D. C., 1984, Developmental Defects of Enamel: Factors Influencing Their Macroscopic Appearance, in: *Tooth Enamel IV* (R. W. Fearnhead and S. Suga, eds.), Elsevier, Amsterdam, pp. 357–362.

Ten Cate, A. R., 1994, *Oral Histology: Development, Structure, and Function*, 4th ed., C. V. Mosby, St. Louis, Missouri.

Thomas, D. H. 1985, *The Archaeology of Hidden Cave, Nevada*, Anthropological Papers 61(1), American Museum of Natural History, New York.

Thomas, D. H. (ed.), 1988, *The Archaeology of Monitor Valley: 3. Survey and Additional Excavations, Anthropological Papers* 66(2), American Museum of Natural History, New York.

Tieszen, L. L., 1991, Natural Variations in the Carbon Isotope Values of Plants: Implications for Archaeology, Ecology, and Paleoecology, *Journal of Archaeological Science* 18:227–248.

Tuohy, D. R., Dansie, A. J., and Haldeman, M. B., 1987, *Final Report on Excavations in the Stillwater Marsch Archaeological District, Nevada, Archaeological Services Reports*, Nevada State Museum, Carson City.

Ubelaker, D. H., 1992, Enamel Hypoplasia in Ancient Ecuador, in: *Recent Contributions to the Study of Enamel Developmental Defects* (A. H. Goodman and L. L. Capasso, eds.), *Journal of Paleopathology Monographic Publications* 2, Associazione Antropologica Abruzzese, Chieti, Italy, pp. 207–217.

Ubelaker, D. H., 1994, The Biological Impact of European Contact in Ecuador, in: *In the Wake of Contact: Biological Responses to Conquest* (C. S. Larsen and G. R. Milner, eds.), Wiley-Liss, New York, pp. 147–160.

Vogel, J. C., Talma, A. S., Hall-Martin, A. J., and Viljoen, P. J., 1990, Carbon and Nitrogen Isotopes in Elephants, *South African Journal of Science* 86:147–150.

Wheat, M. M., 1967, *Survival Arts of the Primitive Paiutes*, University of Nevada Press, Reno.

Whittington, S. L., and Reed, D. M. (eds.), 1997, *Bone of the Maya: Studies of Ancient Skeletons*, Smithsonian Institution Press, Washington, DC.

Williamson, R. F., and Pfeiffer, S. (eds.), 2003, *Bones of the Ancestors: The Archaeology and Osteobiography of the Moatfield Ossuary, Canadian Museum of Civilization Mercury Series* 163, Gatineau, Quebec.

Wing, E. S., and Brown, A. B., 1979, *Paleonutrition: Method and Theory in Prehistoric Foodways*, Academic Press, New York.

Winterhalder, B., 1993, Work, Resources and Population in Foraging Societies, *Man* 28:321–340.

Witkop, C., 1970, Heritable Disorders, in: *Thoma's Oral Pathology*, Volume 2 (R. Gorlin and H. Goldman, eds.), C. V. Mosby, St. Louis, Missouri, pp. 635–637.

Zeannah, D. W., 1996, *Predicting Settlement Patterns and Mobility Strategies: An Optimal Foraging Analysis of Hunter-Gatherer Use of Mountain, Desert, and Wetland Habitats in the Carson Desert*, Ph.D. dissertation, Department of Anthropology, University of Utah, Salt Lake City.

Zeannah, D. W., 2004, Sexual Division of Labor and Central Place Foraging: A Model for the Carson Desert of Western Nevada, *Journal of Archaeological Archaeology* 23:1–32.

Zeannah, D. W., Carter, J. A., Dugas, D. P., Elston, R. G., and Hammett, J. E., 1995, *An Optimal Foraging Model of Hunter-Gatherer Land Use in the Carson Desert*, Report to the U.S. Fish and Wildlife Service and U.S. Department of Navy, Portland, Oregon.

Chapter *10*

Pathoecology of Two Ancestral Pueblo Villages

KARL J. REINHARD

Pathoecology is the study of the biotic, abiotic, and cultural environments of disease (Martinson et al. 2003). A parasitic infection is the result of the pathoecological interaction of host behavior, parasite life cycle, the environment in which both life forms live, the nutritional status of the host, and host physiological responses to all of these factors. Parasites contribute to anemia in many ways. Some, such as hookworm, actually consume blood and cause iron loss through their activities. For other parasites, symptoms such as profuse diarrhea reduce intestinal absorption of nutrients. Others, such as certain fish tapeworms, actually compete for absorption of nutrients necessary for blood cell production. Often these causes of anemia are interdependent. Human behavior is one of the keys to understanding pathoecology (Martinson et al. 2003; Reinhard 1988, 1992a; Reinhard and Buikstra 2003; Reinhard et al. 2003). Thus, for the pathoecologist studying human health, one goal is to identify the human behaviors related to environmental conditions that promote or reduce disease. Relevant aspects of behavior include population density, seasonal movement, excreta disposal patterns, diet, and personal hygiene (Reinhard 1992b). To examine the complexity of pathoecology among the Ancestral Pueblo culture of the Colorado Plateau, I compared two Ancestral Pueblo villages: Salmon Pueblo, New Mexico, (formerly known as Salmon Ruin) and Antelope House, Arizona (Figure 10-1). Parasitism and nutritional deficiency are both causes of anemia. Previous work (Reinhard 1992b) showed that these sites represented the extremes in anemia. Through this comparison, one begins to understand the pathoecology of ancient anemia.

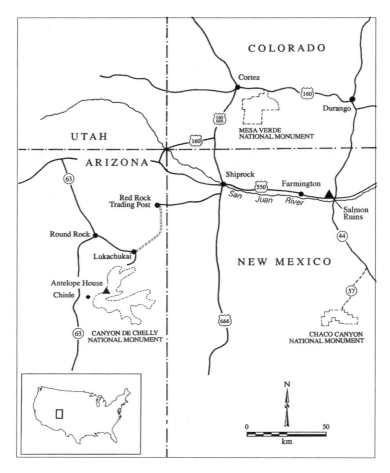

Figure 10-1. Location of Antelope House in Canyon de Chelly, Arizona, and Salmon Pueblo on the San Juan River, New Mexico. Salmon Pueblo is also known as Salmon Ruin.

The two sites have Pueblo III period occupations in very different locations (Figure 10-1). Antelope House is located in a large rock shelter in a canyon bottom in Canyon de Chelly National Monument. Excavation was done by Morris (1986) of the National Park Service between 1970 and 1974. The cave was inhabited by Ancestral Pueblo people between A.D. 500 and 1250. Salmon Pueblo is on the San Juan River. Excavation of Salmon Pueblo was initiated in 1970 by Irwin-Williams (Bohrer and Adams 1977). These excavations ended in 1978. Tree ring evidence indicates that construction of Salmon Pueblo village began in A.D. 1088. It was occupied until it was destroyed in a catastrophic fire between A.D. 1280 and 1290 (Reed 2006:296). At the peak occupation of Salmon Pueblo, it had in excess of 300 inhabitants.

Salmon Pueblo was built on an alluvial fan overlooking the floodplain of the San Juan River. Meander scars show that the river was once close to or in contact with the alluvial fan where the Pueblo was built (Nials and Reed 2006). The location of the Salmon Pueblo village along the river provided ready access to arable land. The flood plain itself was arable and there are two near-by drainages that could have been farmed. Today, the floodplain itself is very moist, with dense vegetation cover. The trash deposits near Salmon Pueblo were washed away at some time in the past.

Likewise, the moist environment of the canyon bottom near Antelope House made it an ideal place for farming. Most trash deposits and an unknown number of rooms were lost to flooding at Antelope House.

MATERIALS AND METHODS

Analysis of the contents of 292 coprolites (paleofeces or dried feces) formed the basis of this study of the pathoecologies at the two sites. Sampling strategies during excavation of these sites optimized diversification. In selecting coprolites, it is ideal to diversify the sample so that defecations by as many different individuals as possible are obtained. Obviously, sites with several latrines and occupied for longer periods of time offer better conditions for diversifying coprolite samples than do single latrine sites from short occupations. Antelope House reflects the former situation and contained several hundred separate latrine deposits. The Antelope House analysis included 180 coprolites selected from 34 separate latrine areas. Differences in texture, coloration, and size were considered to avoid sampling the same defecation twice. Salmon Pueblo reflects the latter situation, with only one large latrine suitable for study. Other large latrines excavated at Salmon Pueblo were burned and the carbonized coprolites in them were not suitable for study. In order to increase heterogeneity, Salmon Pueblo coprolites were recovered from alternate levels in alternate grid squares. Of an estimated 4,000 coprolites collected, 112 were analyzed.

The macrofossil analysis of the coprolites used procedures outlined by Fry (1977). The specimens were cleaned, photographed, sketched, measured, weighed in grams, and described in terms of surface morphology and visible inclusions. Subsamples were removed from each coprolite: 1 g samples from Antelope House and 1.0–2.5 g samples from Salmon Pueblo. These were rehydrated, disaggregated, screened, concentrated by centrifugation, and, in the case of macroscopic components, dried. As a general procedure, the entire coarse fraction was separated into individual components, which were weighed separately. The fine fraction was subsampled by weight, and approximately one third of the fraction was fully separated into the various components. The remainder of the fine fraction was scanned for unique items, such as seeds and ectoparasites, but was not otherwise sorted. The microscopic fraction was examined for parasite remains after Reinhard (1985a) and Reinhard et al. (1988). Parasite identifications are based on morphological similarities to modern comparative materials. Microscopic remains were sedimented by gravity in acetic formalin alcohol. A sample of the uppermost sediments was pipetted onto a glass microscope slide with a drop of glycerol, covered with a cover slip, and sealed. After parasitological analysis, microscopic sediments were processed for pollen (Bryant 1974; Clary 1984; Martin and Sharrock 1964). A minimum of 200 pollen grains was counted for each specimen (Barkley 1934). Dietary interpretations of the pollen data are based on percentage expressions of these counts. Recently, a second analysis of coprolites from Salmon Pueblo was conducted (Reinhard et al. 2006b). This analysis relied on analysis of very small residues less than 0.5 millimeters in size.

Most North Americans are unfamiliar with parasites even though they are important to human health on a global scale. Although less sensational than emergent viral disease such as Ebola and West Nile, parasites nonetheless take a tremendous, chronic toll on humans all over the world. The phylum Platyhelminthes, or flatworms, contains two parasitic classes. One class is the Cestoda which includes the tapeworms. The second parasitic class is the Trematoda which includes the flukes. Of the tapeworms, eggs of the families Hymenolepidae and Taeniidae have been found in Southwest U.S. coprolites. In shape, tapeworms resemble bands of tape or ribbons. The hymenolepidids that parasitized ancient southwesterners were quite small. The dwarf tapeworm, *Hymenolepis nana*, is the smallest tapeworm that infects humans. In contrast, some taeniid species are at least 50 cm long. Flukes are similar to the free-living flatworms (*Planaria* spp.) that are commonly used as experimental animals in biology courses. The exact fluke species that is represented by eggs in prehispanic coprolites is unknown. Roundworms are more common parasites of prehispanic southwesterners. Roundworms are in the phylum Nematoda and are unsegmented worms. The pinworms (*Enterobius vermicularis*), hookworms (family Strongyloidea), and threadworms (*Strongyloides* sp.) are roundworms. Most, if not all, basic biology laboratories use the giant intestinal roundworm of pigs, *Ascaris suum*, as a dissection animal. Students are therefore usually very familiar with the form of roundworm parasites. More recently, coprolites from each site were tested for the flagellated protozoan parasite *Giardia lamblia* (Gonçalves et al. 2002; Wilson et al. 2006). Coprolites from Antelope House have been tested for the amoebid that causes amoebic dysentery, *Entamoeba histolytica* (Gonçalves et al. 2004). Both of these protozoa are common causes of diarrhea, especially in babies and children. High worm infections are accompanied by high protozoal infections in the modern world, and it is almost a certainty that protozoal parasites (*Entamoeba* and *Giardia* species) were more common at Antelope House as well. Therefore, protozoa-caused diarrhea and anemia were probably greater problems for Antelope House inhabitants than for those at Salmon Pueblo.

RESULTS

The two sites had very different levels and types of parasitism. Only 9 (8%) of the Salmon Pueblo coprolites contained eggs of parasitic worms and only one species is represented, the pinworm (*Enterobius vermicularis*) (Figure 10-2). Examination of Salmon Pueblo coprolites for *Giardia* cysts was negative (Wilson et al. 2006).

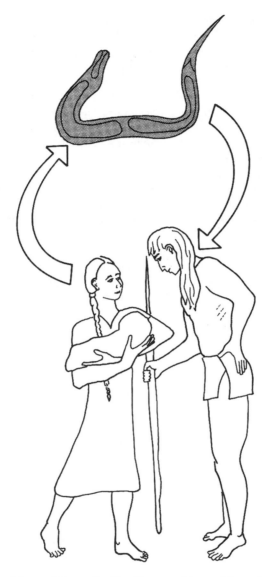

Figure 10-2. Parasitism at Salmon Pueblo. The parasitism evidenced at this site is limited and includes only one parasitic worm, pinworm (*Enterobius vermicularis*). This parasite is transmitted by eggs through the air or by person-to-person contact. In the modern world, children are most often infected, followed by women, probably due to greater contact between women and children.

Of 180 coprolites studied from Antelope House, 52 (29%) contained worm eggs. Five taxa are represented: pinworm, threadworm (*Strongyloides* sp.), tapeworm (family Hymenolepidae), the giant intestinal roundworm of humans (*Ascaris lumbricoides*), and a strongylate species (Figure 10-3). The strongylate worm category refers to the eggs of parasites that could be in either the order Strongyloidea or the order Trichostrongyloidea. The size of these eggs is consistent with hookworms (family Ancylostomidae, *Ancylostoma duodenale, Necator americans*). Other tests show that Antelope House inhabitants were infected by the amoebids that causes amoebic dysentery *G. lamblia* and *E. histolytica.*

The difference in parasitism between the sites had a dramatic impact on health. Pinworms in small numbers are essentially harmless. They are, however, an annoyance. In high numbers, pinworms can be distinct health problems. They can cause minute ulcerations of the intestine with subsequent bacterial infection. Also, the anus can become ulcerated due to scratching, with subsequent bacterial infection. Beyond these rare problems, insomnia, appetite loss, and weight loss can result from pinworm infection. The pinworm prevalence at Antelope House is very high, higher than any modern standards and higher than any other coprolite series studied in the world. Therefore, pinworm was undoubtedly a health problem for Antelope House inhabitants.

Figure 10-3. Parasitism at Antelope House. The parasitism evidenced at this site includes several species. Pinworm (*Enterobius vermicularis*) was transmitted as described in Figure 10-2. As illustrated here, dogs (*Canis familiaris*) were reservoir hosts for threadworms (*Strongyloides stercoralis*), which burrow through the skin of humans. Tapeworms (Hymenolepidae) are commonly transmitted by consumption of grain beetles (*Tribolium* spp. and *Tenebrio* spp.). An unknown strongylate worm is also present at Antelope House, but its exact species identification and therefore its life cycle is currently unknown.

Species of threadworms (*Strongyloides*) are very pathogenic. Third-stage larvae penetrate the skin of humans, enter the bloodstream, and go to the lungs. From the lungs, they migrate up the throat and are swallowed. When they reach the intestine, they mature to the adult stage. The adults burrow through the intestinal lining and cause ulcerations. The eggs hatch in the intestine and the larvae enter the bloodstream through the ulcerations caused by the parents. They then mature in the normal way. This process is called *autoinfection*. Autoinfection can result in a massive population of worms (*hyperinfection*). The blood loss from ulcerations causes anemia. Also, because scar tissue replaces normal intestinal lining at the ulcerations, less food is absorbed through the intestinal lining. Thus, threadworm (*Strongyloides*) infection is dangerous and debilitating.

I cannot determine which species of worm laid the strongylate-type egg. It is probable that hookworm is present. Hookworm is one of the most dangerous human parasites. This possibility is under investigation using DNA within the eggs.

Two other relatively harmless worms are the hymenolepidid tapeworm and the giant intestinal roundworm. The tapeworm found in a Antelope House coprolites rarely causes disease but may cause nausea and other minor symptoms. The giant intestinal roundworm is noteworthy in the numbers of people it infects and the insignificant pathology it causes. In the world today, about 1.5 billion people are infected. Of these only 59,000 become ill, and 10,000 die (Crompton 1999; de Silva et al. 1997).

The results of the parasite analysis show a higher prevalence of parasitism at Antelope House, which undoubtedly caused disease. Fewer parasites were present in the Salmon Pueblo coprolites. As a pathoecologist, I wonder what factors led to greater disease at Antelope House. By looking at other archaeological data, this difference can be understood.

DISCUSSION

Season of Occupation

It is likely that both villages were occupied year-round, although the concentrated habitation of Salmon Pueblo was probably more typical of the late fall, winter, and early spring, when people took refuge from the cold.

On the basis of results from the coprolite analyses (Fry and Hall 1986; Williams-Dean 1986), Morris (1986:55) concluded that Antelope House was occupied throughout the year. Williams-Dean (1986) included macrofossil and palynological data in her analysis of residential patterns. She identified two types of coprolites, designating them "spring-summer" and "four seasons." The first category included coprolites with constituents that could be gathered only in the warm months. Coprolites in the second category contained food items available year-round. Williams-Dean (1986) concluded that year-round occupation of Antelope House was probable, but that winter occupation of the site cannot be empirically demonstrated. Winter occupation can be inferred, however, from the presence of the remains of storable foods and ethnographic analogy for the use of harvested plants. This inference is supported by another study (Sutton and Reinhard 1995), which examined associations of macroscopic plant remains through cluster analysis. This study revealed a difference between foods associated with whole maize (*Zea mays*) kernels and ground maize kernels. The whole kernels were associated with a great diversity of warm-season foods such as prickly pear (platy *Opuntia* sp.) fruits. Ground maize was associated with less variety and foods that reflect cool-season diets. It is likely therefore that Antelope House was occupied year-round. Reinhard et al. (2006a) completed a pollen concentration analysis of coprolites from both sites and concluded that the background pollen was not statistically different between the sites. Both sites were occupied partly in the warm seasons when pollination occurs.

The Salmon Pueblo fire carbonized substantial amounts of stored cultivated and wild plant foods (Bohrer 1980). These foods were harvested or collected during the warm months and stored for winter. The combined data suggest that the site was intensively occupied at least during the winter months when the stores would have been most needed. Adams (1980a; 2006a) interprets the presence of purslane (*Portulaca* sp.) and goosefoot (*Chenopodium* sp.) seeds in the Pueblo as evidence of spring and fall activity.

Resource Exploitation

Environmental parameters are important in the formation of conditions suitable for parasitism. Moist, warm conditions are optimal for the transfer of many parasite species. The analysis of the way in which people at Antelope House and Salmon Pueblo exploited local resources provides information about the ecological conditions to which they were habitually exposed. These data speak to the environments in which Salmon Pueblo and Antelope House inhabitants lived and foraged. They demonstrate that the inhabitants of Antelope House were closely tied to the wetland canyon bottom in which they lived. Conversely, the inhabitants of Salmon Pueblo utilized drier areas for habitation and food collection.

Coprolites from Antelope House indicate that residents there tended to eat plant foods from wet areas (Table 10-1). For example, horsetail (*Equisetum* spp.) pollen is present in 12 of 27 coprolites (Reinhard et al. 2006a), although it is apparently absent in coprolites from other Ancestral Pueblo sites (Clary 1984). Cattail (*Typha* spp.) pollen is found in 17 of 27 Antelope House coprolites. Analysis of pollen and macroscopic remains shows that the pollen and spores of these plants were eaten as a food. The terminal stems of horsetail, which occur just beneath the spore-producing bodies (strobili), are present in macroscopic remains. Cattail pollen is still aggregated in rows and is associated with the fibrous matrix of immature cattail heads in macroscopic remains. Obviously, the strobili of horsetail and the inflorescences of cattail were common food sources for Ancestral Pueblo living at Antelope House. Both plants occur in wetland habitats, often associated with standing water. Horsetail is also found in wet soils of stream beds.

Table 10-1. Number of Coprolites Containing Economic Pollen Types from Antelope House and Salmon Puebo

Taxa	Antelope House	Salmon Pueblo
Parsley family (Apiaceae)	2	6
Sunflower-type (Compositae)	13	12
Mustard family (Brassicaceae)	3	8
Cactus family (Cactaceae)	3	5
Hackberry (*Celtis* spp.)	1	–
Cheno-Am[a] (*Chenopodium-Amaranthus*)	9	9
Beeweed (*Cleome* spp.)	24	28
Squash (*Cucurbita* spp.)	3	3
Horsetail (*Equisetum* spp.)	12	2
Bean family (Fabaceae)	5	4
Lily family (Liliaceae)	1	4
Prickly pear (platy *Opuntia* spp.)	–	2
Grass family[a] (Poaceae)	2	3
Purslane (*Portulaca* spp.)	1	–
Sumac (*Rhus* spp.)	6	3
Cattail (*Typha* spp.)	17	–
Maize (*Zea mays*)	18	25

[a] Only percentages exceeding 5% of a 200 grain-count are included in these categories. Number of coprolites analyzed for pollen: Antelope House $N = 27$, Salmon Pueblo $N = 30$. Data from Reinhard et al. (2006a).

Salmon Pueblo data contrast sharply with those from Antelope House. Wetland-associated plants are absent in the Salmon Pueblo coprolites. In particular, cattail and horsetail were not eaten at Salmon Pueblo (Table 10-1); instead, there was a tendency to consume dry-adapted plants. The pollen evidence from Salmon Pueblo indicates that the flowers of prickly pear or cholla (platy *Opuntia* sp. or cylindro *Opuntia* sp.), beeweed (*Cleome* spp.), and cultivated gourd/squash (*Cucurbita* sp.) were eaten. In addition, pollen of maize, a high-spine composite such as sunflower (*Helianthus* sp.), and perhaps a plant in the parsley family (Apiaceae) are also present. This evidence suggests a more maize-dependent diet for Salmon Pueblo with less diversity in wild plant foods than at Antelope House. The macrofossil remains from the coprolites contrast between the two sites (Table 10-2). There were more wild plant taxa eaten at Antelope House than at any other Ancestral Pueblo site (Reinhard 1992b). Compared to 21 wild plant taxa eaten at Antelope House, just 12 wild plant taxa were eaten at Salmon Pueblo. In the total of 112 coprolites from Salmon Pueblo, juniper (*Juniperus* sp.) fruits are the most common noncultivated macrobotanical component (23 of 112 coprolites). This plant grows in the juniper woodlands overlooking the San Juan River plain. Its presence signals foraging in xeric environments. Other major macrobotanical components are maize, cultivated bean (*Phaseolus vulgaris*), goosefoot, and pigweed (*Amaranthus* spp.). No evidence of wetland plants such as horsetail, hackberry (*Celtis* spp.), or chokecherry (*Prunus virginiana*) was found.

Analyses of food plants, nondietary plants, and vertebrates in noncoprolite contexts provide additional evidence that Antelope House inhabitants utilized aquatic resources to a greater degree than did the inhabitants of Salmon Pueblo

Table 10-2. Percentage of Coprolites Containing Macrofloral Fossil Types from Antelope House and Salmon Pueblo[a]

Common Name and Genus or Family	Antelope House	Salmon Pueblo
Onion (*Allium* spp.)	2	–
Pigweed or amaranth (*Amaranthus* spp.)	10	6
Saltbush (*Atriplex* spp.)	1	–
Cactus (Cactaceae)	42	–
Goosefoot (*Chenopodium* spp.)	4	21
Beeweed (*Cleome* spp.)	16	6
Gourd/squash (*Cucurbita* spp.)	29	4
Ringwing (*Cycloloma* spp.)	–	1
Tansy mustard (*Descurainea* spp.)	–	4
Wild rye (*Elymus* spp.)	1	–
Horsetail (*Equisetum* spp.)	7	–
Hedgehog cactus (*Echinocereus* spp.)	–	4
Cotton (*Gossypium* spp.)	18	–
Sunflower (*Helianthus* spp.)	4	8
Juniper (*Juniperus* spp.)	–	23
Pepperweed (*Lepidium* spp.)	1	–
Prickly pear (platy *Opuntia* spp.)	12	1
Rice grass (*Oryzopsis* spp.)	2	–
Panic grass (*Panicum* spp.)	1	–
Bean (*Phaseolus* spp.)	1	10
Groundcherry (*Physalis* spp.)	16	13
Pine (*Pinus* spp.)	29	6
Grass (Poaceae)	1	–
Purslane (*Portulaca* spp.)	21	6
Sumac (*Rhus* spp.)	7	–
Drop-seed (*Sporobolus* spp.)	1	–
Grape (*Vitis* spp.)	2	–
Yucca (*Yucca* spp.)	1	–
Maize (*Zea mays*)	91	43

[a] Data from Reinhard (1992b) and Reinhard et al. (2006b).

and other Ancestral Pueblo sites in general. The vast majority of bird remains from Antelope House are wading birds (Ardeidae) and ducks (Anatidae), which contrasts with other Ancestral Pueblo collections considered by McKusik (1986). Most of the wild bird species identified prefer standing, shallow pools, further demonstrating that Antelope House inhabitants hunted in moist places. Of plant remains found in noncoprolite contexts, 37% of 78 species from the Antelope House excavations were from "wet places" (Harlin and Dennis 1986). It was concluded from these data that "the canyon bottom, in general, provides more plant species suitable for food than do any of the other [ecological] areas" (Hall and Dennis 1986:139). Other uses of wetland species, particularly woody genera, were noted as well (Morris 1986:548–549). In contrast, Durand and Durand (2006) and Harris (2006) find relatively few wetland species other than fish from the San Juan River at Salmon Pueblo. The faunal remains from Salmon Pueblo were dominated by turkey (*Meleagris gallopavo*) and animals from the desert such as jack rabbit, cottontail rabbit, deer, rodents, and lizards.

Xeric macrobotanical data from Salmon Pueblo contrast with the wetland plants and animals at Antelope House. The most common plants recovered from Salmon Pueblo were maize, cultivated beans, pinyon nut (*Pinus edulis*), gourd/squash, goosefoot, pigweed, wild onion (*Allium* spp.), yucca (*Yucca* spp.), juniper, and purslane (Adams 2006a, b, c, d; Bohrer and Doebly 2006; Doebly 2006). These plants derive from gardening and foraging in xeric areas. Thirteen other plants are listed as common, but only sedge (*Carex* sp.), spikerush (*Eleocharis* sp.), and bulrush (*Scirpus* spp.) are from wet locations (Adams 1980a, b; 2006a, b, d; Bohrer 1980).

Seven burned storerooms were excavated at Salmon Pueblo. When the burning roofs and walls of the rooms collapsed, they smothered the fire, thus preserving carbonized plant remains. Reviewing the contents of these store-rooms provides a unique view of the plants harvested and stored at Salmon Pueblo. All seven of the rooms contained stores of maize. Six contained cultivated beans. Squash seeds and rinds were found in three rooms. Three rooms contained goosefoot seeds and cholla cactus stems, buds, and seeds. Two rooms contained tumble ringwing (*Cycloloma* spp.) seeds as well as prickly pear seeds and pads. Beeweed seed, Indian rice grass (*Oryzopsis hymenoides*) florets, purslane seeds, and pinyon nuts were found in one room. The area did not support pinyon pine, so these nuts were probably collected at some distance from the village. These plant foods probably represent major dietary components. None is specific to wetlands.

The consumption of chokecherry fruit bears on the relative importance of wetland-adapted plants at Salmon Pueblo. Bohrer (1980) discusses the evidence of pinyon pine nut, juniper berry, wild onion bulb, yucca pod, and chokecherry fruit consumption at the site on the basis of botanical data from 53 trash strata. Pinyon nuts and juniper berries occurred in roughly equal numbers of strata, 48 and 43, respectively. Wild onion remains were recovered from 34 trash strata and yucca remains from 26 trash strata. In contrast, chokecherry was found in only 8 strata. Bohrer notes that chokecherry should be expected to produce a reliable crop. In comparison, pinyon pine and yucca are undependable resources. With respect to chokecherry Bohrer (1980:247) states that "the heavy use of pinyon nuts runs counter to its erratic seed production and distance from Salmon Pueblo. The low frequency of chokecherry pits may indicate a lack of popularity, for they seem to be available and reliable." The contrast in utilization of chokecherry with other desert plants suggests that Salmon Pueblo inhabitants preferred foraging in xeric areas even though productive species were available in wetlands near the village.

Sanitation

The patterns of excreta disposal at the two sites reflect drastic differences in general hygiene. At Antelope House, approximately 150 separate fecal deposits were found in the excavation. All these deposits can be considered to be individual latrines. They are located throughout the site, both in plaza areas and in habitation rooms. Thus, the pattern of excreta disposal shows no organized system for segregating human waste from other activity areas.

In contrast, coprolites were recovered from specific, sequestered areas in Salmon Pueblo. Usually, rooms were set aside for the purpose of trash and feces disposal. An example of such a room is 62W, in which a trench was excavated into deep trash deposits next to a narrow masonry bench. Feces were deposited into the trench by people sitting on the bench. Several other rooms were used as latrines, and coprolites were recovered from trash strata as well. The pattern of feces disposal at Salmon Pueblo was one that resulted in the isolation of large amounts of fecal material in specific areas. These rooms were two stories deep and had walls up to 1.5 m thick, so latrines were quite effectively separated from living areas.

Personal hygiene is more difficult to reconstruct. The only information relative to this aspect of life comes from Bohrer (1980) for Salmon Pueblo. She speculates that the inhabitants of Salmon Pueblo made soap from yucca roots. This idea is based on the abundance of yucca leaves, pods, seeds, and hearts at the site, although no yucca roots have been identified. Bohrer suggests that the roots were pulverized to make soap and consequently were destroyed. Soap could have impacted parasitism, especially that of lice (*Pediculus humanus capitis*). Similar information about personal hygiene is not available for Antelope House.

One important aspect of Antelope House was that it was largely enclosed in a cave. The enclosure limited air movement and prevented removal of suspended particles from the air. Because pinworm is commonly transferred by aerial contamination, the lack of air movement at Antelope House could have aggravated pinworm infections.

The pattern of parasite infection coupled with the nature of human habitations can promote or limit infection level (Reinhard 2007). This effect is especially true for pinworms. Pinworms work the night shift. The females crawl out of the anus of the host and desiccate on the skin. Desiccation causes the worms to burst, and the eggs are showered into the air. Now, because pinworm eggs are transmitted through inhalation, areas where the air is still promote infection because the eggs remain in ambient air longer and have a better chance of infecting a new host. Therefore, airborne contamination of food is another source of pinworm infection. Caves, such as the one in which Antelope House was built, are ideal for the transmission of pinworms compared to open sites such as Salmon Pueblo. The open sites are subject to air movement (winds) that blow the tiny eggs out of the village. Therefore, the lower prevalence of pinworm eggs in Salmon Pueblo is due to the fact that it is an open site. Once the pinworms are established via air contamination, other modes of infection exacerbate the problem. When female worms lay their eggs, the eggs are laid in a viscous substance that irritates the perianal skin and causes itching (pruritus). Thus, the human scratches in his or her sleep and transfers the eggs to the hands. The eggs are easily transmitted from person to person in the morning. The eggs must be eaten or inhaled in order to cause infection. If personal hygiene is poor and the perianal folds are not cleaned, the eggs can hatch and the worms wander back into the intestine and mature to adulthood. This process is called *retroinfection*. In addition, the eggs are very light and can be suspended in even the slightest air currents.

Environmental Collapse, Water Contamination, and Starvation

Reinhard and Bryant (2007) present a case that the Pueblo III occupation of Antelope House is the best documented example of an Ancestral Pueblo village suffering from declining health due to contamination of water with parasite eggs and cysts. Morris (1986) summarized the pathoecology of Antelope House water contamination. Towards the end of the occupation, drought affected the Canyon de Chelly region. Water sources outside of the canyon dried up, and people moved into communities such as Antelope House to take advantage of more reliable water within the canyon. Reinhard and Bryant (2007) argue that the increased population and decreased water resulted in contamination. They write "It is also probable that the long-term drought that is recorded in the Antelope House region resulted in an aggregation of human populations around the dwindling water sources. This in turn led to a proliferation of crowd diseases as well as diseases associated with contaminated water and inadequate sanitation. Bioarchaeologically, this phenomenon is commonly expressed in skeletal remains as elevated levels of porotic hyperostosis." Three parasites in Antelope House are associated with such contamination; *E. histolytica*, *G. lamblia*, and hookworm. These cause diarrhea and anemia, especially in children and mothers, which coincides with the increased prevalence of anemia during the Pueblo III occupation of Canyon de Chelly relative to other time periods (El-Naijar 1986; El-Naijar et al. 1976). Thus, there is a relationship between environmental stress, increased parasitism, and skeletal indicators of morbidity in mother and infants.

This infectious pathoecology was compounded by starvation. Early comparative analysis of Ancestral Pueblo coprolites (Reinhard 1992b) suggested that Antelope House inhabitants diversified their subsistence strategy during the Pueblo III occupation and became more reliant on wild plants than any other Ancestral Pueblo Village. Sutton and Reinhard's (1995) component analysis of 180 Antelope House coprolites showed that there was a distinct period when the variety of foods used was low and dominated by yucca and prickly pear. These studies suggested temporary food shortages were suffered at Antelope House. Reinhard and Danielson (2005) examined coprolites for phytolith evidence of reliance on starvation foods at Ancestral Pueblo communities. They discovered that starvation foods (yucca leaf bases and prickly pear pads) were present in nearly all Antelope House coprolites. Reinhard and Danielson (2005) related these starvation foods to the pathoecology of dental disease. Relative to other Anasazi sites, Salmon Pueblo has a moderate prevalence of dental caries and a low prevalence of dental abscesses. In contrast, Antelope has a high level of dental wear relative to other sites. Reinhard and

Danielson (2005) hypothesize that ecological collapse in the region of Antelope House resulted in higher reliance on starvation foods containing high levels of phytoliths which, in turn, abraded dental enamel and cause dental wear and abscesses.

SUMMARY OF PATHOECOLOGY

Several factors relevant to the pathoecology of anemia have been examined, including general environment of the sites, parasite prevalence, parasite diversity, population size, residential seasonality, resource exploitation, and general hygiene. In some aspects, no difference between the sites is apparent. The sites are essentially similar with respect to general environment. Both are built on moist soils adjacent to riparian or wetland environments. The demographic estimates suggest that the populations of the two sites were roughly similar. Both sites were probably year-round habitations.

The major differences are seen in environmental stability, resource exploitation, and hygiene. Antelope House was occupied at a time of drought and environmental collapse. The canyon population became crowded and the water sources were contaminated with protozoa cysts and hookworm eggs. The excreta disposal pattern at Antelope House was erratic in comparison to the defined, isolated latrines at Salmon Pueblo. These factors probably had a pronounced influence on parasitism.

Use of wetlands kept inhabitants of Antelope House in contact with moist soils infested with the facultative threadworm parasite *Strongyloides stercoralis* and hookworms. Dogs (*Canis familiaris*) at Antelope House served as reservoir hosts for threadworm (Reinhard 1985b) and could have maintained threadworm in the soil. Deposits of dog and human feces near and in habitations could have led to human threadworm infection as well. The human crowding throughout the site combined with poor air circulation may have contributed to higher levels of pinworm infection. The considerable difference in pinworm prevalence between the two sites suggests that the inhabitants of Antelope House lived in more crowded conditions than did residents of Salmon Pueblo. This crowding is significant with regard to other diseases. High worm infections are accompanied by high protozoal infections in the modern world, and it is almost a certainty that protozoal parasites of the order Amoebida were more common at Antelope House as well. Therefore, amoebid-caused diarrhea and anemia were probably greater problems for Antelope House inhabitants than for those at Salmon Pueblo.

The health impacts of the combined starvation and parasitism are best seen in skeletal pathology. Stuart-MacAdam and Kent (1992) present a strong case that infectious disease is the main cause of porotic hyperostosis seen in human skeletal remains. Porotic hyperostosis refers to cranial lesions on the parietal and occipital regions that develop as a response to infection-related anemia.

Among Ancestral Pueblo sites that have been examined for porotic hyperostosis in crania and parasites in coprolites, Salmon Pueblo has the lowest prevalence of both. The Canyon de Chelly area, where Antelope House is located, has the highest prevalence of both (Reinhard 1992b). Further, of 33 children from Salmon Pueblo, 14 (45%) exhibit porotic hyperostosis, compared to 15 of 17 children from Canyon de Chelly (88%). Therefore, skeletal pathology at the two sites reflects the health impact of parasitic disease and poor nutrition.

CONCLUSION

Humans have endured parasitic disease throughout their evolution. Before plant domestication, aspects of hunting-gathering life served to limit the prevalence and diversity of parasites in human populations (Figure 10-4). When horticulture emerged, parasitism increased, with debilitating results (Figure 10-5). Only through controlling aspects of behavior could horticultural populations limit the intensity of their parasitic diseases. Through comparison of the coprolites from Salmon Pueblo and Antelope House, one begins to understand the consequences of environmental collapse, sedentism and population increase, as well as the behavioral changes that evolved to limit parasitism in settled villages.

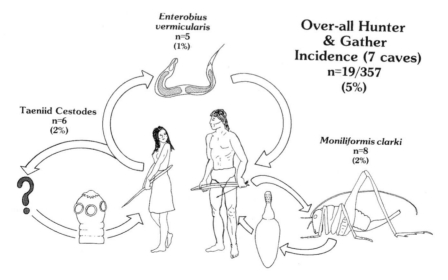

Figure 10-4. Parasitism of southwestern hunter-gatherers. Parasitism in this population was limited by seasonal movement, small band size, diet, and other factors. Studies of coprolites from hunter-gatherer sites show only three parasites. Thorny-headed worm (probably *Moniliformis clarki*) became a human infection when humans ate uncooked insects. Pinworm (*Enterobius vermicularis*) existed in small numbers. An unknown tapeworm (Taeniidae) may be the result of contamination and not actual parasitism of humans. (Data from Reinhard 1988, 1992b).

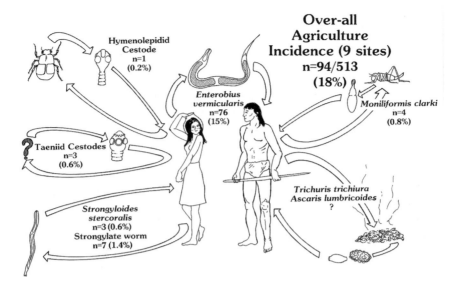

Figure 10-5. Parasitism of horticulturists in the Southwest was more diverse and infections were more common than among hunter-gatherers. Pinworm (*Enterobius vermicularis*) infection increased dramatically. Fecal-borne parasitism emerged (*Trichuris trichiura* and *Ascaris lumbricoides*). Hymenolepidid tapeworms occurred in grain beetles. Threadworm (*Strongyloides* spp.) and an unknown strongylate worm are also evident.

ACKNOWLEDGMENTS

The illustrations were drawn by Jonathan Ham and Debra K. Meir.

REFERENCES

Adams, K. R., 1980a, *Pollen, Parched Seeds, and Prehistory: A Pilot Investigation of Prehistoric Plant Remains from Salmon Ruin, a Chacoan Ruin in Northwest New Mexico, Eastern New Mexico University Contributions to Anthropology* 9, Eastern New Mexico University, Portales.

Adams, K. R., 1980b, Relative Numbers of Native Macrofossils in Strata of Poor Preservation With Emphasis on Flotation, in: V. L. Bohrer (1980), Salmon Ruin Ethnobotanical Report, in: *Investigations at the Salmon Site: The Structure of Chacoan Society in the Northern Southwest* (C. Irwin-Williams and P.H. Shelly, eds.), final report to funding agencies, Part 7, Volume 3, manuscript on file, Department of Anthropology, Texas A & m University, College Station, pp. 163–351.

Adams, K. R., 2006a, Native (Wild) Plants from Salmon Pueblo, in: *Thirty-Five Years of Research at Salmon Ruins, New Mexico* (P. F. Reed, ed.), Center for Desert Archaeology and Salmon Ruins Museum, Bloomfield, pp. 741–757.

Adams, K. R., 2006b, An Archaeobotanical Study of Room 93W at Salmon Pueblo, in: *Thirty-Five Years of Research at Salmon Ruins, New Mexico* (P. F. Reed, ed.), Center for Desert Archaeology and Salmon Ruins Museum, Bloomfield, pp. 785–821.

Adams, K. R., 2006c, Pines and Other Conifers from Salmon Pueblo, in: *Thirty-Five Years of Research at Salmon Ruins, New Mexico* (P. F. Reed, ed.), Center for Desert Archaeology and Salmon Ruins Museum, Bloomfield, pp. 823–851.

Adams, K. R., 2006d, Archaeobotanical Summary and Conclusions, in: *Thirty-Five Years of Research at Salmon Ruins, New Mexico* (P. F. Reed, ed.), Center for Desert Archaeology and Salmon Ruins Museum, Bloomfield, pp. 867–873.

Barkley, F. A., 1934, The Statistical Theory of Pollen Analysis, *Ecology* 15:283–289.

Bohrer, V. L., 1980, Salmon Ruin Ethnobotanical Report, in: *Investigations at the Salmon Site: The Structure of Chacoan Society in the Northern Southwest* (C. Irwin-Williams and P. H. Shelly, eds.), final report to funding agencies, Part 7, Volume 3, manuscript on file, Department of Anthropology, Texas A & m University, College Park, pp. 163–351.

Bohrer, V. L., and Adams, K. R., 1977, *Ethnobotanical Techniques and Approaches at Salmon Ruin, New Mexico, Eastern New Mexico Contributions in Anthropology* 8(1), Eastern New Mexico University, Portales.

Bohrer, V. L., and Doebly, J. F., 2006, Cultivated Plants from Salmon Ruin, in: *Thirty-Five Years of Research at Salmon Ruins, New Mexico* (P. F. Reed, ed.), Center for Desert Archaeology and Salmon Ruins Museum, Bloomfield, pp. 721–739.

Bryant, V. M., Jr., 1974, Prehistoric Diet in Southwest Texas: The Coprolite Evidence, *American Antiquity* 39(3):407–420.

Clary, K. H., 1984, *Prehistoric Coprolite Remains from Chaco Canyon, New Mexico: Inferences for Ancestral Pueblo Diet and Subsistence*, M.S. thesis, Department of Biology, University of New Mexico, Albuquerque.

Crompton, D. W. T., 1999, How Much Human Helminthiasis Is There in the World? *Journal of Parasitology* 85:379–403.

de Silva, N. R., Chan, M. S., and Bundy, D. A. P., 1997, Morbidity and Mortality Due to Ascariasis: Re-estimation and Sensitivity Analysis of Global Numbers at Risk, *Tropical Medicine and International Health* 2:519–528.

Doebly, J. F., 2006, Plant Remains from Trash Deposits at Salmon Pueblo, in: *Thirty-Five Years of Research at Salmon Ruins, New Mexico* (P. F. Reed, ed.), Center for Desert Archaeology and Salmon Ruins Museum, Bloomfield, pp. 759–769.

Durand, K. R., and Durand, S. R., 2006, Variation in Ritual and Economic Fauna at Salmon Ruins, in: *Thirty-Five Years of Research at Salmon Ruins, New Mexico* (P. F. Reed, ed.), Center for Desert Archaeology and Salmon Ruins Museum, Bloomfield, pp. 1079–1100.

El-Naijar, M. Y., 1986, The Biology and Health of the Prehistoric Inhabitants of Canyon de Chelly, in: *Archaeological Investigations at Antelope House* (D. P. Morris, ed.), National Park Service, Washington, DC, pp. 206–220.

El-Naijar, M. Y., Ryan, D. J., Turner, II C. G., and Lozoff, B., 1976, The Etiology of Porotic Hyperostosis among the Prehistoric and Historic Anasazi Indians of the Southwestern United States, *American Journal of Physical Anthropology* 44:447–448.

Fry, G. F., 1977, *Analysis of Prehistoric Coprolites from Utah, University of Utah Anthropological Papers 97*, University of Utah Press, Salt Lake City.

Fry, G. F., and Hall, H. J., 1986, Human Coprolites, in: *Archaeological Investigations at Antelope House* (D. P. Morris, ed.), National Park Service, Washington, DC, pp. 165–188.

Gonçalves, M. L. C., Araújo, A., Duarte, R., Silva, J. P., Reinhard, K. J., Bouchet, F., and Ferreira, L.F., 2004, Detection of *Entamoeba histolitica* Antigen in Coprolites Using a Commercially Available Enzyme Immunoassay, *Transactions of the Royal Society of Tropical Medicine and Hygiene* 98:88–91.

Gonçalves, M. L. C., Araújo, A., Ferreira, L. F., Reinhard, K. J., Bouchet, F., and Duarte, R., 2002, Detection of *Giardia intestinalis* Infection Using a Commercial Immunological Assay in Coprolites, *Transactions of the Royal Society of Tropical Medicine and Hygiene* 96:640–643.

Hall, R. L., and Dennis, A. E., 1986, Cultivated and Gathered Plant Foods, in: *Archaeological Investigations at Antelope House* (D. P. Morris, ed.), National Park Service, Washington, DC, pp. 110–141.

Harlin, A., and Dennis, A. E., 1986, A Preliminary Plant Geography of Canyon de Chelly National Monument, in: *Archaeological Investigations at Antelope House* (D. P. Morris, ed.), National Park Service, Washington, DC, pp. 72–83.

Harris, A. H., 2006, Preliminary Analysis of Faunal Material from Salmon Ruin, in: *Thirty-Five Years of Research at Salmon Ruins, New Mexico* (P. F. Reed, ed.), Center for Desert Archaeology and Salmon Ruins Museum, Bloomfield, pp. 1065–1077.

Martin, P. S., and Sharrock, F. W., 1964, Pollen Analysis of Prehistoric Human Feces: New Approach to Ethnobotany, *American Antiquity* 30:168–180.

Martinson, E., Reinhard, K. J., Buikstra, J. E., and Dittmar, K., 2003, Pathoecology of Chiribaya Parasitism, *Memorias do Instituto do Oswaldo Cruz* 98:195–205.

McKusik, C., 1986, The Avian Remains, in: *Archaeological Investigations at Antelope House* (D. P. Morris, ed.), National Park Service, Washington, DC, pp. 142–158.

Morris, D. P., 1986, *Archaeological Investigations at Antelope House*, National Park Service, Washington, DC.

Nials, F. L., and Reed, P. F., 2006, Salmon's Environmental Setting, in: *Thirty-Five Years of Research at Salmon Ruins, New Mexico* (P. F. Reed, ed.), Center for Desert Archaeology and Salmon Ruins Museum, Bloomfield, pp. 29–39.

Reed, P. F., 2006, Chronology of Salmon Pueblo, in: *Thirty-Five Years of Research at Salmon Ruins, New Mexico* (P. F. Reed, ed.), Center for Desert Archaeology and Salmon Ruins Museum, Bloomfield, pp. 287–297.

Reinhard, K. J., 1985a, Parasitism at Antelope House, A Puebloan Village in Canyon de Chelly, Arizona, in: *Health and Disease in the Prehistoric Southwest* (C. F. Merbs and R. J. Miller, eds.), *Arizona State University Anthropological Research Papers* 34, Tempe, pp. 220–233.

Reinhard, K. J., 1985b, *Strongyloides stercoralis* in the Prehistoric Southwest, in: *Health and Disease in the Prehistoric Southwest* (C. F. Merbs and R. J. Miller, eds.), *Arizona State University Anthropological Research Papers* 34, Tempe, pp. 234–242.

Reinhard, K. J., 1988, Cultural Ecology of Prehistoric Parasitism on the Colorado Plateau as Evidenced by Coprology, *American Journal of Physical Anthropology* 77:355–366.

Reinhard, K. J., 1992a, Parasitology as an Interpretive Tool in Archaeology, *American Antiquity* 57:231–245.

Reinhard, K. J., 1992b, The Impact of Diet and Parasitism on Anemia in the Prehistoric West, in: *Diet, Demography, and Disease: Changing Perspectives on Anemia* (P. Stuart-MacAdam and S. Kent, eds.), Aldine de Gruyter, New York, pp. 219–258.

Reinhard, K. J., 2007, Parasite Pathoecology of Chacoan Great Houses: The Healthiest and Wormiest Ancestral Puebloans, in: *Advances in Chaco Canyon Archaeology* (P. Reed and L. Baker, eds.), University of Utah Press, Salt Lake City, in press.

Reinhard, K. J., and Bryant, V. M., 2007, Pathoecology and the Future of Coprolite Studies in Bioarchaeology, in: *Reanalysis and Reinterpretation in Southwestern Bioarchaeology* (A. W. M. Stodder, ed.), Arizona State University Press, Tempe, in press.

Reinhard, K., and Buikstra, J., 2003, Louse Infestation of the Chiribaya Culture, Southern Peru: Variation in Prevalence by Age and Sex. *Memorias do Instituto do Oswaldo Cruz* 98:173–179.

Reinhard, K., Fink, T. M., and Skiles, J., 2003, A Case of Megacolon in Rio Grande Valley as a Possible Case of Chagas Disease. Memorias do Instituto do Oswaldo Cruz 98: 165–172.

Reinhard, K. J., Confalonieri, U. E., Herrmann, B., Ferreira, L. F., and Araujo, A. J. G., 1988, Recovery of Parasite Eggs from Coprolites and Latrines: Aspects of Paleoparasitological Technique, *Homo* 37:217–239.

Reinhard, K. J., and Danielson, D. R., 2005, Pervasiveness of Phytoliths in Prehistoric Southwestern Diet and Implications for Regional and Temporal Trends for Dental Microwear, *Journal of Archaeological Science* 32:981–988.

Reinhard, K. J., Edwards, S. K., and Meier, D. K., 2006a, Pollen Concentration Analysis of Salmon Ruin and Antelope House: Documenting Anasazi Dietary Variation, *Journal of Palaeogeography, Palaeoclimatology, and Palaeoecology* 237:92–109.

Reinhard, K. J., LeRoy-Toren, S., and Danielson, D. R., 2006b, Coprolite Analysis from Salmon Ruin: The San Juan Occupation, in: *Thirty-Five Years of Research at Salmon Ruins, New Mexico* (P. F. Reed, ed.), Center for Desert Archaeology and Salmon Ruins Museum, Bloomfield, pp. 875–887.

Stuart-MacAdam, P., and Kent, S. (eds.), 1992, *Diet, Demography, and Disease: Changing Perspectives on Anemia*, Aldine de Gruyter, New York.

Sutton, M. Q., and Reinhard, K. J., 1995, Cluster Analysis of the Coprolites from Antelope House: Implications for Ancestral Pueblo Diet and Cuisine, *Journal of Archaeological Science* 22(6):741–750.

Williams-Dean, G. J., 1986, Pollen Analysis of Human Coprolites, in: *Archaeological Investigations at Antelope House* (D. P. Morris, ed.), National Park Service, Washington, DC, pp. 189–205.

Wilson, S. D., Jordan, M. M., and Jordan, M. A., 2006, Coprolite Analysis for *Giardia lamblia* at Salmon Ruins, in: *Thirty-Five Years of Research at Salmon Ruins, New Mexico* (P. F. Reed, ed.), Center for Desert Archaeology and Salmon Ruins Museum, Bloomfield, pp. 889–891.

Chapter **11**

Nutritional Constraints and Mobility Patterns of Hunter-Gatherers in the Northern Chihuahuan Desert

KRISTIN D. SOBOLIK

Residential decisions related to mobility or sedentism are of central importance in cultural evolution. Ascertaining a populations' mobility pattern provides important insights into its society and culture. Mobility "is universal, variable, and multi-dimensional" (Kelly 1992:43) and includes components such as religion, personal choice, trade, and kinship. One of the strongest variables in mobility is subsistence or the acquisition of resources. In the middle-range theory of Binford (1980), the environment, particularly as it relates to acquisition of resources, is the primary mold that shapes hunter-gatherer adaptations and dictates the different combinations of mobility, settlement, and economic and social relations that characterize different cultures. Consequently, although determining patterns of mobility and resource exploitation is important in and of itself, it also aids in understanding other cultural dynamics (Kelly 1992; Price and Brown 1985). By integrating many types of archaeological evidence, it is found that populations in the Lower Pecos region followed a seasonal mobility pattern that allowed them to maximize the dietary resources available in the environment and maintain a nutritionally-sound and stable diet year-round.

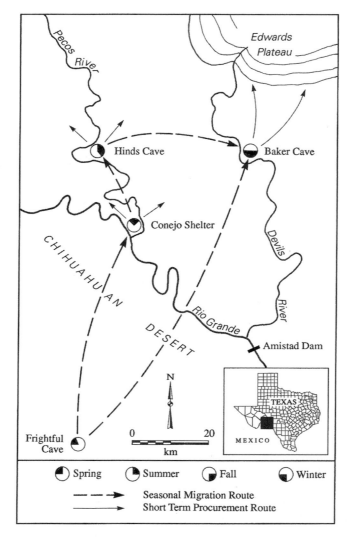

Figure 11-1. Map of the Lower Pecos Region, the Chihuahuan Desert, illustrating the seasonal round hypothesized by Shafer (1986).

This case study considers evidence for the seasonal round of Archaic hunter-gatherers in the Lower Pecos region of the northeastern Chihuahuan Desert (Figure 11-1). These people occupied a territorial range, which they marked with rock art, centered on the Pecos and Devils rivers and the Rio Grande (Shafer 1977). Shafer (1986) postulated that Archaic peoples of the Lower Pecos region followed a cyclical seasonal round based on the availability and quantity of food resources. This seasonal pattern hypothetically included the desert areas and lower canyons (e.g., Frightful Cave and Conejo Shelter) during the spring to early summer, when foods such as flowers, bulbs, and fruits were available. In the late summer, people moved to deeper canyon regions along the Pecos and Rio Grande (e.g., Hinds Cave) to take advantage of aquatic and upland resources. During the fall, they moved to the Devils River drainage (e.g., Baker Cave) to use acorns (*Quercus* spp.), walnuts (*Juglans* spp.), and pecans (*Carya* spp.) abundant in that area at that time. In the winter, when food supplies were low and available plant foods were primarily restricted to desert succulents such as yucca (*Yucca* sp.), agave (*Agave lechuguilla*), sotol (*Dasylirion* sp.), and prickly pear (platy *Opuntia* sp.), people moved to the northern fringes (e.g., Edwards Plateau) and focused their time and energy on the acquisition of upland game (Shafer 1986). From his studies in the Blue Hills area surrounding Hinds Cave, Saunders (1986, 1992) also states that the uplands adjacent to base camps in rockshelters were used for hunting and gathering, whereas the upland areas farther removed from the rockshelters were used mainly for hunting.

The validity and feasibility of such a seasonal round are tested using one of the most important variables in mobility: the acquisition of resources, particularly resources related to diet. This case study is based upon two lines of evidence. One is an assessment of the nutritional adequacy of the diet and health of the population as previously determined by Sobolik (1991a, 1994b), in conjunction with DNA analysis (Poinar et al. 2001). Diet is the food we eat; nutrition is the necessary energy, trace elements, minerals, and vitamins that are provided by the diet and used by the body.

The second line of evidence is an examination of the significant differences in food remains found at some sites as they relate to environment and dietary choice. This analysis provides information on the types of resources used by Lower Pecos populations that can be used to assess the accuracy of the seasonal round model. All dietary and health data available from the area are synthesized, including faunal, botanical, human skeletal, and remains from paleofeces (coprolites), which include macroremains (faunal and botanical) and microremains (pollen and DNA). The reconstructed diet of the population is then evaluated for nutritional adequacy. Significant differences among the food assemblages from each site are compared with the proposed seasonal round. Thus, availability of food resources during the year and at specific seasons in several different environments as well as the nutritional status and health of the population are considered in terms of their relationship to seasonal mobility.

Archaeological data from the Lower Pecos region are useful for exploring the relationship between resource acquisition and population mobility for several reasons. The region has many well-documented, excavated archaeological sites with well-preserved organic materials, many of which represent diet. Numerous studies indicate that there was little if any change in subsistence practices through time, signifying that differences in subsistence patterns among various sites most likely involve factors other than chronology. Moveover, models of population mobility based on seasonal rounds, such as the Shafer (1986) model, have been hypothesized for populations in this region.

BACKGROUND

Region

The Lower Pecos region is on the eastern margin of the Chihuahuan Desert. It includes the Pecos and Devils rivers and their confluence with the Rio Grande (Figure 11-1). The region has a long, highly stratified record of human occupation beginning approximately 10,000–11,000 years ago (Hester 1988; Shafer 1986, 1988). This sequence is broadly divided into Early, Middle, and Late Archaic, and Late Prehistoric periods. These chronological divisions are based upon stylistic changes in rock art, sandal and basketry technology, and the shape of burned rock middens. Although studied extensively, subsistence and the tools related to it do not appear to have changed significantly, with the exception of the introduction of the bow and arrow in the Late Prehistoric period (Alexander 1974; Andrews and Adovasio 1980; Dering 1979; Dibble 1967; Dibble and Lorrain 1968; Epstein 1963; Hester 1988; Irving 1966; Lord 1984; Saunders 1986; Shafer 1988; Sobolik 1991a; Turpin 1984; Williams-Dean 1978).

The area is dominated by eroded limestone canyons and plateaus containing a large number of rockshelters, solution cavities, and overhangs formed by downcutting in the canyon areas (Shafer 1988). The environment is primarily desert, semiarid, and dry, although the three rivers provide a continuous water supply supporting a diverse array of plants and animals. The most variable aspect of the climate is average rainfall, which ranges along a west–east gradient from 305 mm to 560 mm annually (Shafer 1988). Today, the amount of precipitation in the area is insufficient to support dry farming (National Oceanic and Atmospheric Administration 1981).

Although data from a large number of sites are incorporated into this study, data from four sites (Frightful Cave, Conejo Shelter, Hinds Cave, and Baker Cave) are used more extensively (Figure 11-1). Frightful Cave is located at least 80 km from the Rio Grande in the State of Coahuila, Mexico. The area surrounding the cave is the most xeric of the Lower Pecos sites and is more similar to the internal regions of the Chihuahuan Desert. Conejo Shelter, although on the Rio Grande, is located farther south than Baker and Hinds caves and at a greater distance from the mesic Edwards Plateau. The plants and animals near Conejo Shelter are more adapted to arid conditions, and there are fewer trees and shrubs, except along the waterways. Hinds Cave is located on the Pecos River, and Baker Cave is located on a tributary of the Devils River. Both sites are situated in the northern portion of the study area, in a setting more biologically diverse than either Conejo Shelter or Frightful Cave. Baker Cave is in an environment that is the most mesic of the Lower

Pecos sites. It is on the extreme periphery of the Chihuahuan Desert and the southern periphery of the more mesic Edwards Plateau. Edwards Plateau provides a greater amount and variety of trees and shrubs, and the vegetation supports a higher diversity of animals.

Paleonutritional Analysis

Analysis of paleonutritional patterns in the Lower Pecos region reveals that a comprehensive, integrated study of food residue and human skeletal remains greatly enhances the nutritional and dietary information available for a population (Sobolik 1994a). I provide a brief summary of the paleonutrition analysis here; for more comprehensive information, see Sobolik (1991a, 1994b) and Poinar et al. (2001). These studies indicate that the diet and nutrition of hunter-gatherers of the Lower Pecos area were relatively good or adequate throughout the approximately 10,000 years of occupation. The combined nutritional quality of available dietary resources was sufficient to support normal growth and development, assuming each member of the population had equal access to the resources on a year-round basis. Dietary staples provided good sources of energy: suitable fats, carbohydrates, and protein. In addition, the diversity of dietary items provided essential minerals, vitamins, and trace elements, and the river water was a good source of calcium and electrolytes, as indicated by an analysis of the modern constituents (Sobolik 1991a: Table 41).

Human skeletal remains from the region indicate that the population was generally healthy. The most frequently observed problems were dental diseases: caries (50% of 22 samples), abscesses (55% of 22 samples), antemortem tooth loss (86% of 22 samples), and wear (64% of 22 samples) (Hartnady 1988; Reinhard et al. 1989). Caries and abscesses indicate that the population probably had a high-carbohydrate diet. Paleofecal and macrobotanical evidence shows that agave, mesquite (*Prosopis* spp.), and nuts were prevalent dietary items, and all contain high amounts of carbohydrates. Antemortem tooth loss most likely resulted from the high incidence of caries and abscesses, and tooth wear probably reflects the abrasive character of the diet. Numerous food items could have severely worn the teeth, including silicate and calcium-oxalate crystals in the plant foods. Nutshells, hard seed coats and seeds, and bone found in the paleofeces indicate that people chewed and ingested these items, which also abraded their teeth.

Short term stress was indicated in some of the skeletal remains by the presence of Harris lines (50% of 2 samples) and enamel hypoplasias (86% of 7 samples). These growth-arrest lines occur only when a person's growth has halted or slowed for a short period of time and then resumed. A person who has severe stress such as chronic malnutrition will not exhibit growth-arrest lines. Growth-arrest lines could be the result of a number of causes; nutritional stress is only one possibility. Although the nutritional analysis of consumed food items indicates that the population had access to all the nutrients necessary for a healthy existence, such analysis is insufficient to demonstrate that every individual had the appropriate array of nutrients or that these nutrients were available year-round. If Harris lines and enamel hypoplasias were actually caused by nutritional stress, then the source of the stress may have been experienced by only a few individuals in the population. Nutritional stress also may have been experienced seasonally rather than throughout the year.

In the Lower Pecos region, seasonal stress would most likely have occurred in the winter months, when the range of available foods narrowed as a result of natural factors. During the winter, people would have had to rely on those few dietary items available year-round or on those available only in the winter. As a comprehensive nutritional analysis indicates (Sobolik 1991a, 1994b), such items do not provide the diversity of amino acids, vitamins, trace elements, and minerals necessary for a healthy existence. Therefore, if the growth-arrest lines observed on some of the human skeletal material are a result of nutritional stress, that stress may have occurred during the winter season, when nutritional variety was limited. The lack of dietary diversity during this season, rather than the quality of the diet, may have produced nutritional stress in the Lower Pecos populations.

METHODS

Only data from four sites are presented here, although the original comprehensive analysis included data from all analyzed Lower Pecos sites. Zooarchaeological, archaeobotanical, and paleofecal samples from Baker Cave (Douglas 1970; Hester 1983, 1986; Sobolik 1991a, 1991b), Hinds Cave (Dering 1979; Lord 1984; Poinar et al. 2001; Reinhard 1992; Sobolik 1991a; Stock 1983; Williams-Dean 1978), Conejo Shelter (Alexander 1974; Bryant 1969, 1974), and Frightful Cave (Bryant 1975; Fry 1975) are considered here. These sites were selected because of the quality of excavation techniques, suitability of materials analyzed, environmental differences related to site location, and comparability of quantification methods.

The food items recovered at each site are compared to test the feasibility of a seasonal round as a way to obtain resources. The data were compared according to site location using chi-square and Yate's correction for continuity. The chi-square test was used to assess the significance of the apparent relationships among the food remains according to site location. Yate's correction for continuity was calculated for each chi-square to adjust the chi-square statistic to small sample sizes.

Chi-square analysis requires that food remains from each site be quantified in a similar manner. Unfortunately, the macrobotanical remains from these sites are difficult to compare and quantify because archaeobotanists lack a uniform quantification technique. To compensate for this discrepancy, overall comparisons were based on the presence/absence data of botanical remains at each site. This means that botanical data could not be evaluated with chi-square analyses.

Although a wide variety of quantification techniques are available for pale-ofecal studies, these coprolite data are assessed using relative frequencies of samples in which dietary remains were observed. This type of analysis does not take into consideration the observed frequency of each dietary item in each sample, but compares the actual number of samples in which the dietary item appears from each study. Analyses of DNA from Hinds Cave paleofeces help define a broader range of dietary diversity, but cannot be considered in the statistical analyses due to quantification questions.

Lower Pecos vertebrate remains are more precisely and frequently quantified than archaeobotanical or paleofecal samples. Quantification and comparison of Lower Pecos faunal remains use presence/absence, Minimum Number of Individuals (MNI), and Number of Identified Specimens (NISP). NISP was used in the chi-square analyses because faunal material from sites analyzed for this study were all quantified using NISP.

RESULTS

Chi-square analyses were conducted on the dietary data from each site to ascertain whether significant differences existed in the resources obtained and used at each location. Because numerous, extensive studies on food remains from the region have revealed no subsistence change through time (Alexander 1974; Dering 1979; Irving 1966; Lord 1984; Saunders 1986; Sobolik 1991a), any differences noted probably reflect environmental conditions among sites and the types of dietary items that people occupying each site obtained, rather than chronology. Any contrasts and patterns will help ascertain the probability that any or all of the populations have an established seasonal round that can be considered in terms of diet, health, and nutrition.

Archaeobotanical Evidence

The archaeobotanical data (Table 11-1) indicate that Lower Pecos occupants ate a wide variety of plant foods. Despite this variety, people apparently relied on a few significant dietary staples. Agave was an important dietary component, judging from the large amount of quids and cut leaf bases recovered from sites as well as the large numbers of burned rock middens used for roasting agave. The relatively high frequency of agave remains may in part reflect the large number of nonfood uses for agave. Agave processing for both food and nonfood purposes produces a large amount of unusable residue. Dering (1979:68) states that processing an adequate amount (of agave) for a meal creates huge amounts of waste. When the by-products of basket, mat, and sandal

Table 11-1. Botanical Macroremains[a]

Seeds, nuts, and pods	Hinds Cave	Conejo Shelter	Baker Cave
Acacia (*Acacia* spp.)	X		
Guajillo (*A. berlandieri*)	X		X
Catclaw acacia (*A. greggii*)		X	
Blackbrush acacia (*A. rigidula*)	X	X	
Lechuguilla (*Agave lechuguilla*)	X	X	X
Agarito (*Berberis trifoliata*)	X		X
Hackberry (*Celtis* spp.)		X	
Netleaf hackberry (*C. reticulata*)			X
Goosefoot (*Chenopodium* spp.)	X		
Texas colubrina (*Colubrina texensis*)	X		
Condalia (*Condalia* spp.)	X		
Buffalo gourd (*Cucurbita foetidissima*)	X		
Sotol (*Dasylirion* spp.)	X		
Jimsonweed (*Datura* spp.)	X		
Texas persimmon (*Diospyros texana*)	X	X	X
Texas kidney-wood (*Eysenhardtia texana*)	X		
Ocotillo (*Fouquieria splendens*)	X		
Tanglehead (*Heteropogon contortus*)	X		
Leatherstem (*Jatropha dioica*)	X		
Walnut (*Juglans* spp.)		X	
Little walnut (*J. microcarpa*)	X		X
Juniper (*Juniperus* spp.)	X		
Red-berry juniper (*J. pinchotii*)	X		
Coyotillo (*Karwinskia* spp.)	X	X	X
Littleleaf leadtree (*Leucana retusa*)		X	
Sacahuiste (*Nolina* spp.)	X		
Prickly pear (platy *Opuntia* spp.)	X	X	X
Grasses (Poaceae)	X		
Mesquite (*Prosopis* spp.)		X	
Honey mesquite (*P. glandulosa*)	X		X
Plum, cherry (*Prunus* spp.)	X		
Oak (*Quercus* spp.)		X	X
Vasey shin oak (*Q. pungens*)	X		
Plateau oak (*Q. virginiana*)	X		
Littleleaf sumac (*Rhus microphylla*)	X		X
Capul (*Schaefferia cuneifolia*)	X		
Bristlegrass (*Setaria leucopila*)	X		
Plains bristlegrass (*S. lutescens*)			X
Mountain laurel (*Sophora* spp.)		X	
Texas mountain laurel (*S. secundiflora*)	X		X
Buckeye (*Ungnadia* spp.)		X	
Mexican buckeye (*U. speciosa*)			X
Skeleton-leaf goldeneye (*Viqueria stenoloba*)	X		
Grape (*Vitis* spp.)	X		
Canyon grape (*V. arizonica*)			X
Yucca (*Yucca* spp.)	X	X	

[a] Data from Hinds Cave, Block A (Dering 1979) and Block B (Sobolik 1991a), Conejo Shelter (Alexander 1974), and Baker Cave (Chadderdon 1983), condensed from Sobolik (1991a).

manufacture are added to the waste from food processing, the amount of semi-succulent plant remains becomes overwhelming in these dry rockshelters.

Remains of other succulents, particularly yucca and sotol, are prevalent. Prickly pear cacti, particularly the pads, were important to human subsistence, as indicated by their ubiquity in the sites. In combination, desert succulents and prickly pear were extremely important because they are available for food and other purposes year-round. As a result, they probably were dietary staples, particularly when other food resources were unavailable or during times of stress.

Nuts and legumes (Fabaceae) also are present in the Lower Pecos sites. Nuts, including Texas walnut (*Juglans microcarpa*), acorns, Texas mountain laurel (*Sophora secundiflora*), and Mexican buckeye (*Ungnadia speciosa*), were important because of the quantities of food they provided during the fall. On the basis of the large quantities of pods and seeds found, mesquite, and acacia (*Acacia* sp.) were the most important leguminous plants. As with nuts, legumes were probably most important for nutrition during the fall.

Baker and Hinds caves occupy very similar environmental settings and offer the most direct environmental contrast with other Lower Pecos sites. Both caves are situated in an upland region where rainfall is more frequent than in southern areas. Their common environmental setting is reflected in the food remains. For example, seeds of grape (*Vitis* sp.), agarito (*Berberis trifoliata*), and littleleaf sumac (*Rhus microphylla*) were observed exclusively in samples from these two sites. Agarito and littleleaf sumac are shrubby trees that were probably more prevalent in northern and north-eastern portions of the region because floristic diversity of the area includes more plant species originating in the Edwards Plateau. Grapes are observed in marginally desertic regions, although they tend to grow mainly in areas with more moisture. Seeds of juniper (*Juniperus* sp.), plum or cherry (*Prunus* sp.), goosefoot (*Chenopodium* sp.), and sacahuiste (*Nolina* sp.) were observed only in samples from Hinds Cave. Although sacahuiste tends to be associated with xeric regions, it requires more rainfall for proper growth than do other strictly desertic plants.

Faunal Evidence

The faunal evidence from Lower Pecos sites (Tables 11-2 and 11-3) indicates that similar vertebrates were exploited at all sites, although there are significant differences in the relative frequencies of these resources. The faunal remains listed in Table 11-2 are those taxa used for chi-square analyses and are a condensed version of the overall array of faunal material from the sites. For a complete list of faunal remains identified from the region, see Sobolik (1991a).

The most frequent faunal resources used were small animals, particularly rabbits, rodents, and fishes. The relative frequency of rabbit and fish remains was significantly higher in the Baker Cave sample compared to those from either Hinds Cave or Conejo Shelter. Rodents were important food items, showing the least significant difference among the sites. The difference in rodent assemblages is important because it indicates that the presence of

Table 11-2. Fauna Used for Chi-Square Analysis[a]

Taxa	NISP/total		
	Conejo Shelter	Hinds Cave	Baker Cave
Rabbits	491/1251	1509/17943	737/4177
		1509/12986[b]	
Rodents	67/1251	5756/17943	205/4177
		779/12986[b]	
Carnivores	25/1251	331/17943	9/4177
		331/12986[b]	
Deer/antelope	57/1251	260/17943	321/4177
		260/12986[b]	
Birds	19/1251	567/17943	108/4177
		567/12986[b]	
Reptiles	26/1251	5756/17943	205/4177
		5756/12986[b]	
Fish	132/1251	512/17943	935/4177
		512/12986[b]	

[a] Data from Conejo Shelter (Alexander 1974), Hinds Cave (Lord 1984), and Baker Cave (Sobolik 1991a), condensed from Sobolik (1991a). NISP = Number of Identified Specimens.
[b] Rodents from Hinds Cave Analysis Units 7 and 8 were removed due to their probable noncultural origin.

rodents in these samples remains similar even after obvious nonculturally accumulated rodent remains are excluded from the analysis (i.e., the basal levels of Hinds Cave and throughout the excavated levels of Baker Cave) (see Table 11-2). This finding indicates that rodents were extremely important to the diet of people at all Lower Pecos sites.

Although small game animals were exploited frequently, nutritional input (particularly protein) from large animals, i.e., one of two species of deer (*Odocoileus* spp.) or antelope (*Antilocapra americana*), or both, was most likely extremely important to the diet of the people. Although deer were not as prevalent in the diet as smaller game, deer probably provided a large portion of the protein. It would take a large number of rabbits to equal the amount of protein provided by one deer.

Deer and rabbits are more plentiful near the brushlands of the Edwards Plateau than in the open areas of the desert; not unexpectedly, the faunal remains attest to the greater use of these animal food resources near the Edwards Plateau. Baker Cave contained proportionately more fish. Since many of the sites examined are near rivers, the high frequencies of fish in Baker Cave assemblages may be a reflection of site location or function.

Paleofecal Evidence

Paleofeces provide dietary information on a different scale than either the faunal or macrobotanical remains because they directly reflect ingested food items. In addition, DNA analysis of Hinds Cave paleofeces provides evidence of dietary

Table 11-3. Chi-Square Analysis of Faunal Remains[a]

Taxon/site	Chi-square	p	Significant at $\alpha = 0.05$
Rabbits			
Conejo–Hinds	1188.2	0.00	Yes
Hinds–Baker	315.7	0.00	Yes
Conejo–Baker	255.5	0.00	Yes
Conejo–Hinds[b]	719.1	0.00	Yes
Hinds–Baker[b]	100.3	0.00	Yes
Rodents			
Conejo–Hinds	3052.0	0.00	Yes
Hinds–Baker	6323.9	0.00	Yes
Conejo–Baker	0.31	0.57	No
Conejo–Hinds[b]	1.13	0.287	No
Hinds–Baker[b]	8.67	0.00	Yes
Carnivores			
Conejo–Hinds	0.08	0.78	No
Hinds–Baker	58.36	0.00	Yes
Conejo–Baker	46.34	0.00	Yes
Conejo–Hinds[b]	1.2	0.27	No
Hinds–Baker[b]	87.4	0.00	Yes
Deer/antelope			
Conejo–Hinds	67.6	0.00	Yes
Hinds–Baker	512.7	0.00	Yes
Conejo–Baker	14.06	0.00	Yes
Conejo–Hinds[b]	33.03	0.00	Yes
Hinds–Baker[b]	310.3	0.00	Yes
Birds			
Conejo–Hinds	10.10	0.00	Yes
Hinds–Baker	3.59	0.06	No
Conejo–Baker	4.34	0.04	Yes
Conejo–Hinds[b]	22.73	0.00	Yes
Hinds–Baker[b]	26.05	0.00	No
Reptiles			
Conejo–Hinds	498.6	0.00	Yes
Hinds–Baker	1269.3	0.00	Yes
Conejo–Baker	18.23	0.00	Yes
Conejo–Hinds[b]	842.6	0.00	Yes
Hinds–Baker[b]	2164.4	0.00	Yes
Fish			
Conejo–Hinds	211.4	0.00	Yes
Hinds–Baker	2110.0	0.00	Yes
Conejo–Baker	84.6	0.00	Yes
Conejo–Hinds[b]	113.87	0.00	Yes
Hinds–Baker[b]	310.3	0.00	Yes

[a] Degree of freedom = 1 in all cases.
[b] Data calculated without rodents from Hinds Cave Analysis Units 7 and 8.

Table 11-4. Dietary Remains Identified Through DNA Analysis of Hinds Cave Paleofeces

Dietary Item Common name (Family)	Paleofecal sample (number of clones assigned)[a]
Sunflower (Asteraceae)	I (7); II (3); III (4)
Ocotillo (Fouquieriaceae)	III (1)
Legumes (Fabaceae)	III (7)
Oak (Fagaceae)	II (7); III (1)
Lily (Liliaceae)	I (10); II (4)
Buckthorn (Rhamnaceae)	III (10)
Nightshade (Solanaceae)	II (6)
Elm, hackberry (Ulmaceae)	I (8); II (12); III (5)
Pronghorn (Antilocapridae)	I (2)
Sheep/goat (Caprinae)	III (11)
Cottontail (Leporidae)	I (1)

[a] Data from Poinar et al. (2001).

components not observed through normal macro- and micro- reconstruction (Table 11-4). Chi-square analysis of Lower Pecos paleofeces (Tables 11-5–11-8) indicates that there were significant differences among samples from each site. The paleofeces indicate that people ate a large amount of fiber (Tables 11-4 and 11-5); prickly pear and onion (*Allium drummondii*) bulbs are the most prevalent fiber sources. Agave, yucca, and sotol are present, as are seeds and nuts (Table 11-5). This prevalence demonstrates the importance of prickly pear fruit as well as of grass as dietary items. Texas persimmon (*Diospyros texana*), mesquite, and hackberry (*Celtis* sp.) were important dietary items, judging from their frequencies. The paleofeces contain a wide variety of economic pollen, indicating probable flower ingestion (Sobolik 1994b).

Rodents and fishes were the most frequent faunal remains in the paleofeces (Tables 11-4, 11-7, and 11-8), although reptiles, birds, and rabbits appear. Evidence for the consumption of large animals, Antilocapridae and Bovidae, is provided only through the DNA analysis of Hinds Cave samples (Table 11-4). However, faunal remains from the sites and in the macro- and microremains from paleofeces, indicate that small game was most likely more frequently eaten than larger game. The importance of rodents to the diet is also revealed by their sheer frequency and variety in the paleofeces (Sobolik 1993).

The spatial comparisons reveal that differences exist among sites for most of the food items observed in the paleofeces (Tables 11-6 and 11-8). The most noteworthy differences include: the absence of nut hulls in Frightful Cave and Conejo Shelter samples, the absence of fish remains in Frightful Cave samples, the absence of birds in Conejo Shelter samples, the low prevalence of onion fiber at Frightful Cave, and prickly pear fiber in the greatest number of samples from Frightful Cave and Conejo Shelter.

Table 11-5. Botanical Remains from Paleofeces

Botanical remains	Percentage of total sample that contained remains[a]									
	A	B	C	D	E	F	G	H	I	J
Fiber/epidermis										
Fiber/epidermis	100	100	98	100	100	–	18	100	63	100
Agave (*Agave* spp.)	–	–	–	–	51	–	5	–	–	13
Onion (*Allium* spp.)	55	17	58	–	40	8	3	–	55	29
Sotol (*Dasylirion* spp.)	–	75	–	–	7	–	3	–	–	24
Cactus (Cactaceae)	–	75	–	100	–	–	–	90	–	–
Prickly pear (platy *Opuntia* spp.)	28	–	8	–	70	42	5	–	93	50
Christmas cactus (*O. leptocaulis*)	38	–	19	–	46	–	–	–	–	–
Yucca (*Yucca* spp.)	–	–	–	–	–	12	–	–	–	11
Seeds/nuts										
Cactus (Cactaceae)	–	8	–	10	–	–	–	30	–	–
Hackberry (*Celtis* spp.)	–	8	–	20	4	8	10	30	–	3
Goosefoot (*Chenopodium* spp.)	–	–	7	10	–	–	–	20	5	3
Texas persimmon (*Diospyros texana*)	10	–	31	20	14	4	10	20	–	–
Sunflower (*Helianthus* spp.)	–	8	–	20	–	–	–	30	–	–
Walnut (*Juglans* spp.)	–	–	8	–	23	4	–	–	–	5
Juniper (*Juniperus* spp.)	–	–	–	–	–	–	–	–	–	5
Prickly pear (platy *Opuntia* spp.)	35	75	85	10	74	65	59	60	40	16
Grass (Poaceae)	10	8	27	50	2	23	1	20	5	–
Mesquite (*Prosopis* spp.)	–	33	8	–	13	4	–	20	–	3
Acorn (*Quercus* spp.)	–	–	–	–	–	–	–	–	–	3
Drop-seed (*Sporobolus* spp.)	7	–	42	–	42	–	–	–	–	–

[a] (A) 29 samples from Hinds Cave (Stock 1983); (B) 12 samples from Frightful Cave (Fry 1975); (C) 26 samples from Hinds Cave (Stock 1983); (D) 10 samples from Frightful Cave (Fry 1975); (E) 100 samples from Hinds Cave (Williams-Dean 1978); (F) 25 samples from Hinds Cave (Reinhard 1992); (G) 40 samples from Hinds Cave (Edwards 1990); (H) 10 samples from Frightful Cave (Fry 1975); (I) 43 samples from Conejo Shelter (Bryant 1974); (J) 38 samples from Baker Cave (Sobolik 1991b). Data condensed from Sobolik (1991a).

Table 11-6. Chi-Square Analysis of Botanical Remains from Paleofeces

Botanical item	Site	Chi-square	df	p	Significant at $\alpha = 0.05$
Fiber	All sites combined	27.99	3	0.000	Yes
	Frightful–Conejo	14.22	1	0.000	Yes
	Conejo–Hinds	1.43	1	0.231	No
	Hinds–Baker	11.736	1	0.001	Yes
Prickly pear fiber	All sites combined	48.130	3	0.000	Yes
	Frightful–Conejo	0.170	1	0.680	No
	Frightful–Hinds	18.521	1	0.000	Yes
	Hinds–Baker	0.156	1	0.693	No
	Baker–Conejo	16.762	1	0.000	Yes
Onion fiber	All sites combined	20.580	3	0.000	Yes
	Frightful–Conejo	17.77	1	0.000	Yes
	Conejo–Hinds	6.031	1	0.014	Yes
	Hinds–Baker	0.238	1	0.625	No
	Baker–Conejo	4.889	1	0.027	Yes
	Frightful–Baker	4.512	1	0.034	Yes
	Frightful–Hinds	9.183	1	0.002	Yes
Seeds	All sites combined	73.629	3	0.000	Yes
	Frightful–Hinds	3.738	1	0.053	Yes[a]
	Hinds–Conejo	33.068	1	0.000	Yes
	Frightful–Conejo	22.235	1	0.000	Yes
	Hinds–Baker	40.822	1	0.000	Yes
	Frightful–Baker	26.168	1	0.000	Yes
Prickly pear seeds	All sites combined	60.87	3	0.000	Yes
	Frightful–Conejo	0.446	1	0.504	No
	Frightful–Hinds	7.437	1	0.006	Yes
	Frightful–Baker	7.913	1	0.000	Yes
	Conejo–Baker	4.487	1	0.034	Yes
	Hinds–Conejo	19.60	1	0.000	Yes
	Baker–Hinds	48.214	1	0.000	Yes

(Continued)

Table 11-6. (*Continued*)

Botanical item	Site	Chi-square	df	p	Significant at $\alpha = 0.05$
Grass seeds	All sites combined	24.135	3	0.000	Yes
	Frightful–Conejo	4.93	1	0.026	Yes[a]
	Frightful–Hinds	0.047	1	0.828	No
	Hinds–Conejo	9.870	1	0.002	Yes
Nuts	All sites combined	9.712	3	0.021	Yes[a]
	Hinds–Baker	0.002	1	0.964	No[a]
	Frightful–Conejo–Hinds/Baker	7.198	1	0.007	Yes

[a] At least one cell contains fewer than 5 samples.

Table 11-7. Animal Remains from Paleofeces

Animal remains	Percentage of total sample that contained remains[a]									
	A	B	C	D	E	F	G	H	I	J
Bone	83	57	96	50	97	65	74	40	41	53
Fur/hair	45	100	73	100	79	23	5	100	–	45
Bats (Chiroptera)	–	–	–	–	–	3	–	–	–	–
Rabbits (Leporidae)	7	–	8	–	13	–	10	–	–	3
Rodents	14	–	15	–	55	4	10	–	23	32
White-footed mouse (Peromyscus leucopus)	–	–	–	2	–	–	–	18	–	–
Packrat (Neotoma spp.)	3	–	–	19	–	–	–	8	–	–
Hispid cotton rat (Sigmodon hispidus)	–	–	4	–	13	–	–	–	3	–
Golden mouse (Ochrotomys nuttalli)	–	–	–	–	1	–	–	–	–	–
Harvest mouse (Reithrodontomys spp.)	–	–	–	–	1	–	–	–	–	–
Ground squirrel (Spermophilus spp.)	–	–	–	–	3	–	–	–	–	–
Birds	–	–	–	10	10	–	3	20	–	18
Bobwhite (Colinus spp.)	–	–	–	–	2	–	–	–	–	–
Mourning dove (Zenaida macroura)	–	–	–	–	1	–	–	–	–	–
Reptiles	14	8	–	–	–	17	–	12	–	–
Lizards	–	–	–	–	5	16	–	–	18	–
Spiny lizard (Sceloporus spp.)	–	–	–	–	1	–	–	–	–	–
Snakes	7	–	–	3	–	–	–	3	–	–
Fishes	3	–	23	–	12	8	–	–	9	37
Insects	52	100	85	70	–	58	69	80	11	47
Shell	–	–	2	–	19	19	28	–	–	7

[a] (A) 29 samples from Hinds Cave (Stock 1983); (B) 12 samples from Frightful Cave (Fry 1975); (C) 26 samples from Hinds Cave (Stock 1983); (D) 10 samples from Frightful Cave (Fry 1975); (E) 100 samples from Hinds Cave (Williams-Dean 1978); (F) 25 samples from Hinds Cave (Reinhard 1992); (G) 40 samples from Hinds Cave (Edwards 1990); (H) 10 samples from Frightful Cave (Fry 1975); (I) 43 samples from Conejo Shelter (Bryant 1974); (J) 38 samples from Baker Cave (Sobolik 1991b). Data condensed from Sobolik (1991a).

Table 11-8. Chi-Square Analysis of Animal Remains from Paleofeces

Animal item	Site	Chi-square	df	p	Significant at α = 0.05
Bone	All sites combined	61.63	3	0.000	Yes
	Frightful–Conejo	0.217	1	0.641	No
	Frightful–Hinds	24.41	1	0.000	Yes
	Frightful–Baker	0.000	1	1.00	No
	Baker–Hinds	24.23	1	0.000	Yes
	Conejo–Hinds	43.32	1	0.000	Yes
	Baker–Conejo	0.557	1	0.455	No
Fur/hair	All sites combined	71.73	3	0.000	Yes
	Frightful–Conejo	63.34	1	0.000	Yes
	Frightful–Hinds	14.57	1	0.000	Yes
	Frightful–Baker	23.15	1	0.000	Yes
	Baker–Hinds	1.45	1	0.228	No
Rodents	All sites combined	12.76	3	0.005	Yes
	Conejo–Hinds	0.676	1	0.411	Yes
	Conejo–Baker	0.348	1	0.555	No
	Baker–Hinds	0.000	1	1.00	No
Birds	All sites combined	12.509	3	0.006	Yes[a]
	Baker–Hinds	6.198	1	0.013	Yes[a]
	Frightful–Hinds	0.226	1	0.634	No[a]
	Baker–Frightful	0.054	1	0.463	No
Reptiles	All sites combined	79.03	3	0.048	Yes[a]
	Frightful–Conejo	0.832	1	0.362	No[a]
	Frightful–Hinds	0.456	1	0.499	No[a]
	Frightful–Baker	3.51	1	0.61	No[a]
	Conejo–Hinds	0.171	1	0.679	No[a]
	Conejo–Baker	0.722	1	0.395	No
	Baker–Hinds	4.088	1	0.043	Yes[a]

(Continued)

Table 11-8. (Continued)

Animal item	Site	Chi-square	df	p	Significant at $\alpha = 0.05$
Fishes	All sites combined	28.69	3	0.000	Yes[a]
	Frightful–Conejo	1.572	1	0.210	No[a]
	Frightful–Hinds	2.199	1	0.138	No[a]
	Frightful–Baker	11.756	1	0.001	Yes
	Conejo–Hinds	0.000	1	1.000	No[a]
	Conejo–Baker	7.33	1	0.007	Yes
	Hinds/Baker	18.328	1	0.000	Yes

[a] At least one cell contains fewer than 5 samples.

DISCUSSION

Significant differences are noted among most of the faunal, botanical, and pale-ofecal data. The differences are most likely a result of environmental variations associated with site locations, which are reflected in the foods collected. Environmental diversity is high in the Lower Pecos region. A variety of factors, mainly the location of sites near water sources and the location of the sites along a southwest–northeast environmental gradient defined by temperature and rainwater, are reflected in the varied zooarchaeological and archaeobotanical assemblages.

Many of the botanical resources found in Lower Pecos samples flower or set fruit along this southwest–northeast gradient. For example, prickly pear fruits begin to ripen in the summer in the southwestern portion of the region. The "wave" of ripening continues throughout the Lower Pecos until, finally, plants in the northeastern portion bear ripe fruit in the late summer and early fall. Thus, there is a significant period in which prickly pear flowers, fruits, and seeds are available in the region, but that availability varies for specific locations depending upon the season. Sites located in the southwestern portion of the Lower Pecos region should contain different relative frequencies of dietary items than sites located in the northeastern portion.

Frightful Cave is located in the extreme southwestern portion of the region in an environment with little rainfall and no permanent water source. The absence of nut hulls and fish in samples from this cave reflects the lack of water. Similarly, the low frequency of onion fiber from Frightful Cave reflects the scarcity of water, particularly from rainfall. Prickly pear pads, a food staple that is generally available, were observed in more samples from Frightful Cave and Conejo Shelter, possibly indicating that the inhabitants of these sites did not have a large variety of food items to choose from and thus most commonly ate prickly pear.

Conejo Shelter is located in the central portion of the Lower Pecos region to the south of Baker and Hinds caves. Although the site is situated on the Rio Grande, it is a great distance from the mesic Edwards Plateau. Due to Conejo Shelter's location in a more xeric environment, it is not surprising that samples from the rockshelter contain few nut hulls, seeds, or bird remains. Although some birds live in and near desert scrubs and cacti, the majority of birds live in and near trees or grasslands. Since Conejo Shelter is not near stands of trees, and trees are observed only rarely along the rivers and canyons of the Lower Pecos region, the absence of birds is probably related to environmental variables.

The location of Hinds and Baker caves in the more mesic northeastern portion of the region explains the significant differences observed between the dietary array found at these caves and that found at the southern sites, Frightful Cave and Conejo Shelter. These differences include the highest number of samples from Baker Cave containing nut hulls, the highest number of samples from Baker Cave containing fish, bird remains observed most frequently in samples from Hinds and Baker caves, prickly pear fiber observed in fewer samples from Hinds and Baker caves; and bone observed more frequently in samples from Hinds Cave.

In the seasonal round hypothesized by Shafer (1986), the Lower Pecos populations inhabited the Frightful Cave region in the spring, Conejo Shelter in the late spring to early summer, Hinds Cave in the late summer and early fall, and Baker Cave in the fall and possibly into the winter (Figure 11-1). Dietary remains (animals, plants, and paleofeces) indicate that there were distinct differences in the food items collected by people at each site and that these differences reflect environmental conditions specific to each site location. Archaic populations of the Lower Pecos region were therefore choosing dietary resources found in the local environment at each home base.

If the Lower Pecos population(s) followed the seasonal round postulated here, they would have obtained a variety of dietary resources reflecting seasonal availability and plant growth cycles. The variety of nutrients in their diet would be maximized by this cycle, and their health, as a population, would have benefited. This assessment of their nutritional status is indicated by the paleonutritional analysis, which found that the population was relatively healthy and that the people had access to the nutrients and food resources necessary for a healthy existence.

Short-term nutritional stress was observed in the population, most likely related to seasonal stress incurred during the lean winter months when dietary resources were restricted. At present, there is no evidence of storage in the region, indicating that food storage for use during other seasons was not probable. Food resources in all areas of the region were scarce in the winter, although a greater number of plant and animal resources were available nearer the Edwards Plateau and upper canyon areas. These few plants and animals, as well as the dietary staples (agave, yucca, and prickly pear), provided the population with sustenance but may not have been nutritionally adequate for everyone in the population through the winter months.

CONCLUSIONS

To test models of seasonal mobility, it is important to combine identifications of dietary resources obtained from sites located in diverse environments within the hypothesized annual cycle with paleonutritional analyses of the diet and health assessments. This combination played an important role in testing the seasonal round hypothesized for Archaic populations in the Lower Pecos region. These analyses indicate that the population was relatively fit and obtained the nutrients and food resources necessary for a healthy existence during most of the year. The seasonal round corresponds to the nutritional information and allowed the population to maximize the food resources available in the region by following the seasonal availability and growth cycles of resources.

This study provides evidence of the important role that seasonal availability of food resources played in population mobility. While the importance of the environment and food resources to mobility patterns is reported in many ethnographic and modern examples (Belovsky 1988; Chatters 1986; Speth 1990), applying such correlates to the archaeological record is more difficult. Only through integrating many types of archaeological evidence could it be observed that populations in the Lower Pecos region followed a seasonal mobility pattern that allowed them to maximize the dietary resources available in the environment. By maximizing seasonal resources, the population maintained a nutritionally sound and stable diet year-round.

ACKNOWLEDGMENTS

I want to thank Elizabeth Reitz, Lee Newsom, Daniel Sandweiss, and Sylvia Scudder for their insightful comments on the contents of this chapter. The original research for this chapter benefited from the comments and guidance of D. Gentry Steele, Vaughn M. Bryant, Jr., and Harry J. Shafer. I thank Harry J. Shafer for hypothesizing a seasonal round that could be tested, but any errors in interpretation are my own. I also thank Hendrik Poinar for agreeing to analyze DNA from paleofeces. The research was funded by the National Science Foundation (BNS-90004064) and Sigma Xi. Stephen Bicknell drafted Figure 11-1.

REFERENCES

Alexander, R. K., 1974, *The Archaeology of Conejo Shelter: A Study of Cultural Stability at an Archaic Rockshelter Site in Southwestern Texas*, Ph.D. dissertation, Department of Anthropology, University of Texas, Austin.

Andrews, P., and Adovasio, J., 1980, Perishable Industries from Hinds Cave, Val Verde County, Texas, *Ethnology Monographs* 5, Department of Anthropology, University of Pittsburgh, Pittsburgh, Pennsylvania.

Belovsky, G. E., 1988, An Optimal Foraging-Based Model of Hunter-Gatherer Population Dynamics, *Journal of Anthropological Archaeology* 7:329–372.

Binford, L. R., 1980, Willow Smoke and Dogs' Tails: Hunter–Gatherer Settlement Systems and Archaeological Site Formation, *American Antiquity* 45:4–20.

Bryant, V. M., Jr., 1969, *Late Full-Glacial and Post-Glacial Pollen Analysis of Texas Sediments*, Ph.D. dissertation, Department of Botany, University of Texas, Austin.

Bryant, V. M., Jr., 1974, Prehistoric Diet in Southwest Texas: The Coprolite Evidence, *American Antiquity* 39(3):407–420.

Bryant, V. M., Jr., 1975, Pollen as an Indicator of Prehistoric Diets in Coahuila, Mexico, *Bulletin of the Texas Archaeological Society* 46:87–106.

Chadderdon, M. F., 1983, Baker Cave, Val Verde County, Texas: The 1976 Excavations, *Center for Archaeological Research Special Report* 13, University of Texas at San Antonio, San Antonio.

Chatters, J. C., 1986, Hunter-Gatherer Adaptations and Assemblage Structure, *Journal of Anthropological Archaeology* 6:336–375.

Dering, J. P., 1979, *Pollen and Plant Macrofossil Vegetation Record Recovered from Hinds Cave, Val Verde County, Texas*, M.S. thesis, Department of Anthropology, Texas A & M University, College Station.

Dibble, D. S., 1967, Excavations at Arenosa Shelter, 1965–66, report submitted to the National Park Service by the Texas Archaeological Salvage Project.

Dibble, D. S., and Lorrain, E. R., 1968, Survey and Test Excavations at Amistad Reservoir, 1964–65, *Texas Archaeological Salvage Project Survey Report* 3, Austin.

Douglas, C. L., 1970, Analysis of the Faunal Remains, in: Excavations at Baker Cave, Val Verde County, Texas (J. H. Word and C. L. Douglas, eds.), *Bulletin of the Texas Memorial Museum* 16, University of Texas, Austin, pp. 113–151.

Edwards, S. K., 1990, *Investigations of Late Archaic Coprolites: Pollen and Macrofossil Remains from Hinds Cave (41VV456), Val Verde County, Texas*, M.A. thesis, Department of Anthropology, Texas A & M University, College Station.

Epstein, J. F., 1963, Centipede and Damp Caves: Excavations in Val Verde County, Texas, 1958, *Bulletin of the Texas Archaeological Society* 33:1–130.

Fry, G. F., 1975, Human Coprolites from Frightful Cave (CM 68) Coahuila, Mexico, paper presented at the 40th Annual Meeting of the Society for American Archaeology, Dallas, Texas.

Hartnady, P. W., 1988, *Premature Molar Tooth Loss in the Archaic Trans-Pecos Region of South Texas*, M.A. thesis, Department of Anthropology, University of Arkansas, Fayetteville.

Hester, T. R., 1983, Late Paleo-Indian Occupations at Baker Cave, Southwestern Texas, *Bulletin of the Texas Archaeological Society* 53:101–119.

Hester, T. R., 1986, Baker Cave: A Rich Archaeological Record, in: *Ancient Texas: Rock Art and Lifeways along the Lower Pecos* (H. J. Shafer, ed.), Texas Monthly Press, Austin, pp. 84–87.

Hester, T. R., 1988, Chronological Framework for Lower Pecos Prehistory, *Bulletin of the Texas Archaeological Society* 59:53–64.

Irving, R. S., 1966, A Preliminary Analysis of Plant Remains from Six Amistad Reservoir Sites, in: *A Preliminary Study of the Paleoecology of the Amistad Reservoir Area* (D. A. Story and V. M. Bryant, Jr., comps.), final report of research under the auspices of the National Science Foundation (GS-667).

Kelly, R. L., 1992, Mobility/Sedentism: Concepts, Archaeological Measures, and Effects, *Annual Review of Anthropology* 21:43–66.

Lord, K. J., 1984, *The Zooarchaeology of Hinds Cave (41V456)*, Ph.D. dissertation, Department of Anthropology, University of Texas, Austin.

National Oceanic and Atmospheric Administration, 1981, *Local Climatological Data*, Del Rio, Texas.

Poinar, H. N., Kuch, M., Sobolik, K. D., Barnes, I., Stankiewicz, A. B., Kuder, T., Spaulding, W. G., Bryant, Jr. V. M., Cooper, A., and Pääbo, S., 2001, A Molecular Analysis of Dietary Diversity for Three Archaic Native Americans, *Proceedings of the National Academy of Sciences* 98(8):4317–4322.

Price, T. D., and Brown, J. A., 1985, Aspects of Hunter–Gatherer Complexity, in: *Prehistoric Hunter-Gatherers: The Emergence of Cultural Complexity* (T. D. Price and J. A. Brown, eds.), Academic Press, New York, pp. 3–20.

Reinhard, K. J., 1992, Patterns of Diet, Parasitism and Anemia in the Prehistoric West, in: *Diet, Demography, and Disease: Changing Perspectives of Anemia* (P. Stuart-MacAdam and S. Kent, eds.), Aldine de Gruyter, New York, pp. 219–258.

Reinhard, K. J., Olive, B. W., and Steele, D. G., 1989, Bioarchaeological Synthesis, in: *From the Gulf to the Rio Grande: Human Adaptation in Central, South, and Lower Pecos, Texas* (T. R. Hester, S. L. Black, S. G. Steel, B. W. Olive, A. A. Fox, K. J. Reinhard and L. Bement, eds.), *Arkansas Archaeological Survey Research Series* 33:93–114.

Saunders, J. W., 1986, *The Economy of Hinds Cave*, Ph.D. dissertation, Department of Anthropology, Southern Methodist University, Dallas, Texas.

Saunders, J. W., 1992, Plant and Animal Procurement Sites in the Lower Pecos Region, Texas, *Journal of Field Archaeology* 19:335–349.

Shafer, H. J., 1977, Art and Territoriality in the Lower Pecos Archaic, *Plains Anthropologist* 22(75):13–22.

Shafer, H. J., 1986, *Ancient Texas: Rock Art and Lifeways along the Lower Pecos*, Texas Monthly Press, Austin.

Shafer, H. J., 1988, The Prehistoric Legacy of the Lower Pecos Region of Texas, *Bulletin of the Texas Archaeological Society* 59:23–52.

Sobolik, K. D., 1991a, *Paleonutrition of the Lower Pecos Region of the Chihuahan Desert*, Ph.D. dissertation, Department of Anthropology, Texas A & M University, College Station.

Sobolik, K. D., 1991b, The Prehistoric Diet and Subsistence of the Lower Pecos Verde County, Texas, *Studies in Archaeology Series* 7, Texas Archaeological Research Lab, University of Texas, Austin.

Sobolik, K. D., 1993, Direct Evidence for the Importance of Small Animals to Prehistoric Diets: A Review of Coprolite Studies, *North American Archaeologist* 14(3):227–244.

Sobolik, K. D., 1994a, Introduction, in: *Paleonutrition: The Diet and Health of Prehistoric Americans* (K. D. Sobolik, ed.), *Center for Archaeological Investigations, Occasional Paper* 22, Southern Illinois University at Carbondale, Carbondale, pp. 1–18.

Sobolik, K. D., 1994b, Paleonutrition of the Lower Pecos Region of the Chihuahuan Desert, in: *Paleonutrition: The Diet and Health of Prehistoric Americans* (K. D. Sobolik, ed.), *Center for Archaeological Investigations, Occasional Paper* 22, Southern Illinois University at Carbondale, Carbondale, pp. 247–264.

Speth, J. D., 1990, Seasonality, Resource Stress, and Food Sharing in So-Called "Egalitarian" Foraging Societies, *Journal of Anthropological Archaeology* 9:148–188.

Stock, J. A., 1983, *The Prehistoric Diet of Hinds Cave (41VV456), Val Verde County, Texas: The Coprolite Evidence*, M.A. thesis, Department of Anthropology, Texas A & M University, College Station.

Turpin, S., 1984, The Red Linear Style Pictographs of the Lower Pecos River Region, Texas, *Plains Anthropologist* 29:181–198.

Williams-Dean, G. J., 1978, *Ethnobotany and Cultural Ecology of Prehistoric Man in Southwest Texas*, Ph.D. dissertation, Department of Biology, Texas A & M University, College Station.

Chapter *12*

Developing Models of Settlement for the Florida Gulf Coast

MICHAEL RUSSO AND IRVY R. QUITMYER

One puzzle with which archaeologists struggle is determining when people began to be less mobile and more sedentary. In many parts of the world sedentism is linked to plant domestication. Residential sedentism without domestic plants is less frequently reported. However, Calusa communities on the southwestern Gulf coast of Florida in the 16th century A.D. exhibited many features considered hallmarks of sedentary societies: permanent structures, storage facilities, large villages, and large-scale public works in the form of shell and earthen mounds. The Calusa political system maintained hegemony over the entire southern tip of the peninsula (Widmer 1988). Although the Calusa lived in permanently-occupied villages, domesticated plants were not a significant food source (Scarry and Newsom 1992). Instead, Calusa economies were based on the bountiful, non-domesticated plants and animals of Florida's coastal estuaries, marshes, and swamps. When did this tradition of sedentism in the absence of a domestic food base appear and how was it supported? This study elaborates upon evidence for early sedentism on the southwestern Gulf coast of Florida and reveals two distinct patterns of bay scallop (*Argopecten irradians*) collection.

EVIDENCE FOR SEASON OF USE

Traditional markers of sedentism present difficulties for interpreting archaeological sites in southwestern Florida (Figure 12-1). "Permanent structures" made of leaves and branches degraded quickly and subterranean storage pits were less

235

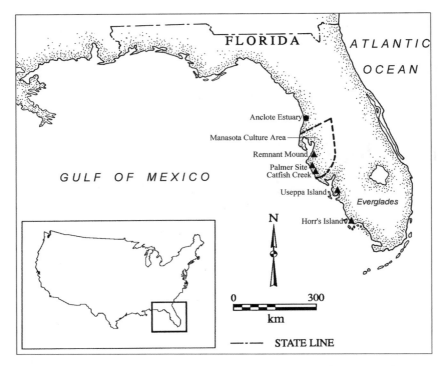

Figure 12-1. Archaeological and other locations in south Florida mentioned in text.

frequently used in Florida's damp soils than in drier areas of the Southeast (Milner 2004:33, 37, 183; Smith 1986:17, 54; Steponaitis 1986:375, 381). Millennia of shellfish accumulations obscure site boundaries, making it difficult to distinguish between refuse of brief camping episodes and that of large, permanent villages. Large-scale public works such as ceremonial mounds, hallmarks of the sedentary and hierarchically-organized Mississippian farming societies of the interior Southeast, traditionally are interpreted on the Florida Gulf coast as refuse rather than architecture (e.g., Bullen and Bullen 1976:20; Houck 1996:32; McMichael 1982; Russo 2004:59–63). Consequently, distinguishing permanent settlements from semipermanent base camps or frequently revisited collection camps is problematic.

Fortunately, many of the animal remains at these sites contain information about the season in which they were harvested, providing a basis for understanding a community's residential pattern. If the remains provide evidence that a site was occupied during more than one season, then the likelihood that the site was a single-season camp is reduced. The more collection seasons identified, the stronger the support for an interpretation of year-round occupation (i.e., permanent settlement), particularly if such evidence is derived from several organisms.

Assessing season of death most commonly is based on correlations of time with body size, incremental grown structures, and chemical properties in skeletal remains (e.g., Hudson et al. 1976; Rhoads and Lutz 1980). Quahogs (*Mercenaria* spp.; Quitmyer et al. 1997), coquina (*Donax variabilis*; Jones et al. 2004; Russo and Ste. Claire 1992), bay scallop (Russo and Quitmyer 1996), and the oyster parasite, the impressed odostome (*Boonea impressa*; Russo 1991b) often are used to estimate seasonal collection episodes at Florida sites.

Seasonal site occupation and seasonal patterns of resource procurement (e.g., shellfish collection) represent two different aspects of human behavior (Deith 1983:423; Quitmyer et al. 1997). Humans are broad-based omnivores who use many plants and animals seasonally. A single or multiple seasonal indicator may be sufficient to identify when a specific site was occupied in the annual cycle, but may not be sufficient to characterize that site's full role within the annual cycle or within the larger, regional settlement pattern. Redundant sampling and repeated analyses of collections representing the full range of site types are necessary to fully resolve the question of seasonal site occupation.

BAY SCALLOP SEASONALITY

Season of death for animals that grow throughout their lives may be estimated by correlating growth increments or the sizes of archaeological specimens with those of modern populations at specific times of the year (e.g., Claassen 1982; Clark 1979; Clark and Lutz 1982; Quitmyer et al. 1997). The size-correlation approach assumes that most individuals of the taxon reproduce during a specific season and grow at a predictable rate that can be equated with size at a specific time in the annual cycle. Animals most easily studied in this way are those that live a single year.

One such species is the bay scallop, which in Florida waters generally lives only one year (Austin and Russo 1989; Quitmyer 1992, 1998; Russo 1998). The annual cycle commences with fall spawning, but little or no growth occurs during the winter. Winter bay scallops average less than 20 mm in height (Figure 12-2, see methods). Rapid growth begins as water temperatures warm in early spring (March) and continues through June, by which time average shell height reaches 50 mm. Growth continues through September until spawning occurs in October, when shell height reaches an average of 60 mm. After spawning most bay scallops die. Mass mortality of year-old bay scallops in the fall means that few larger bay scallops are available during winter months. At a given time, then, the mean size of most bay scallops is predictable. By calculating the mean size of bay scallops from an archaeological deposit and comparing it to known seasonal size classes, the time of year that the bay scallops were collected can be estimated.

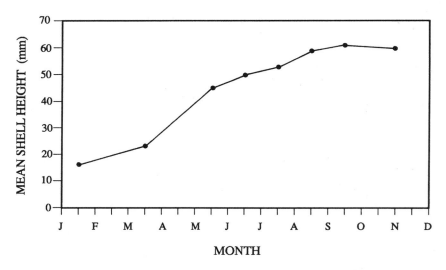

Figure 12-2. Monthly mean shell heights of modern bay scallops (*Argopecten irradians*) collected from Anclote estuary, Florida (after Barber and Blake 1983). The sample size is unknown.

METHODS

In this study, bay scallop (Russo and Quitmyer 1996) and quahog (Quitmyer et al. 1997) season of death estimates are combined with the sizes of fishes, the presence/absence of loons (*Gavia immer*), and nonmetric markers of seasonality and settlement permanency to examine the character of year-round coastal settlement achieved in southwest Florida. The assessment of seasonal periodicity derived from these organisms is elaborated upon by reference to capture technologies, settlement patterns, population increases, and political boundaries.

Valve height is measured from the umbo to the ventral margin (dorsoventrally). Occasionally, height and length (the anterior to posterior dimension) are confused in the literature (see Broom 1976; Perlman 1973), but the two measures are nearly identical on a given valve (Bushnell 1965). When an archaeological sample did not provide a sufficient number of whole valves to be statistically reliable (e.g., 30 or more), the hinge heights of fragmentary bay scallops were measured and appropriate allometric formulae applied to estimate shell size (Reitz et al. 1987; Russo 1991a:506). When selecting random samples from larger collections, measurable hinges from both whole and broken valves were selected so no biases resulted from measuring only whole valves. These methods are described in greater detail elsewhere (Quitmyer 1998; Quitmyer et al. 1997; Russo 1991b, 1998).

Four cultural periods represented by five Gulf coast archaeological sites are considered in this case study. The Preceramic Middle Archaic (5000–2000 B.C.) and the Late (Ceramic) Archaic (2000–500 B.C.) periods are collectively termed the Archaic. Their discussion is followed by the Manasota (500 B.C.–A.D. 900) and Safety Harbor (A.D. 900–1500) periods. The Archaic sites are located on the lower Gulf coast of the Florida peninsula, whereas the Manasota and Safety Harbor sites are located in the Manasota culture area to the north (Figure 12-1).

ARCHAIC SEASONALITY AND SETTLEMENT

Rising sea levels during the mid-Holocene submerged and destroyed many, if not most, Gulf coast Archaic sites (e.g., Dunbar et al. 1992; Faught and Donoghue 1997; Faught and Latvis 1999) and few archaeological sites occupied before 2500 B.C. are found along Florida's present shoreline. For years, archaeologists interpreted the few surviving Archaic sites as special-purpose or seasonal encampments of mobile hunter-gatherers who otherwise occupied the interior non-coastal environments of the peninsula (Cockrell 1970; Milanich and Fairbanks 1980:29; Widmer 1988:211). Widmer (1988) hypothesized that prior to 2500 B.C., when the rising Gulf stabilized near its current level, the coastal zone was not sufficiently stable to support even "semipermanent" settlements. In the absence of biological evidence, archaeologists speculated that shellfish must have been a cold-weather resource (e.g., Bullen and Bullen 1961; Milanich and Fairbanks 1980:150). Shellfish were presumed to be a poor nutritional source or otherwise inedible during the summer because of diseases and parasites. Consequently, they were interpreted as a hard-luck food eaten in what were presumed to be the lean months of winter when presumably preferred foods were unavailable. Thus, the few Archaic archaeological shell deposits on Florida's Gulf coast were interpreted as seasonal camps left by people from the interior of Florida incapable of living on the coast on a year-round basis. The invention of pottery in the Late Archaic period, it was thought, facilitated permanent settlement on the coast by providing vessels for cooking and storing seafood, thus accounting for an increase in sites at the end of the Archaic period (Sassaman 1993:216–217).

The large sizes and seasonal signatures of some Archaic sites (Russo 1991a, b, 1992, 1998, 2004; Russo and Heide 2001, 2004; Russo and Saunders 1999; Saunders 2004) require reconsideration of this model of Archaic coastal settlement and subsistence patterns. Instead of seasonal migrations from the interior, this evidence indicates that the productive estuaries were exploited from large, permanently-occupied coastal villages as well as from smaller logistical foraging camps during the Archaic period (*sensu* Binford 1980; Quitmyer and Massaro 1999; Russo 1991a, 1998; Torrence 1999). Evidence for such a pattern is found at one such village on Horr's Island.

Horr's Island: A Permanent Archaic Village

The Horr's Island site (8Cr208) is a Preceramic Late Archaic village and ceremonial mound complex occupied circa 3000–2400 B.C. (Figure 12-1). The village consists of a large (150 m long) U-shaped ring of shell up to 5 m high.
Associated with the ring are three conical, ceremonial shell/sand mounds, the
tallest rising over 6 m above the surrounding village (Russo 1994, 1998). This
village/mound complex is over a kilometer long and has yielded evidence (post
molds) of numerous domestic structures and hearths. The site has an elaborate
shell tool assemblage in contrast to the typical stone assemblages associated
with interior groups; stone resources are not available on the coast. The monumental architecture and abundant utilitarian artifacts indicate diversified,
long-term maintenance activities typical of permanently-occupied villages.

The site was originally interpreted as a series of winter encampments on the
assumption that shellfish were a cold-weather resource (McMichael 1982). To
test whether the site was a seasonal encampment or year-round village,
seasonal indicators were examined. Although scallops were recovered
throughout the site, only one feature produced sufficient numbers of valves to
assess season of collection. An average height of 59 mm indicates that most of
these scallops were collected in the late summer. Because scallops of this size
are strictly a warm-weather resource, they could not, in and of themselves, be
used to differentiate between a single season of occupation or year-round
occupation. When the scallop data are combined with seasonal data for quahog, oyster (*Crassostrea virginica*), and several species of fish (Figure 12-3),
however, it appears that animals collected from every season of the year are
represented. Quahogs were predominately collected in the spring and early
summer and, to a lesser extent, in the fall and winter; oysters and scallops were
collected in the late summer and early fall. The small sizes estimated for some
fishes indicate that hardhead catfish (*Ariopsis felis*) and pinfish (*Lagodon
rhomboides*) were caught in the fall and summer and thread herrings
(*Opisthonema oglinum*) were caught in the winter months (Russo 1998).

These animals were not the only ones consumed at the site. The species list
from Horr's Island is the richest of any Archaic faunal collection from Florida
(Russo 1991a). This richness, combined with data from the seasonal marker
species, provides a compelling case for the year-round occupation of Horr's
Island. The case for sedentism is supported by plant remains collected during
multiple seasons (Newsom 1991), ceremonial mounds, numerous residential
structures, and extensive village size. Together the seasonal measures in support
of a year-round interpretation correct the previous interpretation that the Horr's
Island site was a winter encampment and shellfish were a winter resource.

Useppa Island: A Seasonal Archaic Camp

Useppa Island is a small barrier island north of Horr's Island (Figure 12-1). It
contains four shell middens (8LL51) ranging in size from 50 m to over 150 m

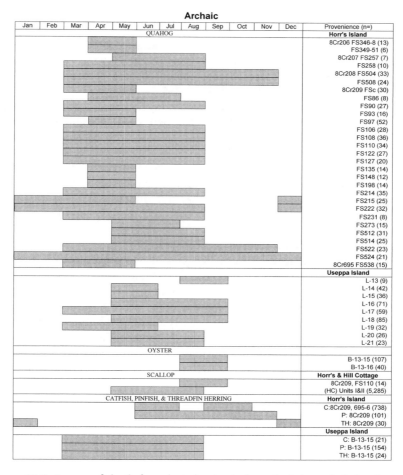

Figure 12-3. Season of death for selected animals from Archaic period sites in south Florida (Horr's Island [Russo 1991a]; Useppa [Quitmyer and Massaro 1999]) and the Manasota culture area (Palmer Site-Hill Cottage Midden [Bushnell 1965]).

deposited during both the Preceramic and Ceramic Archaic periods (Marquardt 1999:77). Based largely on a preponderance of whelk (*Busycon* spp.) remains in all stages of reduction and evidence that the quahogs and fishes were taken in the spring and summer (Figure 12-3), the site is interpreted as an Archaic seasonal camp where whelks were collected and made into tools (Quitmyer and Massaro 1999:122–123; Torrence 1999:73–74). The site may have been a logistical foraging camp related to the permanent settlement at Horr's Island (*sensu* Binford 1980; Russo 1991a:487; Torrence 1999:72).

Palmer Site-Hill Cottage Midden: An Archaic Sedentary Village?

The Hill Cottage midden component of the Palmer site (8SO2) (Figure 12-1) is a Late Archaic shell midden occupied between 2150 and 1400 B.C. (uncorrected) (Bullen and Bullen 1976). The midden is 150 m long and forms a U-shaped ring similar to that at the Horr's Island village.

Bushnell (1965) found significant changes in the size of scallops recovered from Hill Cottage by Bullen (Bullen and Bullen 1976). Bushnell interpreted changes in mean scallop size to correlate with rising sea levels and variability in temperature. Bullen and Bullen (1976) subsequently used the changes in scallop sizes to support their idea that the Ceramic Archaic period in Florida began during a cool, dry period (Bullen and Bullen 1961). Apparently, neither Bushnell nor the Bullens were aware that size differences in scallops result from seasonal growth. In consequence, they overlooked the more parsimonious interpretation that the differences they observed represented different times of the year in which scallops were collected.

Bushnell measured scallops from 11 levels of two test units from both the Preceramic and the Ceramic Archaic middens (Figure 12-4). A seasonal reappraisal of his measurements reveals that the scallops averaged between 42 and 54 mm in height, indicating that they were collected in most stages of warm-weather growth beginning in late spring/early summer and followed by collection of larger scallops later in the season up to the fall (Figure 12-3). With the arrival of a subsequent spring, the spring-through-summer collection pattern was repeated. The nearly parallel growth phases in each level indicate either

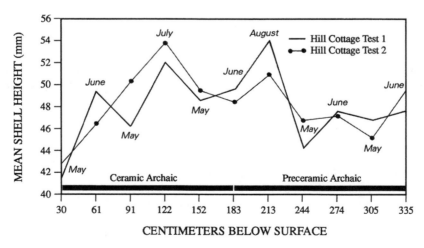

Figure 12-4. Comparison of mean bay scallop valve heights from two column samples from Hill Cottage midden (after Bushnell 1965). The number of specimens was 764 for Test 1 and 4,521 for Test 2.

that the two test units intersected the same unique seasonal deposits or, more likely, a similar seasonal pattern of collection was practiced during both the Preceramic and the Ceramic Archaic periods across the site.

Despite the introduction of pottery (at 183 cm below the surface in Figure 12-4), the use of scallops continued with no apparent effects on the generalized collection strategy. Bullen and Bullen (1976:6–20) observed that despite slight changes in the relative amounts of shellfish throughout the midden, the kinds of shellfish and other subsistence remains persisted unchanged throughout the Preceramic and Ceramic Archaic deposits. Scallops, oysters, quahogs, whelks, and conchs (*Melongena corona, Strombus* spp.), as well as fishes, were identified in both Preceramic and Ceramic levels. They noted that the introduction of pottery did not alter the kinds of tools used in the procurement and processing of food. These observations argue against the idea that the coast was not permanently settled until ceramics allowed larger populations to be sustained for longer periods.

The Bullens never directly approached the question of seasonal versus year-round settlement at Hill Cottage. They did contrast the site with nearby small sites, which they interpreted as periodic camps (Bullen and Bullen 1976:35), implying that Hill Cottage midden was occupied for periods longer than those associated with a seasonal camp. The presence of burials, post molds for domestic structures and drying racks, hearths, large numbers of utilitarian artifacts, and numerous decorative items, as well as the U-shape of the midden, support the interpretation of Hill Cottage midden as a large, permanently-occupied site (Bullen and Bullen 1976; Russo 2006). To date, however, only the season of death for the scallops is available for Hill Cottage midden, and more evidence is needed to confirm that the site represents a sedentary occupation.

MANASOTA SEASONALITY AND SETTLEMENT

Safety Harbor and Manasota period cultures are characterized by shell middens and mound sites along coastal estuaries presumed to be permanent villages (Luer and Almy 1982). Smaller sites have been identified in the interior as well as among the estuaries. Regardless of location, small sites are interpreted as seasonal collection stations (Almy 1988; Austin and Russo 1989; Luer and Almy 1982). Some 25 years after the model was put forward, few Manasota sites have yielded seasonality data that have been or could be used to test the model (e.g., Austin 1995:220; Austin and Russo 1989; Quitmyer 1998). Typically sites are fit into the bipartite settlement typology (camp versus village) based solely on their relative size, complexity, and artifact diversity (e.g., Ardren et al. 2003:53; Schwadron 2002:208). That is, sites are identified as villages or camps in the absence of seasonality determinations. Below we present seasonality data to test the model of Manasota-area settlement patterns.

Palmer Site-Shell Ridge Midden: A Permanent Manasota and Safety Harbor Village

The Shell Ridge midden (8SO1) component of the Palmer site was occupied during the Manasota and Safety Harbor periods. It was first excavated by Bullen and Bullen (1976) and subsequently by Torrence, who reopened the Bullens' excavation units H and I (Quitmyer 1992, 1998; Torrence 1992). Shell Ridge midden consists of shell deposits over a kilometer long and a burial mound. The size and complexity of the site suggest it was a permanently-settled coastal village (Luer and Almy 1982; Milanich 1994:225). The combination of several lines of biological evidence indicates multi-season occupations during the Manasota and Safety Harbor periods. Season of death for quahogs indicates spring through summer occupations; the quahogs and loon data combined suggest fall and winter occupations during the Manasota period (Figure 12-5; Quitmyer 1992:56; 1998:203). In addition, wild plants attest to a spring through fall occupation (Newsom 1992, 1998). The quahog, loon, and scallop data verify multi-season occupation of the site during the Safety Harbor period as well (Figure 12-5).

Although the Shell Ridge midden data demonstrate that the site was occupied during multiple, if not all, seasons of the year, scallops were, of course, a seasonal resource. Quitmyer (1998) measured 1,143 scallop valves from 14 Safety Harbor proveniences. The mean valve heights fell between 38 mm and 43 mm (except for one sample mean of 49 mm), indicating a spring collection period. This contrasts with the warm-weather, late-season scallop harvest at the Archaic-period Hill Cottage component (8SO2); the narrower size range of the Shell Ridge scallops suggests a different collection strategy. During the Safety Harbor period, scallops were collected at the Palmer site in the spring, early in their growth cycle when they were relatively small.

Supporting evidence for such seasonal exploitation may be found in other aspects of the assemblage. The small scallops indicate that a mass capture technique was used to collect them, a technique that would incidentally gather benthic shell hash. The Shell Ridge midden contains large amounts of very fine shell debris typical of benthic hash (Quitmyer 1998). Collection techniques such as drag-netting and basket scooping might capture large numbers of scallops regardless of their size but limit the number of scallops available later in the season. In contrast, Archaic scallop assemblages contained few small individuals and lacked the fine shell debris, indicating a more selective collection strategy, probably simple hand collection.

Catfish Creek Site: A Seasonal Manasota Camp

The Catfish Creek site (8SO608), which was occupied around A.D. 700, is less than a kilometer northeast of the Palmer site on a small freshwater creek (Figure 12-1). The site consists of numerous small shell middens or "scatters"

Manasota

Jan	Feb	Mar	Apr	May	Jun	Jul	Aug	Sep	Oct	Nov	Dec	Provenience (n=)
				QUAHOG								**SR/RM/CC**
												Shell Ridge IV (42)
												Remnant Md. Zone C (26)
												Remnant Md. Zone D (30)
												Catfish Creek (57)
				LOON								**Shell Ridge**
												313-340 (present)
												Test E (present)
												Test F (present)
												Test G (present)
				SCALLOP								**Remnant Mound**
												Eu2, Zn A (31)
												Eu2, Zn B (32)
												Eu2, Zn C (24)
												Eu2, Zn D (13)
												Catfish Creek
												FS20 (70)
												FS22 (78)
				OYSTER								**Remnant Mound**
												Zone C, FS179 (157)
												Zone D, FS181 (138)
			PINFISH AND PIGFISH									
												Zone A, FS177 (16)
												Zone B, FS178 (20)
												Zone C, FS179 (57)
												Zone D, FS181 (52)

Safety Harbor

Jan	Feb	Mar	Apr	May	Jun	Jul	Aug	Sep	Oct	Nov	Dec	Provenience (n=)
				QUAHOG								**Shell Ridge**
												IA (47)
												IB (126)
												ID (42)
												IE (72)
												IIA (19)
												III (25)
				LOON								
												IA (present)
												39-50 (present)
												50 (present)
												140 (present)
												110-155 (present)
												136-150 (present)
												156 (present)
												166 (present)
												173-181 (present)
				SCALLOP								
												IA (614)
												IB (388)
												IIA (69)

Figure 12-5. Season of death for selected animals from Manasota and Safety Harbor period sites in the Manasota culture area; (Catfish Creek [Austin and Russo 1989]; Remnant Mound [Russo and Powell 2005] and Palmer Site-Shell Ridge Midden [Quitmyer 1998]).

with pit features. Season of occupation estimates are based on quahogs and scallops (Figure 12-5; Austin and Russo 1989). The mean sizes of the scallops were 43 mm and 46 mm, indicating an early spring death. The quahogs also were collected in early spring. These data, combined with a limited assemblage of utilitarian tools, suggest the area served as a seasonal camp. No evidence of diverse, long-term maintenance activities, storage, structures, mounds, or elaborate ceremonialism is present. The site probably articulated with the larger Shell Ridge midden, likely functioning as a logistical place from which to collect resources for the larger village (Austin and Russo 1989:77). While at the camp, collectors processed and consumed other estuarine resources, whose remains are found in the Catfish Creek midden.

Differences in Archaic and Manasota collecting strategies at Shell Ridge and Catfish Creek may be linked to human population increase. Milanich (1991) noted an increase in the number of Manasota-period sites from the previous Archaic period and speculated that population had increased significantly. The apparent increase in seasonal encampments during the Manasota period may reflect a need for an expanded resource base. The intensive exploitation of scallops at Catfish Creek and Shell Ridge during a brief season reflects increased pressure on the resource. Intensive harvesting of scallops early in the season might have reduced the number of scallops available for harvest later in the season.

Remnant Mound: A Permanent Manasota Village

One other Manasota village site has yielded seasonal data (Figure 12-1). The Remnant Mound at DeSoto National Memorial is part of the Shaw Point site (8Ma7), which consists of a massive, complex array of shell mounds, ridges, ramps, and burial mounds (Schwadron 2002). Evidence for season of death for quahogs, scallops, oysters, pinfishes, and pigfishes (*Orthopristis chrysoptera*) was considered (Quitmyer 2002:188; Russo and Powell 2005). The oysters, pinfishes, and pigfishes probably died from the late summer through fall, possibly into early winter. Based on four scallop samples, it appears that scallops were collected during the early spring (Russo and Powell 2005) in a pattern similar to that at the Catfish Creek and Shell Ridge middens (Figure 12-6). Two quahog samples indicated a late winter/spring period of collection (Russo and Powell 2005). In combination, these data demonstrate that this large village was used throughout the year.

DISCUSSION

Seasonality analyses demonstrate that Archaic peoples inhabited the coastal zone throughout the annual cycle. These data challenge the notion that Archaic peoples were hunter-gatherers who wandered the interior forests of peninsular Florida, only occasionally making seasonal forays to the coast. At large shell ring sites such as Horr's Island and Hill Cottage, Archaic peoples lived on the Gulf coast year round. Without the benefit of cultivated plants or pottery, in some cases, Archaic peoples became sedentary through their strategies of exploiting the extensive, highly-productive coastal fisheries. Gulf coast estuaries offered abundant resources, and peoples using these were not

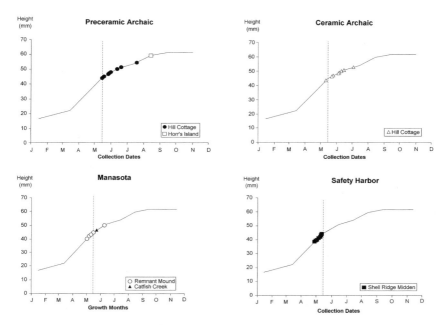

Figure 12-6. Comparison of seasons of scallop collection among Safety Harbor, Manasota, and Archaic periods within the Manasota culture area. Proveniences provided in legends (see Bushnell 1965; Quitmyer 1998; Russo 1991a; Russo and Powell 2005). The curve follows the monthly mean shell heights of modern bay scallops in Figure 12-2.

compelled to engage in a seasonal round in search of less abundant interior terrestrial foods. Thus they were free of the migratory constraints typically considered limiting factors for interior Archaic hunter-gatherer population growth and social organization (Smith 1986; Steponaitis 1986).

A much larger, more sedentary, Archaic population lived on the southwestern Gulf coast practicing more varied residential and subsistence patterns than previously thought. Both large village/ceremonial shell rings (Dickel 1992; Russo and Heide 2004) and smaller logistical sites are found along the coast (Lee et al. 1998; Torrence 1999; Wheeler et al. 1997). It remains to be seen if the smaller coastal sites were logistical foraging stations for the large, permanent coastal villages; or if the smaller sites were year-round habitations that interacted with the larger settlements in other ways.

The presence of both large and small Archaic coastal settlements does not mean that coastal peoples did not interact with interior resources and peoples. Archaic sites in the interior of southern Florida indicate that the coastal zone was not the only place with intensive Archaic occupations. Dense Archaic middens occupied between 1000 and 3000 B.C. are found throughout the freshwater Everglades (Beiter 2001; Masson et al. 1988; Mowers and Williams 1972; Newman 1993; Russo 2005; Schwadron 2005). The sites presently known are situated on relatively small tree islands (circa 50 m across), which probably limited the scope of the communities living on them. These interior Archaic sites refute the idea that interior Archaic peoples were seasonal hunters and collectors. At least some interior groups were freshwater fisher-folk, intensively exploiting the interior marshes for fishes, turtles, and other aquatic reptiles. The occasional, but ubiquitous, presence of coastal mollusks indicates that these people were connected to coastal settlements in some way.

How can these inland sites be shown to be different cultures or polities from coastal sites if they have the same artifacts? It is difficult to determine Archaic cultural relationships largely due to the absence of differences in diagnostic arti-facts between the interior and coast. Specifically, are the interior Everglades sites logistical foraging camps for coastal groups or home to a separate and culturally-distinct people (Griffin 2002:287–289)? Although the question may not be answerable with material culture, ultimately seasonality studies may determine if inland sites were occupied throughout the year or only seasonally. Testing whether interior and coastal Archaic sites were occupied during the same seasons or year-round will enable us to determine if they represent separate and distinct cultural entities or different aspects of a seasonal round. This will be a pivotal test for the model of seasonal logistic foraging.

In addition to raising questions about settlement patterns during the Archaic period, this study reveals two distinct patterns of scallop collection heretofore unrecognized: the Archaic strategy of collecting from summer to early fall and the Manasota pattern of collecting in the spring (Figure 12-6). In mod-ern Florida, the scallop season is closed between April and July to allow the scallops to reach reproductive maturity (Bowles 1989). The Manasota cultures did not practice this conservation strategy. Rather, they collected scallops as soon as the majority of the scallops reached collectible size, precisely during the period when modern authorities believe it is necessary to avoid exploitation in order to ensure the long-term productivity of the species. The question, then, is whether the Manasota strategy impacted the scallop populations so adversely that their numbers were reduced later in the year or, perhaps, in sub-sequent years.

Typically, changes in the size of animals recovered from archaeological sites are equated with environmental change or overexploitation (e.g., Broughton 1997; Bushnell 1965; Cumbaa 1976; Jackson et al. 2001; Jones et al. 2004; Quitmyer 1998; Quitmyer and Jones 2000; Wing 2001). More rarely are such changes linked to human choice (see Quitmyer 1998; Quitmyer and Jones 2000:165; Sandweiss 1996). Despite the fact that there appears to be a reduc-tion in size of individual scallops from the Preceramic Archaic into the Safety Harbor period (Figure 12-6), the slight differences in size among samples within each period can be attributed to differences in season of collection rather than to environmental change (e.g., Bushnell 1965). This suggests that the collection strategy did not result in overexploitation, at least to any irreparable degree. The amounts and sizes of scallops in the assemblages are more simply understood in terms of subsistence behavior associated with seasonality than in terms of overexploitation.

We have suggested that the early season scallop collection during the Manasota and Safety Harbor periods limited peoples' options. Such a strategy may have reduced the scallop populations to a point where late season collection was not practical. But that suggestion does not provide insight into the basis for this decision. The question remains, why did the Manasota people not wait until later in the year to collect the larger scallops as did the Archaic people? A myriad of scenarios could account for early-season targeting of scallops. Spring may have been an otherwise lean time requiring such intensification, making longer-term implications irrelevant. Or, summer and fall, when scallops were larger, may have been a time when other resources were bountiful and scallops were not an attractive resource. Or spring may been a time when the logistically foraging fishers left the village to obtain other resources elsewhere, leaving fishing and shellfish collection in the hands of those (e.g., children and older individuals) who could easily manage the shallow, warm waters where scallops were collected.

By suggesting that each strategy was a choice that resulted in long-term, sustainable patterns of scallop collection at each site, we are not concluding that the collectors were imbued with a conservation ethic. Other shellfish taxa were collected at these sites and some shellfish did decrease in size during the Archaic and Manasota periods, suggesting the possibility of over-harvesting (Quitmyer 1998:204). More likely, the choices in scallop-collection strategy had to do with other aspects of social and subsistence behavior.

CONCLUSION

Oysters are usually the dominate shellfish in coastal shell middens of the Gulf and Atlantic coasts of the southeastern United States. Scallops, although abundant, just happen to show us a more easily measured seasonal signature than do oysters and the other plant and animal remains. In either case, it may be misleading to characterize shell-bearing sites as scallop-collection sites or whelk-reduction stations or some other functional aspect of shellfish processing endeavors. Certainly, activities related to shellfish and subsistence occurred at these sites, but undoubtedly so did other social behaviors such as raising children, visiting neighbors, celebrating events, and holding rituals, the common occurrences of daily human life. These social functions may more usefully describe these sites, e.g., households, villages, ceremonial centers, and lend insight into the broader settlement and social organization of the greater society. But to arrive at these interpretations, minimally we have to determine when and for how long each site was occupied during the annual cycle.

Our seasonality data provide evidence of year-round occupation of the coastal region in southwest Florida in general, and at some sites in particular, beginning as early as the Preceramic Middle Archaic. It is clear that the tradition of sedentism supported by the rich estuarine environment but without a domestic food base appeared very early and continued into the 16th century A.D. in ever-changing subsistence strategies. In this study, we have tested earlier models of settlement which were based on no seasonal data whatsoever. The patterns identified here await the next challenge, and hopefully even more rigorous testing of seasonality.

ACKNOWLEDGMENTS

We wish to thank William Marquardt, director of the Southwest Florida Project, and Key Marco Developments, which funded the research on Horr's Island. We thank Piper Archaeological Research, Inc., and Bob Austin for allowing Russo the creative leeway to explore new seasonality techniques at Catfish Creek. Thanks also go to the Gulf Coast Heritage Foundation, the Florida Department of State, and marion Almy for providing support for work undertaken on the Shell Ridge midden. Thanks to the many archaeologists whose models of subsistence and settlement have offered students the many opportunities to test and expand. We give special thanks to Elizabeth Wing, whose many works set the examples and provided the inspiration for this and all our other zooarchaeological investigations.

REFERENCES

Almy, M., 1988, Phase I Archaeological Reconnaissance of Area "A" Oscar Scherer State Recreation Area, Sarasota County, Florida, manuscript on file, Florida Bureau of Archaeological Research, Tallahassee.

Ardren, T., Burger, B., and Sullivan, B. K., 2003, The Mystery River Point Site, Englewood, Sarasota County, Florida, *Florida Anthropologist* 56(1):47–55.

Austin, R. J., 1995, Life in a Coastal Hamlet: A Summary of What We Have Learned, in: *Yat Kitischee: A Prehistoric Coastal Hamlet, 100 B.C.–A.D. 1200* (R. J. Austin, ed.), Pinellas County Planning Department, Clearwater, Florida, pp. 219–229.

Austin, R. J., and Russo, M., 1989, Limited Excavations at the Catfish Creek Site (8SO608), Sarasota County, Florida, report submitted to Palmer Venture, Inc., by Piper Archaeological Research, Inc., St. Petersburg, Florida.

Barber, B. J., and Blake, N. J., 1983, Growth and Reproduction of the Bay Scallop, *Argopecten irradians*, at its Southern Distributional Limit, *Journal of Experimental Marine Biology and Ecology* 66:247–256.

Beiter, G. N., 2001, Salvage and Excavation of Bamboo Mound (8DA94), Dade County, Florida: a Multi-component Site, *Florida Anthropologist* 54(1):30–48.

Binford, L. R., 1980, Willow Smoke and Dogs' Tails: Hunter-Gatherer Settlement Systems and Archaeological Site Formation, *American Antiquity* 45(1):4–20.

Bowles, R., 1989, Scallops Lure Harvesters, *The Gainesville Sun*, August 11.

Broom, M. J., 1976, Synopsis of Biological Data on Scallops, *Food and Agriculture Organization of the United Nations Fisheries Synopsis* 114, Rome, Italy.

Broughton, J. M., 1997, Widening Diet Breadth, Declining Foraging Efficiency, and Prehistoric Harvest Pressure: Ichthyofaunal Evidence from the Emeryville Shellmound, California, *Antiquity* 71:845–862.

Bullen, R. P., and Bullen, A. K., 1961, The Summer Haven Site, *Florida Anthropologist* 14(1–2):1–16.

Bullen, R. P., and Bullen, A. K., 1976, The Palmer Site, *Florida Anthropological Society Publication* 8.

Bushnell, F. F., 1965, Preliminary Report on Ecology of Little Sarasota Bay Based on Shell Samples from the Palmer Site, manuscript on file, Florida Museum of Natural History, University of Florida, Gainesville.

Claassen, C. P., 1982, *Shellfishing Patterns: An Analytical Study of Prehistoric Shell from North Carolina Coastal Middens*, Ph.D. dissertation, Department of Anthropology, Harvard University, University Microfilms, Ann Arbor, Michigan.

Clark, G. R., II, 1979, Seasonal Growth Variations in the Shells of Recent and Prehistoric Specimens of *Mercenaria mercenaria* from St. Catherines Island, Georgia, *Anthropological Papers of the American Museum of Natural History* 51:161–179.

Clark, G. R., II, and Lutz, R. A., 1982, Seasonal Patterns in Shell Microstructure of *Mercenaria mercenaria* along the U.S. Atlantic Coast, *Geological Society of America Abstracts* 14:464.

Cockrell, W. A., 1970, *Glades I and Pre-Glades Settlement and Subsistence Patterns on Marco Island (Collier County, Florida)*, M.A. thesis, Department of Anthropology, Florida State University, Tallahassee.

Cumbaa, S. L., 1976, A Reconsideration of Freshwater Shellfish Exploitation in the Florida Archaic, *Florida Anthropologist* 29(1):49–59.

Deith, M. R., 1983, Molluscan Calendars: The Use of Growth-line Analysis to Establish Seasonality of Shellfish Collection at the Mesolithic Site of Morton, Fife, *Journal of Archaeological Science* 10:423–440.

Dickel, D., 1992, An Archaeological and Historical Survey of Bonita Spings, Parcel Three, Lee County, Florida, Archaeological and Historical Conservancy Technical Report 43, Miami, manuscript on file, Florida Bureau of Archaeological Research, Tallahassee, Florida.

Dunbar, J. S., Webb, S. D., and Faught, M. K., 1992, Archaeological Sites in the Drowned Tertiary Karst Region of the Eastern Gulf of Mexico, in: *Paleo-shorelines and Prehistory: An Investigation in Method* (L. L. Johnson, ed.), CRC Press, Boca Raton, Florida, pp. 117–146.

Faught, M. K., and Donoghue, J. F., 1997, Marine Inundated Archaeological Sites and Paleofluvial Systems: Examples from a Karst-controlled Continental Shelf Setting in Apalachee Bay, *Geoarchaeology* 12(5):417–458.

Faught, M. K., and Latvis, J., 1999, *Report of the 1999 Field Operations: PaleoAucilla Prehistory Project*, Underwater Prehistoric Archaeology in Apalachee Bay, Research Reports 1, Program in Underwater Archaeology, Florida State University, Tallahassee.

Griffin, J. W., 2002, *Archaeology of the Everglades*, University Press of Florida, Gainesville.

Houck, B. A., 1996, Archaeological Excavations at 8LL717, Bonita Springs, Lee County, Florida, Archaeological and Historical Conservancy Technical Report 78 (revised), Miami, manuscript on file, Florida Bureau of Archaeological Research, Tallahassee, Florida.

Hudson, J. H., Shinn, E., Halley, R., and Lidz, B., 1976, Sclerochronology: A New Tool for Interpreting Past Environments, *Geology* 4:361–364.

Jackson, J. B. C., Kirby, M. X., Berger, W. H., Bjorndal, K. A., Botsford, L. W., Bourque, B. J., Bradbury, R. H., Cooke, R., Erlandson, J., Estes, J. A., Hughes, T. P., Kidwell, S., Lange, C. B., Lenihan, H. S., Pandolfi, J. M., Peterson, C. H., Steneck, R. S., Tegner, M. J., and Warner, R. R., 2001, Historical Overfishing and the Recent Collapse of Coastal Ecosystems, *Science* 293:629–638.

Jones, D. S., Quitmyer, I. R., and Andrus, C. F. T., 2004, Seasonal Shell Growth and Longevity in *Donax variabilis* from Northeastern Florida: Evidence from Oxygen Isotopes, *Journal of Shellfish Research* 23(3):707–714.

Lee, A. R., Beriault, J. G., Belknap, J., Buschman, W. M., Thompson, J. W., and Johnson, C. B., 1998, Heineken Hammock, 8Cr231: A Late Archaic Corridor Site in Collier County, *Florida Anthropologist* 51(4):223–239.

Luer, G., and Almy, M., 1982, A Definition of the Manasota Culture, *Florida Anthropologist* 35(1):34–58.

Marquardt, W. H., 1999, Useppa Island in the Archaic and Caloosahatchee Periods, in: *The Archaeology of Useppa Island* (W. H. Marquardt, ed.), *University of Florida Institute of Archaeology and Paleoenvironmental Studies Monograph* 3, University of Florida, Gainesville, pp. 77–98.

Masson, M., Carr, R. S., and Sandler, D., 1988, The Taylor's Head Site (8Bd74): Sampling a Prehistoric Midden on an Everglades Tree Island, *Florida Anthropologist* 41(3):336–350.

McMichael, A. E., 1982, *A Cultural Resource Assessment of Horr's Island, Collier County, Florida*, M.A. thesis, Department of Anthropology, University of Florida, Gainesville.

Milanich, J. T., 1991, Central Peninsular Gulf Coast: 500 B.C.–A.D. 900, in: *Florida's Comprehensive Historic Preservation Plan* (C. Payne and J. T. Milanich, eds.), Florida Bureau of Archaeological Research, Tallahassee, pp. 89–95.

Milanich, J. T., 1994, *Archaeology of Precolumbian Florida*, University Press of Florida, Gainesville.

Milanich, J. T., and Fairbanks, C. H., 1980, *Florida Archaeology*, Academic Press, New York.

Milner, G. R., 2004, *The Moundbuilders: Ancient Peoples of Eastern North America*, Thames & Hudson, London.

Mowers, B., and Williams, W. B., 1972, The Peace Camp Site, Broward County, Florida, *Florida Anthropologist* 25(1):1–20.

Newman, C., 1993, The Cheetum Site: An Archaic Burial Site in Dade County, Florida, *Florida Anthropologist* 46(1):1–37.

Newsom, L. A., 1991, Horr's Island Archaeobotanical Research, in: *Final Report on Horr's Island: The Archaeology of Archaic and Glades Settlement and Subsistence Patterns* (M. Russo, ed.), University of Florida Institute of Archaeology and Paleoenvironmental Studies, Gainesville, pp. 591–644.

Newsom, L. A., 1992, Research with Prehistoric Plant Remains, in: *A Window to the Past: An Archaeological Discovery at Historic Spanish Point Sarasota County, Florida* (W. H. Marquardt, ed.), Archaeological Consultants, Sarasota, Florida, pp. 83–100.

Newsom, L. A., 1998, Archaeobotanical Research at Shell Ridge Midden, Palmer Site (8SO2), Sarasota County, Florida, *Florida Anthropologist* 51(4):207–222.

Perlman, S. M., 1973, *Pecten irradians Growth Patterns: Their Application to Archaeological Economic Analysis*, M.A. thesis, Department of Anthropology, University of Massachusetts, Boston.

Quitmyer, I. R., 1992, Determination of Habitat Exploitation and Seasonality from Selected Zoological Indicators, in: *A Window to the Past: An Archaeological Discovery at Historic Spanish Point Sarasota County, Florida* (W. H. Marquardt, ed.), Archaeological Consultants, Sarasota, Florida, pp. 35–80.

Quitmyer, I. R., 1998, Zoological Indicators of Habitat Exploitation and Seasonality from the Shell Ridge Midden, Palmer Site (8SO2), Osprey, Florida, *Florida Anthropologist* 51(4):193–205.

Quitmyer, I. R., 2002, Faunal Analysis of Remnant Mound Column Samples, in: *Archaeological Investigation of DeSoto National Memorial* (M. Schwadron, ed.), *National Park Service Southeast Archeological Center Technical Reports* 8, Tallahassee, Florida, pp. 158–202.

Quitmyer, I. R., and Jones, D. S., 2000, The Over-exploitation of Hard Clams (*Mercenaria* spp.) from Five Archaeological Sites in the Southeastern United States, *Florida Anthropologist* 53(2–3):160–167.

Quitmyer, I. R., Jones, D. S., and Arnold, W. S., 1997, The Sclerochronology of Hard Clams, *Mercenaria* spp., from the South-Eastern U.S.A: A Method of Elucidating the Zooarchaeological Records of Seasonal Resource Procurement and Seasonality in Prehistoric Shell Middens, *Journal of Archaeological Science* 24:825–840.

Quitmyer, I. R., and Massaro, M. S., 1999, Seasonality and Subsistence in a Southwest Florida Estuary: A Faunal Analysis of Precolumbian Useppa Island, in: *The Archaeology of Useppa Island* (W. H. Marquardt, ed.), *University of Florida Institute of Archaeology and Paleoenvironmental Studies Monograph 3*, University of Florida, Gainesville, pp. 99–128.

Reitz, E. J., Quitmyer, I. R., Hale, H. S., Scudder, S. J., and Wing, E. S., 1987, Application of Allometry to Zooarchaeology, *American Antiquity* 52:304–317.

Rhoads, D. C., and Lutz, R. A., 1980, *Skeletal Growth of Aquatic Organisms*, Plenum Press, New York.

Russo, M., 1991a, *Archaic Sedentism on the Florida Coast: A Case Study from Horr's Island*, Ph.D. dissertation, Department of Anthropology, University of Florida, Gainesville.

Russo, M., 1991b, A Method for the Measurement of Season and Duration of Oyster Collection: Two Case Studies from the Prehistoric South-east U.S. Coast, *Journal of Archaeological Science* 18(2):205–231.

Russo, M., 1992, Chronologies and Cultures of the St. Mary's Region of Northeast Florida and Southeast Georgia, *Florida Anthropologist* 45(2):107–126.

Russo, M., 1994, Why We Don't Believe in Archaic Ceremonial Mounds and Why We Should: The Case from Florida, *Southeastern Archaeology* 13:93–109.

Russo, M., 1998, Measuring Sedentism with Fauna: Archaic Cultures Along the Southwest Florida Coast, in: *Seasonality and Sedentism: Archaeological Perspectives from Old and New World Sites* (T. R. Rocek and O. Bar-Yosef, eds.), Peabody Museum of Archaeology and Ethnology, Harvard University, Cambridge, Massachusetts, pp. 143–164.

Russo, M., 2004, Measuring Shell Rings for Social Inequality, in: *Signs of Power: The Rise of Cultural Complexity in the Southeast* (J. L. Gibson and P. J. Carr, eds.), University of Alabama Press, Tuscaloosa, pp. 26–70.

Russo, M., 2005, In Search of Missing Ancestors, paper presented at the symposium "Florida's 'Fierce' People: Tracing the Calusa Indians from Dominance to Demise," the Florida Humanities Council and the National Endowment for the Humanities, Naples, Florida.

Russo, M., 2006, *Archaic Shell Rings of the Southeast U.S., National Historic Landmarks Historic Centers*, National Park Service, Southeast Archeological Center, Tallahassee, Florida.

Russo, M., and Heide, G., 2001, Shell Rings in the Southeast US, *Antiquity* 75(289):491–492.

Russo, M., and Heide, G., 2004, The Emergence of Pottery in South Florida, in: *Early Pottery: Technology, Style, and Interaction in the Lower Southeast* (R. Saunders and C. Hays, eds.), University of Alabama Press, Tuscaloosa, pp. 105–128.

Russo, M., and Powell, A., 2005, Seasonality of Shellfish from Remnant Mound, DeSoto National Monument, manuscript on file, National Park Service Southeast Archeological Center, Tallahassee.

Russo, M., and Quitmyer, I. R., 1996, Sedentism in Coastal Populations of South Florida, in: *Case Studies in Environmental Archaeology*, 1st ed. (E. J. Reitz, L. A. Newsom, and S. J. Scudder, eds.), Plenum Press, New York, pp. 215–231.

Russo, M., and Saunders, R., 1999, *Identifying the Early Use of a Coastal Fisheries and the Rise of Social Complexity in Shell Rings and Arcuate Middens on Florida's Northeast Coast*, final grant report, National Geographic Society, copy on file at National Park Service, Southeast Archeological Center, Tallahassee, Florida.

Russo, M., and Ste. Claire, D., 1992, Tomoka Stone: Archaic Period Coastal Settlement in East Florida, *Florida Anthropologist* 45(2):336–346.

Sandweiss, D. H., 1996, Environmental Change and Its Consequences for Human Society on the Central Andean Coast: A Malacological Perspective, in: *Case Studies in Environmental Archaeology*, 1st ed. (E. J. Reitz, L. A. Newsom, and S. J. Scudder, eds.), Plenum Press, New York, pp. 127–146.

Sassaman, K. E., 1993, *Early Pottery in the Southeast: Tradition and Innovation in Cooking Technology*, University of Alabama Press, Tuscaloosa.

Saunders, R., 2004, The Stratigraphic Sequence at Rollins Shell Ring: Implications for Ring Function, *Florida Anthropologist* 57(4):249–270.

Scarry, C. M., and Newsom, L. A., 1992, Archaeobotanical Research in the Calusa Heartland, in: *Culture and Environment in the Domain of the Calusa* (W. H. Marquardt, ed.), *University of Florida Institute of Archaeology and Paleoenvironmental Studies Monograph* 1, University of Florida, Gainesville, pp. 375–401.

Schwadron, M., 2002, *Archeological Investigation of DeSoto National Memorial, Southeastern Archeological Center Technical Reports* 8, Tallahassee, Florida.

Schwadron, M., 2005, Archeological Investigation of the Eastern Everglades Tree Islands Sites, Everglades National Park, paper presented at the 57th Annual Meeting of the Florida Anthropological Society, Gainesville.

Smith, B. D., 1986, The Archaeology of the Southeastern United States: From Dalton to DeSoto, 10,500 B.P. to 500 B.P., *Advances in World Archaeology* 5:1–92.

Steponaitis, V., 1986, Prehistoric Archaeology in the Southeastern United States 1970–1985, *Annual Review of Anthropology* 15:363–404.

Torrence, C. McP., 1992, More than Midden: Opening the Shell around Late Archaic Lifeways, paper presented at the 49th Annual Meeting of the Southeastern Archaeological Conference, Little Rock, Arkansas.

Torrence, C. McP., 1999, The Archaic Period on Useppa Island: Excavations on Calusa Ridge, in: *The Archaeology of Useppa Island* (W. H. Marquardt, ed.), *University of Florida Institute of Archaeology and Paleoenvironmental Studies Monograph* 3, University of Florida, Gainesville, pp. 23–76.

Wheeler, S., Snapp, A., and Lewis, S., 1997, Phase II Report: Archaeological Investigations of the Scheurich Midden Site (8PB9261) Jupiter, Palm Beach County, Florida, S. P. Lewis Historic Preservation Services, Fort Lauderdale, Florida, manuscript on file, Florida Bureau of Archaeological Research, Tallahassee, Florida.

Widmer, R. J., 1988, *The Evolution of the Calusa: A Nonagricultural Chiefdom on the Southwest Florida Coast*, University of Alabama Press, Tuscaloosa.

Wing, E. S., 2001, The Sustainability of Resource Use by Native Americans on Four Caribbean Islands, *International Journal of Osteoarchaeology* 11:112–126.

Chapter **13**

Reconstructing Subsistence in the Lowland Tropics: A Case Study from the Jama River Valley, Manabí, Ecuador

DEBORAH M. PEARSALL

One of the archaeological challenges in the tropical lowlands is recovering biological remains. Without the remains of plants and animals utilized by past cultures, many important issues concerning human life in the tropics cannot be addressed. One such issue is the origin and dispersal of domesticated plants and animals (Piperno and Pearsall 1998). To address, for example, the questions of when maize (*Zea mays*) was first introduced from Mesoamerica into the lowlands of Central America and northern South America and when it became important in subsistence requires not only evidence of maize, but some way to evaluate its importance in relation to resources such as root crops and tree fruits. Perhaps the most striking point revealed by this study is that the richness of the resource base remained very consistent through time. Maintaining a broad subsistence base, in the lowland forest environment and early subsistence system, would be highly adaptive in a setting subjected to catastrophic events. The more crops and other resources used, the more likely some would survive.

Environmental archaeologists face special challenges in the lowland tropics. Some are related to the assumption that infertile soils precluded the development of intensive agriculture and limited cultural evolution (see Meggers

255

1954, 1971; Roosevelt 1991:103–105; Steward 1948, 1949). Many of the tropical lowland cultural traits were interpreted as diffusions from the Andes or the circum-Caribbean basin that had degenerated in eastern South America. While Meggers (1971) acknowledged that the extensive alluvial soils of the lowlands were capable of supporting dense populations with complex social organization, most advanced cultural traits were considered derived from elsewhere, rather than the result of local innovation. This view precludes the discovery of complex cultural practices in the lowlands. When archaeologists find evidence for agriculture, extensive landscape modification, or sedentism in a tropical lowland site that predates similar sites in more arid regions, they must battle the perception that their results are wrong or misdated simply because the site is in the lowlands.

The tropical lowland setting itself also presents challenges (Pearsall 1995a; Stahl 1995). It is difficult to locate archaeological sites, especially early ones, to recover biological remains from those sites, and to identify and interpret the remains that are available. The lowland tropics are not a uniform environment (Piperno and Pearsall 1998). Rainfall, soils, and vegetation vary dramatically. Researchers interested in early tropical agriculture must often search in the most difficult conditions an environmental archaeologist can imagine: sites that are buried under meters of alluvium, that can be tested only with small excavations, that have experienced thousands of years of wetting and drying cycles to break up materials, and that are located in soils with high clay content that refuse to release remains. Finally, the analyst must identify fragments from highly diverse biological communities. The key to reconstructing subsistence is to utilize multiple lines of evidence. These lines include quantitative data, when such data are available, and qualitative data (what taxa are present and in what combinations) that place biological remains in their cultural and environmental contexts. The last point means that the environmental archaeologist must understand the archaeology of the region in addition to its environment.

A case study from the tropical lowlands of western Ecuador illustrates this approach, focusing on the role of maize in subsistence in the Jama River valley (Figure 13-1). Macrobotanical remains and phytoliths are compared and combined; the importance of studying on-site and off-site samples is illustrated, and the botanical data are placed in a context of regional site sampling and archaeofaunal recovery programs. This study does not consider pollen because it is not well preserved in Jama archaeological contexts (Zimmerman 1994).

Archaeologists' views on the role of maize in Ecuadorian subsistence have changed over the past 40 years. It was once assumed that maize was introduced during the Late Formative period (circa 1000 B.C.) (Table 13-1) and became the mainstay of diets thereafter (Meggers 1966). With further research, it was found that maize was introduced as early as 6000–4600 B.C. Nonetheless, it was proposed that the crop was relatively unimportant in coastal Ecuadorian subsistence until the Late Formative or thereafter (Pearsall 1994b, 1995b, 1999, 2003).

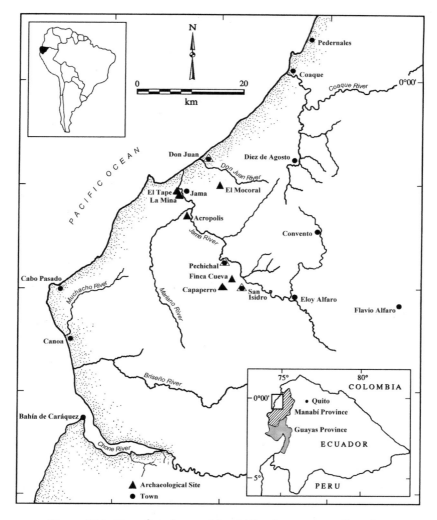

Figure 13-1. Map of study area, Manabí Province, western Ecuador.

This model proposes that the circa 1700 B.C. expansion of populations from southwest Guayas province, where sites with early maize are found, was fueled by a search for new agricultural lands, or in response to agricultural failures, by peoples increasingly dependent on domesticated plants. The limited studies of carbon isotopes in human bone support this scenario by documenting that maize had little dietary impact during the Early Formative. In addition, charred maize macroremains do not become common at sites until after circa 1800 B.C. (Pearsall 2003). However, the transition of maize from low levels of use to a dietary staple is poorly documented. One of the goals of the Jama

Table 13-1. Cultural Chronology of the Jama River Valley

Culture	Phase	Date[a]	Period
Campace (?)	Muchique 5	A.D. 1430–1640	Colonial
	Spanish Conquest (A.D. 1532)		
Jama-Coaque II	Muchique 4	A.D. 1290–1430	Integration
Jama-Coaque II	Muchique 3	A.D. 880–1260	Integration
Jama-Coaque II	Muchique 2	A.D. 420–790	Integration
	Tephra III (immediately prior to A.D. 420)		
Jama-Coaque I	Muchique 1	240 B.C.–A.D. 90[b]	Regional Developmental
	Hiatus (lasts 250–500 years)		
	Tephra II (750 B.C.)		
Chorrera	Tabuchila	1300–750 B.C.	Late Formative
	Hiatus (lasts circa 500 years)		
Valdivia	Late Piquigua	ends 1880 B.C.	Early Formative
	Tephra I		
	Early Piquigua	begins 2030 B.C.	Early Formative

[a] Dates are modal age ranges cal. B.C./A.D. based on 37 radiocarbon dates (Zeidler et al. 1998).
[b] Zeidler et al. (1998) suggest, on stratigraphic grounds, that Muchique 1 may be longer by 200 years.

Valley Archaeological–Paleoethnobotanical Project was to gather data relevant to the evolution of coastal subsistence from circa 2000 B.C. to Spanish colonization in A.D. 1532 through valley-wide survey and site testing (Zeidler and Pearsall 1994).

METHODS

Macrobotanical Remains

The botanical data are of two types. One is charred macrobotanical remains recovered by water flotation from archaeological excavations (described in Pearsall 1994a, c). Charred botanical remains were analyzed at the University of Missouri Paleoethnobotany Lab (MU) following standard procedures (Hastorf and Popper 1988; Pearsall 2000). Wood count per 10 liter of floated soil is used to examine the overall abundance of material by phase. This measure can reveal a taphonomic phenomenon: whether abundance declines with depth. Ratios of each class of food remains (maize, common bean [*Phaseolus vulgaris*], tree fruits, and tuber-roots) per 10 liters of floated soil (maize/10 liters, tree fruit/10 liters, bean/10 liters, tuber-root/10 liters) allow individual remains to be tracked over time, revealing relative abundances.

The ratios of maize to wood by phase (maize:wood), fruit/bean/tuber to wood by phase (other:wood), and maize to other foods by phase (maize:other food) serve two purposes. First, preservation differences caused by deposit depth or frequency of burning are adjusted for by quantifying food remains relative to wood abundance. In other words, if a context has very high wood remains, indicating frequent burning, preservation of all food remains is more likely than in contexts with less frequent burning. Second, since the focus of this discussion is on maize, it is useful to contrast its occurrence to that of all other food remains as a group.

Finally, the percentage presence (ubiquity) of each food class is examined by phase. This simple measure is the percentage of all samples in which a food occurs (i.e., if beans are in 2 of 4 samples examined from Late Muchique 3, then the percentage presence is 50%).

To make a preliminary assessment of subsistence patterning, remains are classified into four groups: (1) maize; (2) domesticated common bean; (3) tree fruit, including palm and other wild/tended tree taxa; and (4) tuber-root. The tuber-root class includes rootlike fragments not identified to taxon. There are at least four types of tubers (arrowroot [*Maranta arundinacea*], llerén, [*Calathea allouia*], achira [*Canna edulis*], and manioc [*Manihot esculenta*]), and at least 17 kinds of tree fruits (e.g., taxa in the palm [Arecaceae], sapote [Sapotaceae], and guanábana [Annonaceae] families, among others) (Pearsall 2004). Small seeds are relatively rare, and none of the species recovered appear to be significant food sources.

Differences in durability among botanical materials is a problem. Tuber-root remains have a fragile edible portion and little nonedible residue and are the least likely to survive. Tree fruits, with dense edible "meats" and woody hulls, are the most likely to survive. Beans (dense seed, ephemeral inedible fruit) and maize (fragile kernel, relatively sturdy inedible cob) fall in between. In broad terms, tuber-root and bean remains were probably more important than suggested by the quantity of remains recovered. Maize and tree fruits were perhaps less so, especially if hulls and cobs were burned as fuel.

The Jama macrobotanical data have limitations. Because the more ancient deposits often are buried under meters of subsequent occupation, data for the early time periods are smaller in numbers of sites, in areas of sites tested, and in abundance of botanical remains. Thus, Formative samples, i.e., Piquigua and Tabuchila, cannot be compared quantitatively to the later, better-preserved samples of Muchique 1–4 (Table 13-1). The latter materials appear to be much less impacted by taphonomic factors. Thus, the Jama macrobotanical data do not provide a direct means of testing whether the Late Formative marks a transition to agriculture in the coastal lowlands. Patterning within the Muchique sequence, however, does provide insight into the process.

Phytoliths

The other type of botanical remains presented here is phytoliths, microscopic opal silica bodies found in plants (e.g., Meunier and Colin 2001; Pearsall 2000; Pearsall and Piperno 1993; Pinilla et al. 1997; Piperno 1988, 2006; Rapp and Mulholland 1992). Following standard sampling procedures, soil samples were taken from profiles in archaeological sites and from the Río Grande profile, a natural stratigraphic profile in the Jama River bank near San Isidro (Pearsall 2000). Identifications were based on comparisons with specimens collected during the Jama Valley Project (Pearsall 2004; for illustrations see the MU phytolith web page: HTTP://web.missouri.edu/~umcasphyto) and maize was identified using the cross-body size and shape method developed by Pearsall (1979, 1982); Piperno (1984). The Río Grande profile samples were analyzed by Veintimilla (1998, 2000), who used absolute counts to express phytolith abundance. All archaeological profiles were analyzed by relative counts, with a sum of 500 phytoliths used, or until an entire slide was scanned.

The numbers of phytoliths produced by plant species vary. Plant families such as the grasses and palms produce abundant phytoliths, and others many fewer; therefore, numbers of phytoliths do not correspond directly with the abundance of those plants in the environment or in an archaeological deposit. Some plants will be consistently over-represented and some under-represented. Data are interpreted by comparing the phytolith types and abundances in an ancient stratum to phytolith assemblages produced by known types of vegetation (e.g., grassland, dry tropical forest) or activities (e.g., storing food), or by using the presence of indicator species to signal the presence of a vegetation formation or activity.

DESCRIPTION OF THE STUDY AREA AND SITES

The project was initiated by Zeidler and Pearsall in the mid-1980s as an inter-disciplinary investigation of long-term sociopolitical change in northern Manabí province (Pearsall 2004; Zeidler and Pearsall 1994). A major focus of the project was investigation of agricultural intensification in the lowland tropics. The ultimate goal is to test models of agricultural evolution and cropping intensification by using biological data recovered from sites in the valley.

The Jama drainage is one of several river valleys on the Pacific coast of Ecuador (Figure 13-1). The Jama's headwaters are in the low hills of the coastal cordillera; the main channel extends 75 km inland. The valley occupies a transitional area from dry coastal to wetter inland conditions. Although the valley is largely deforested today, the natural vegetation cover was dry tropical forest along the coast and humid pre-montane tropical forest farther inland (Zeidler and Kennedy 1994). The landscapes of the valley include the mangrove estuary at the mouth, the floodplains of the Jama and its major tributaries, and the low coastal hills. Floodplain soils are typically deep deposits of silt and silty clay loam favored for agriculture, and the soils of the uplands are more variable in composition and depth, with mollisols near San Isidro as productive for maize as floodplain soils (Pearsall 2004).

An important feature of the chronology (Table 13-1) is the presence throughout the valley of three discrete tephra layers: thick, redeposited layers of wind-transported volcanic ash and pumice. These tephra deposits originated from volcanic eruptions in the western Andes and were blown onto the coastal plain by the prevailing easterly winds (Isaacson 1994). Originally blanketing the landscape, the tephras were redeposited in coastal watersheds by erosion from hillslopes into valley bottoms, where deposits up to 2 m thick accumulated (Donahue and Harbert 1994). Ash deposition of this magnitude has a severe impact, damaging forests and destroying crops, and can lead to long-term disruption of agricultural systems based on cultivation of alluvial lands, where ash deposits are thickest (Isaacson 1994; Pearsall 2004).

In the case of the Jama valley, understanding the impact of tephra deposition is essential for understanding human adaptation in the valley. Although the small Piquigua populations survived the first tephra fall (Tephra I), the valley was abandoned within 100 years of the event and remained unoccupied for some 550 years. Tephra II ended the Tabuchila occupation of the valley, resulting in a major cultural discontinuity that is seen in the Jama valley and elsewhere in northwest Ecuador (Isaacson 1994; Isaacson and Zeidler 1999; Zeidler and Isaacson 2003). The valley was reoccupied during the Muchique 1 phase. Tephra III created the deepest deposits but had relatively less impact on cultural continuity in the valley than Tephra II. A short abandonment of perhaps a generation occurred after Tephra III, followed by reoccupation in the Muchique 2 phase.

The macrobotanical data discussed here are from seven chronological phases or subphases at five sites (Table 13-2). The Muchique 1 phase is represented by materials from the Don Juan site, located at the mouth of the Don Juan River in the lower valley. Feature 7 from this site is examined separately, since it represents the earliest Muchique 1 occupation. An occupation floor and a deep storage pit provide data for the Muchique 2 phase. The El Tape feature is about 200 years younger than the pit from the Pechichal site, providing an opportunity to look at change within the Muchique 2 phase. Both features have abundant remains. The El Tape site is in the lower valley, Pechichal in the upper. Three sites provide

Table 13-2. Macrobotanical Data Base, in Chronological Order

Site	Strata	Phase	Date[a]	Liters	Wood[b]	Maize[b]	Other food[b]
Don Juan	Dep. 1, 2	Post-Muchique 3		302.5	853	208	57
Acropolis		Late Muchique 3	A.D. 1150	94	395	265	44
Don Juan	Dep. 3	Late Muchique 3		151	592	162	24
Capaperro	F4	Early Muchique 3	A.D. 755	51	138	162	48
Pechichal	F5	Muchique 2	A.D. 545	831.5	2702	3568	1922
El Tape	F4	Early Muchique 2	A.D. 340	184.5	8863	647	43
		Tephra III					
Don Juan	Dep. 4	Muchique 1		190	331	85	11
Don Juan	F7	Early Muchique 1	480 B.C.	18	55	9	–

[a] Calibrated dates provided by J. Zeidler (1993, personal communication).
[b] Data for wood, maize, and other food are presented as counts.

Muchique 3 materials: the Capaperro site in the upper valley dates to the early portion of this phase; Don Juan and the Acropolis, both in the lower valley, are later and roughly contemporary. No samples clearly attributable to Muchique 4 were available for this analysis, so the upper two deposits at the Don Juan site that postdate Muchique 3 (Strata 1 and 2) were combined to give some indication of later subsistence patterns.

Phytolith data are from the Río Grande profile (the natural stratigraphic cut in the Jama River bank near San Isidro) and from cultural deposits in San Isidro itself. Both natural and cultural profiles extend into Formative period deposits: Piquigua, in the case of the Río Grande natural profile, and Tabuchila for the San Isidro site. From the perspective of investigating human uses of plants, the advantages of sampling an off-site area such as the Jama River bank are twofold. First, such a profile shows which phytoliths were deposited by the action of water and in situ decay of vegetation. Comparing such a profile with an archaeological deposit permits the analyst to separate naturally occurring taxa from those brought into the site and used by humans. Second, because humans often alter the landscape (e.g., preparation of agricultural fields involves felling trees), an off-site stratigraphic cut may capture vegetation changes related to human presence in the landscape.

RESULTS

Macrobotanical Remains

Looking first at the overall pattern of abundance of charred wood by phase (Figure 13-2A), it is clear that the Early Muchique 2 sample is unique in terms of the great quantity of wood preserved. If Early Muchique 2 is omitted from the graph showing wood count/10 liters of soil, variability in wood abundance during the other phases is easier to assess (Figure 13-2B). Abundance is not a simple product of the age of sediments, suggesting that taphonomic phenomena have not biased the overall plant assemblage to a great extent. Muchique 1, however, clearly has less material than the later phases, suggesting that sample bias may be greatest in this phase. Under-representation of more fragile remains, especially tubers, is the likely result, as is lower richness or numbers of kinds of taxa.

Figure 13-3 illustrates the abundance of each class of food remains per 10 liters of floated soil. The most striking patterns are that maize is the most abundant food in all phases (Figure 13-3A) and that absolute abundances of food remains are highly variable among the phases (Figure 13-3B). Abundances of beans and tuber-root fragments vary by phase, with tubers being more abundant than beans in the Late Muchique 3 sample from the Acropolis. Because tuber remains are extremely fragile, this finding is significant.

Given the variability in burning activity shown in wood abundance by phase (Figure 13-2), a reliable method of examining change over time in the

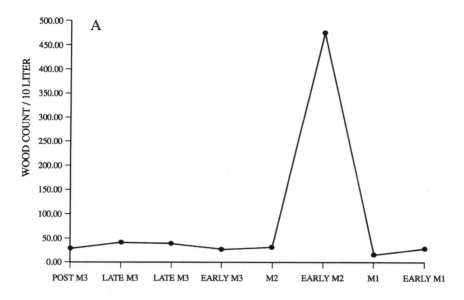

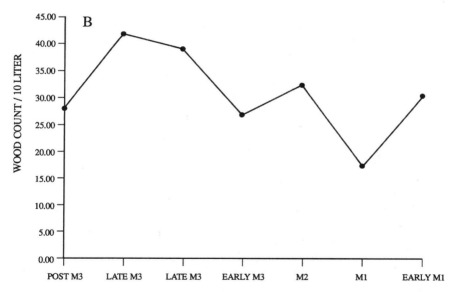

Figure 13-2. Wood count (A) and wood count excluding the Early Muchique 2 phase (B), per 10-liter flotation sample.

relative importance of maize and other foods is to compare the data using maize:wood, other food:wood ratios (Figure 13-4A), and maize:other food (Figure 13-4B) ratios. At first glance, Figure 13-4A seems to document a dramatic increase in the abundance of maize between the Early and Late

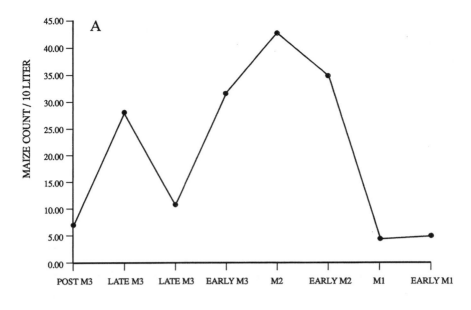

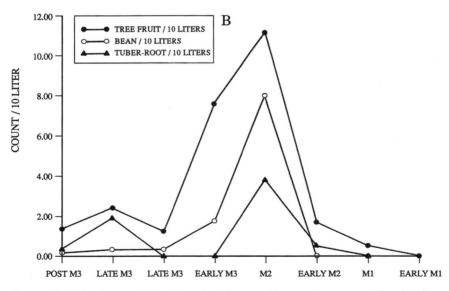

Figure 13-3. Maize count (A) and tree fruit, bean, and root-tuber counts (B) per 10-liter flotation sample.

Muchique 2 phases, as reflected in the contents of these two features with excellent preservation. However, comparing maize and other foods directly (maize:other foods Figure 13-4B) suggests that maize abundance declines relative to that of other foods between these phases.

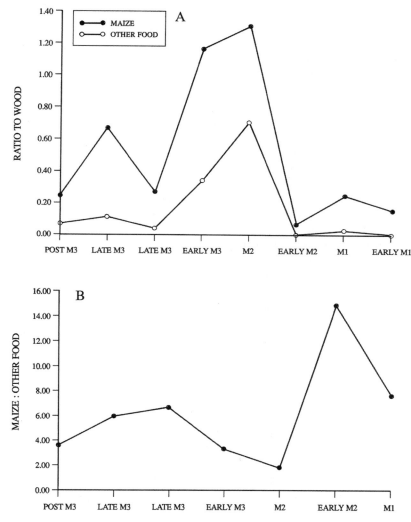

Figure 13-4. Ratios of maize to wood and all other foods to wood (A) and ratios of maize to all other foods (B), by counts.

What does a closer look at the data reveal about this time period? First, food remains are a high percentage of charred materials in the Late Muchique 2 sample (67% foods, 33% wood), and the Early Muchique 2 sample is mostly wood (7% foods, 93% wood). In comparing these two phases, then, all food remains increased markedly. The question becomes: which shows the greatest increase, maize or other foods? Although maize increases by a factor of 1.2 when measured per 10-liter sample (Figure 13-3A), there is a 10-fold increase in other foods (Figure 13-3B). Similarly, the change in the other food:wood ratio is much greater than Figure 13-4A suggests: from 0.005 to 0.71, a 142-fold increase, as compared to a 19-fold increase for maize. It is this difference that produces the declining maize:other food ratio (Figure 13-4B). Although maize remains important in Late Muchique 2 (65% of all food remains recovered), the contribution of tree fruits, and to a lesser extent tuber-root foods and beans, increases significantly. The abundance of all food remains in the Late Muchique 2 sample suggests an overall intensification in food production/procurement. It is significant that the large storage feature tested at the Pechichal site was only one of a number of similar features at this locality.

A further indication that Muchique 2 is a critical period for understanding subsistence change in the valley is the finding that maize apparently increased in abundance from Muchique 1 to Early Muchique 2. Maize abundance per 10 liters of soil increases 7.8-fold, in comparison to a 4-fold increase in other foods, and the ratio of maize:other foods doubles, from 7.73 to 15.05. Because the Muchique 1 sample may be biased by poorer preservation, however, this result must be confirmed by further tests.

Finally, it is useful to consider the same set of data using a very conservative approach, ubiquity (Figure 13-5). Maize is ubiquitous in all phases. Some maize occurred in each sample examined, with the exception that 10% of the Muchique 1 samples lacked maize. This finding demonstrates that maize production and use was widespread in the valley throughout the Muchique phase. This is not the case for the other food remains; tree fruits, although less

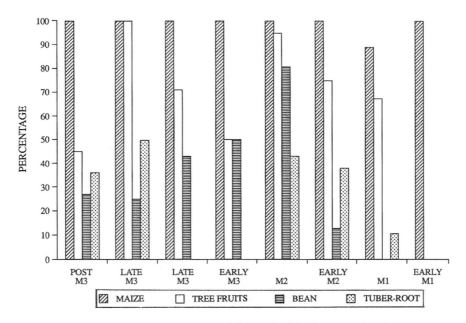

Figure 13-5. Percentage presence (ubiquity) of food remains by phase.

subject to postdepositional destruction, are far from ubiquitous. This finding is another indicator that the importance of tree fruits is less than is indicated by the preserved remains.

Phytoliths

Figure 13-6, a phytolith diagram from area V/A1 at the San Isidro site, illustrates one way in which phytolith data are displayed. The percentage occurrences of each identified taxon based on relative counts are displayed as a series of histograms. Data also may be presented in tabular form; Table 13-3 shows the relative abundances based on absolute counts of forest, open area, grass, and biogenic indicator phytoliths from the Río Grande natural stratigraphic profile (adapted from Veintimilla 1998).

A number of vegetation changes are documented in the Río Grande profile (Table 13-3; Veintimilla 1998, 2000). First, there is a clear pattern of decline in forest cover over time. This decline appears to begin before people settled in the valley. The percentage of phytoliths from forest taxa declines from 71% to 59% in the pre-occupation levels, then remains steady throughout the strata contemporary with the first occupation of the valley by Piquigua peoples. It is possible that our understanding of when people first occupied the valley is incorrect and that earlier sites are buried under river alluvium. Significantly, this reduction in forest cover predates the first tephra fall. Forest cover declines again in Tabuchila times, and still further in Muchique 1. Only 25% of phytoliths recovered from the lower Muchique 1–age stratum (MU Lab No. 871) came from forest taxa, compared to 59–71% in the pre-occupation strata. This decline represents significant landscape modification. Some limited forest regeneration

Table 13-3. Phytolith Data from Selected Strata in the Río Grande Profile[a]

Associated phase[b]	Percentages of phytoliths by type			
	Forest	Open	Grass	Biogenic
Postcontact[c] (877)	44.4	0.8	19.2	35.6
Muchique 2 and after[d] (875)	37.2	3.0	54.5	5.4
Tephra III (873)	77.0	–	23.0	–
Muchique 1 (872)	34.2	2.0	57.3	5.1
Muchique 1 (871)	25.3	0.7	74.0	–
Tephra II (not sampled)	–	–	–	–
Tabuchila (870)	39.1	6.6	40.0	14.3
Tephra I (not sampled)	–	–	–	–
Piquigua (869)	59.6	9.9	–	13.9
Piquigua (867)	55.4	32.1	6.8	5.6
Pre-occupation[e] (865)	59.0	13.7	7.9	19.4
Pre-occupation[e] (863)	71.4	8.2	10.2	10.2

[a] Adapted from Veintimilla (1998).
[b] The numbers in parentheses are the laboratory sample numbers.
[c] Historic–modern in date.
[d] This stratum is the only one in the profile with evidence of cultural activity.
[e] These strata predate the earliest known human occupation in the valley.

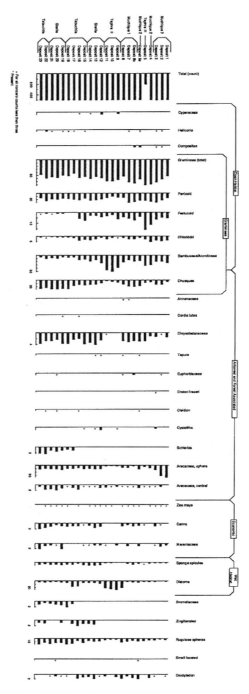

Figure 13-6. Phytolith profile from Sector V/A1, San Isidro site. Cultural affiliation of deposit from Zeidler (1994).

occurs later in time; this regeneration is suggested by an increase in palm phytoliths. Over the course of 3500 years the Jama Valley was converted from a dry tropical forest to a mosaic of open habitats and remnant trees.

The second noteworthy change is an increase in grass phytoliths, especially panicoid grasses, that is associated with the low in forest taxa during Muchique 1. Panicoid grasses include maize. Grass phytoliths, however, including those from panicoid taxa, begin to increase in the earlier, Tabuchila-age stratum. Phytoliths from archaeological samples, discussed below, document that maize was present in the valley from the first occupation by Piquigua peoples. Phytoliths produced by maize occur in the first Piquigua-age stratum in the Río Grande profile, demonstrating that maize was being grown in the Jama watershed as early as 2000 B.C.

The final point of interest in the Río Grande profile is that moisture conditions at the sampling locality declined until the Spanish conquest, after which conditions become locally moist. This change is indicated by the patterning of biogenic silica diatoms and sponge spicules. Possible causes for the vegetation changes in the Río Grande profile are discussed below. The period covered by Deposits 1–6, circa 2800 B.C. to A.D. 1200, correspond to a drier period documented in a number of regions in the Neotropics.

What the Río Grande profile documents at the landscape scale, microfauna from the Feature 5 pit at Pechichal confirm (Stahl 2000). When the natural histories of the animals recovered from this natural trap are considered along with their representation, valuable information about local ecology is revealed. The faunal data indicate that during the time the pit was open, the Muchique 2 phase, it was located close to the modified edge of a forest fragment. Among the entrapped fauna are an abundance of hardy generalists, such as rice rat (*Oryzomys* spp.) and spiny rat (*Proechimys* spp.), animals adapted to forest edges who readily invade forest fragments and are at home in secondary growth and fallow fields.

Understanding vegetation change in the valley makes it easier to interpret phytolith data from habitation sites (Figure 13-6). The earliest cultural strata in this profile represent the Tabuchila phase (Chorrera) occupation at the site (Zeidler 1994). Forest indicators decline in abundance between Tabuchila and Muchique 1 levels. Grasses go up slightly, after fluctuating in abundance during the Tabuchila occupation, and continue to rise in Muchique 2 and Early Muchique 3. This pattern reverses in the uppermost two levels, Late Muchique 3, with grasses declining and palms increasing significantly. Maize occurs throughout the profile, as do achira and Marantaceae (probably arrowroot), both of which produce edible roots.

Without the insights gained from the Río Grande profile, the general increase in grasses seen between Tabuchila and Muchique 1 might have been interpreted as an increase in the use of grasses in house construction or similar functions related to the use of maize documented independently by the presence of large Variant 1 cross-shaped phytoliths (for illustrations of phytolith shapes, see Pearsall [2000] and Piperno [1988, 2006]. Instead, it is more likely that much of the grass component of the assemblage is the result of deposition from the "background" vegetation, which is characterized by declining forest cover and increased open area plants, including grasses. In a similar vein, palm phytoliths decline very little between these cultural levels, and the Río Grande profile documents a more substantial reduction of palms in the watershed (Veintimilla 1998, 2000). This finding suggests that people were important agents of palm deposition at the site. The consistent occurrence of maize at the site, as compared to its sporadic presence in the natural profile, again attests to human agency in deposition. It will be interesting to see in additional studies of natural soil profiles what vegetation changes occurred during the later part of the Muchique sequence (these levels are culturally influenced in the Río Grande profile), i.e., whether the decline in grasses and upsurge in forest indicators documented at the San Isidro site are due to forest regeneration or to a change in subsistence practices such as increased root and tree fruit use relative to maize.

It is unfortunate that Formative (Piquigua, Tabuchila) macrobotanical data cannot be compared quantitatively to those of the Muchique sequence because of preservation and sampling problems. The best information on plant use during the Formative period is from phytoliths. It is difficult to determine the relative importance of crops from phytolith data, however, since phytolith production varies widely among species (Pearsall 2000; Piperno 1988, 2006) and it is difficult to filter out the contribution of background phytoliths. The strength of phytolith data lies in documenting plants that are poorly preserved in the charred botanical record, especially for early time periods.

Phytolith data from the San Isidro site document the presence of a suite of cultivated and utilized plants that remains much the same from Piquigua through Muchique 4. Maize, Marantaceae, and achira are present throughout the sequence. Palms and sedges (Cyperaceae), either cultivated or encouraged, were consistently present, as were useful wild grasses (Poaceae), especially bamboos, such as *Chusquea* sp. Gourd/squash (Cucurbitaceae) phytoliths occur only in Formative samples (Pearsall 1992; Pearsall and Zeidler 1994). Palm and achira phytoliths are especially abundant late in the Muchique phase in some sampling localities; whether this is a shift in subsistence or a change in background vegetation is unclear. Tabuchila deposits from two smaller habitation sites, El Mocoral (M3B4–031) and Finca Cueva (M3D2–009), essentially duplicate the plant suite found at San Isidro.

Phytolith data from cultural contexts contribute the following: (1) confirmation that maize was present from the beginning of the sequence; (2) identification of two root crops, arrowroot and achira; (3) suggestion of an enhanced role for root crops and palms in the latter part of the sequence; and (4) documentation that the richness (variety) of resources utilized was high throughout the sequence. The last point does not negate the possibility that reliance on maize increased during Muchique 2, as suggested by the macrobotanical data, but indicates that other resources were not abandoned.

Phytolith analysis of the Río Grande natural stratigraphic profile provides an opportunity to document vegetation change without the complicating factors of human behavior. What explanations exist for the pattern of changes presented above? Two causal agents stand out in this setting: volcanic ash fall and clearing forests for agriculture (Pearsall 2004). Distinguishing between these agents is critical for evaluating the nature of human-environment interactions in the Jama valley. It is difficult to do so using the data currently available, however.

The effects of volcanic ash fall on vegetation have been studied by Eggler (1948, 1963) for the 1943–1952 activity of Paricutín volcano in the State of Michoacán, Mexico. This case provides useful comparisons for the Jama situation. Before volcanic activity began, 75% of the land in the Paricutín region was pine–oak forest, with agriculture (primarily maize cultivation) confined to flat and gently sloping lands. Hill slopes were generally forested. Rainfall was highly seasonal and caused rapid soil erosion on deforested slopes.

Study of the area in 1945, following 2 years of active lava flow and ash fall, revealed that ash eroded very rapidly on hill slopes: a third of the ash deposited on steep slopes had already eroded into valleys. This reduction in the depth of ash on slopes occurred even as ash was still falling. Erosion during the rainy season was the main avenue for redeposition of ash; during heavy rains, ash flowed as sheets of mud. In areas of ash fall greater than 2.0 m, where all vegetation was buried and killed, once ash eroded away, trees, shrubs, and grasses had begun to regenerate from roots and plant bases by 1959. In areas of ash fall of about 50 cm or less, buried plants regenerated through the ash within a few years. New plants were not established on primary ash deposits, however, unless soil was improved by animal droppings or mixture with organic matter. In some cases, this occurred when ash was redeposited and mixed with soil and plant debris. Seedling germination was apparently inhibited by chemicals in the ash; as little as 10 cm of ash produced this effect. Maize seeds planted in ash did grow, however, as long as plant roots were able to reach the old ground surface (i.e., in ash up to 30 cm deep).

Ash from Paricutín volcano killed all above-ground portions of plants when it accumulated to a minimum depth of 0.7–2.0 m. Small trees (<12 cm diameter) were affected by shallower deposits, dying with accumulations of 20–30 cm of ash. Large trees (>30 cm diameter) were killed by deposits of 30–60 cm, primarily because branches broke under the weight of ash; medium trees (12–30 cm diameter) fared best, surviving ash accumulation of up to 2.0 m because their branches were more flexible. Breakage and starvation, because ash either covered leaves or caused defoliation, had greater impact on tree mortality than depth of ash around the base. If ash was carried away by rain as fast as it fell, trees recovered quickly.

According to data available at this time, most surviving volcanic ash deposits in the Jama valley are tephras (Isaacson 1994). We have discovered few in situ deposits of areal ash in the valley uplands, making it difficult to determine how deep original ash accumulations were before being redeposited as tephras. Based on the Paricutín analogy, erosion of ash from steep terrain could have been rapid. Even as little as 20–30 cm of ash would affect small trees, shrubs, and herbaceous vegetation throughout the valley, at least to the extent of killing them back to the ground. Regeneration of perennials would have been rapid, with slow reappearance of annuals, until original ground surfaces were exposed again. Tephra deposits in valley bottoms were deep enough (1–2 m is common [Donahue and Harbert 1994]) to kill most vegetation, including medium and large trees, and make regrowth of perennials through the tephra difficult. In some areas, however, the mixture of soil and organic matter with ash may have hastened the reestablishment of plants from seed.

One tephra deposit, Tephra III, was examined in the Río Grande off-site profile (Table 13-3, MU Lab No. 873). Phytoliths in tephra deposits presumably represent the remains of plants killed or defoliated by areal ash fall that were then mixed with the ash and redeposited by erosion. Forest indicators are abundant in this stratum. This suggests that ash fall may have affected forest taxa to a greater extent than other plants. Overall abundance of phytoliths is very low in the tephra deposit (Veintimilla 1998, 2000), however, a phenomenon also noted in tephra deposits at archaeological sites. This scarcity indicates that redeposited ash accumulated very quickly, with many fewer phytoliths deposited per volume of sediment than occurs as the result of "normal" sedimentation rates for alluvial deposition. Rapid buildup fits the Paricutín analogy; low phytolith counts suggest a lesser impact on vegetation.

If we compare the two Río Grande samples contemporaneous with Muchique 1, there is evidence for an increase in forest in the watershed between the sample lying immediately above the Tephra II ash fall (MU Lab No. 871) and the later Muchique 1 level (MU Lab No. 872). One interpretation of this change would be that the forest regenerated after Tephra II. Unfortunately, only one Tabuchila-age stratum was sampled following Tephra I, and all samples analyzed above Tephra III are influenced by cultural activities, making it impossible to determine whether forest regrowth occurred. Strata deposited prior to Tephra I document a reduction in forest cover unrelated to ash fall; unfortunately, human presence in the valley at this time cannot be confirmed by independent means, leaving the cause of this change uncertain. Given the contradictory indicators concerning the impact of ash fall versus human activity in forest reduction, it seems best to assume that both were involved and to seek new data to resolve this issue.

Because we know that cultivated plants were grown in the valley from its initial occupation, and that human population, as measured by numbers and sizes of sites, rose throughout the sequence, it is likely that people resumed agricultural activities and cleared forest, dead or alive if needed, after the Tephra II and Tephra III events. Fertility of alluvial lands may have been reduced depending on depth of tephra and degree of admixture with soil and organic matter, leading to increased clearing in the forested uplands. Upland areas not used for agriculture would have recovered rapidly from ash fall; the limited increase in forest cover over the Muchique 1 period may represent such regrowth. Further research is clearly required, however, to delineate the interplay of ash fall and agricultural intensification in transforming the dry tropical forest of the Jama valley into a mosaic of open habitats, fallow fields, and secondary growth. Phytolith analysis of additional tephra deposits might allow more precise modeling of the taxa destroyed by ash fall; analysis of wood samples from sites may yield information on forest clearing.

CONCLUSION

This case study focused on reconstructing subsistence in lowland South America. To investigate this topic, or any topic related to subsistence in the tropical lowlands, requires a well-designed and carefully executed botanical recovery program, including not only charred macrobotanical remains, but also phytoliths, which preserve under conditions where charred materials may be destroyed. Because it is critical to understand the quality of preservation in order to apply quantitative methods to macrobotanical data and to know when not to apply such methods, these issues have been considered at length.

Perhaps the most striking point revealed by this study in the Jama River valley is that the richness (i.e., number of utilized taxa) of the resource base remained very consistent through time. This consistency is the major contribution of the phytolith data, which show continuity of resources present from Piquigua through Muchique. Maintaining a broad subsistence base that made use of the richness of the lowland forest environment, as was characteristic of early subsistence systems, was highly adaptive in a setting subject to catastrophic events like tephra fall. The greater variety of crops and other resources used, the more likely some would be to survive.

Agricultural technology, in this case floodplain farming versus shifting cultivation in uplands, plays a role in successful human adaptation in such settings (Pearsall 2004). Even given a subsistence system characterized by richness, people in the Jama valley were severely impacted by tephra events, with long episodes of abandonment after Tephra I and II. One likely contributing factor early in the cultural sequence was concentration of settlements and fields on the river alluvium, which was buried deeply by tephra. Archaeological surveys show no Piquigua-age settlements off the alluvium. Settlement begins in the uplands during Tabuchila and continues in Muchique 1, with appreciable expansion of settlement in the uplands occurring during Muchique 2, after the Tephra III ash fall (Pearsall and Zeidler 1994). There is no lengthy hiatus after Tephra III, perhaps because people had access to fields in the uplands, and were able to weather the effects of the tephra. Valley bottom villages, including ceremonial centers, were quickly reoccupied, indicating that people occupying these locations were not solely dependent on cultivating the alluvial lands around them, but had access to upland resources through kinship or redistribution networks (Pearsall 2004).

It is in the Muchique 2 phase that changes occur in the apparent importance of maize and other food plants. Maize increases in abundance relative to tree fruits, beans, and tuber-root foods from Muchique 1 to Early Muchique 2, then declines during the last part of the Muchique 2 phase. Following Muchique 2, maize abundance rises again, but not to the levels seen previously. The relationship of these changes to the Tephra III event is suggestive. Of the available crops, perhaps maize was best adapted to the altered growing conditions faced by Muchique 2 populations recolonizing the valley. Would the same pattern of increased use of maize be seen in Tabuchila samples unimpacted by destruction of macrobotanical remains? Very deep tephra deposits on alluvial soils necessitated use of fields in the uplands in both cases. Reestablishing a mixed cropping strategy, as suggested by the Late Muchique 2 data, may have contributed to long-term stability of subsistence in the valley.

One of the goals of the project was to determine when maize became important in subsistence in western Ecuador, specifically, to test whether the Late Formative period marked the beginning of maize-based subsistence systems in this region. Our results suggest that although all the elements of the subsistence system observed in A.D. 1532 were in place during the Formative period, innovation in the system, specifically an increased role for maize relative to root crops and perennials, may have occurred during the Muchique phase, rather than the Formative.

ACKNOWLEDGMENTS

Pearsall and Zeidler's research in the Jama River valley was supported by grants from the National Science Foundation.

REFERENCES

Donahue, J., and Harbert, W., 1994, Fluvial History of the Jama River Drainage Basin, in: *Regional Archaeology in Northern Manabí, Ecuador*, Volume 1. *Environment, Cultural Chronology, and Prehistoric Subsistence in the Jama River Valley* (J. A. Zeidler and D. M. Pearsall, eds.), *University of Pittsburgh Memoirs in Latin American Archaeology* 8:43–57.

Eggler, W. A., 1948, Plant Communities in Vicinity of the Volcano Paricutín, Mexico, after 2 Years of Eruption, *Ecology* 29:415–436.

Eggler, W. A., 1963, Plant Life of Paricutín Volcano, Mexico, 8 Years after Activity Ceased, *American Midland Naturalist* 69:38–68.

Hastorf, C., and Popper, V. (eds.), 1988, *Current Paleoethnobotany*, University of Chicago Press, Chicago, Illinois.

Isaacson, J. S., 1994, Volcanic Sediments in Archaeological Contexts from Western Ecuador, in: *Regional Archaeology in Northern Manabí, Ecuador*, Volume 1. *Environment, Cultural Chronology, and Prehistoric Subsistence in the Jama River Valley* (J. A. Zeidler and D. M. Pearsall, eds.), *University of Pittsburgh Memoirs in Latin American Archaeology* 8:131–140.

Isaacson, J. S., and Zeidler, J. A., 1999, Accidental History: Volcanic Activity and the End of the Formative in Northwestern Ecuador, in: *Actividad Volcánica y Pueblos Precolombianos en el Ecuador* (P. Mothes, ed.), Abyayala, Quito, Ecuador, pp. 41—72.

Meunier, J. D., and Colin, F. (eds.), 2001, *Phytoliths: Applications in Earth Sciences and Human History*, A. A. Balkema Publishers, Lisse.

Meggers, B. J., 1954, Environmental Limitations on the Development of Culture, *American Anthropologist* 58:801–824.

Meggers, B. J., 1966, *Ecuador*, Praeger, New York.

Meggers, B. J., 1971, *Amazonia: Man and Culture in a Counterfeit Paradise*, AHM Publishing, Arlington Heights, Illinois.

Pearsall, D. M., 1979, *The Application of Ethnobotanical Techniques to the Problem of Subsistence in the Ecuadorian Formative*, Ph.D. dissertation, University of Illinois, Urbana-Champaign, University Microfilms, Ann Arbor, Michigan.

Pearsall, D. M., 1982, Phytolith Analysis: Applications of a New Paleoethnobotanical Technique in Archaeology, *American Anthropologist* 84:862–871.

Pearsall, D. M., 1992, Prehistoric Subsistence and Agricultural Evolution in the Jama River Valley, Manabí Province, Ecuador, *Journal of the Steward Anthropological Society* 20:181–207.

Pearsall, D. M., 1994a, Ethnobiological Field Research, in: *Regional Archaeology in Northern Manabí, Ecuador,* Volume 1. *Environment, Cultural Chronology, and Prehistoric Subsistence in the Jama River Valley* (J. A. Zeidler and D. M. Pearsall, eds.), *University of Pittsburgh Memoirs in Latin American Archaeology* 8:145–148.

Pearsall, D. M., 1994b, Issues in the Analysis and Interpretation of Archaeological Maize in South America, in: *Corn and Culture in the Prehistoric New World* (S. Johannessen and C. A. Hastorf, eds.), Westview Press, Boulder, Colorado, pp. 245–272.

Pearsall, D. M., 1994c, Macrobotanical Analysis, in: *Regional Archaeology in Northern Manabí, Ecuador,* Volume 1. *Environment, Cultural Chronology, and Prehistoric Subsistence in the Jama River Valley* (J. A. Zeidler and D. M. Pearsall, eds.), *University of Pittsburgh Memoirs in Latin American Archaeology* 8:149–159.

Pearsall, D. M., 1995a, "Doing" Paleoethnobotany in the Tropical Lowlands: Adaptation and Innovation in Methodology, in: *Archaeology in the Lowland American Tropics: Current Analytical Methods and Recent Applications* (P. Stahl, ed.), Cambridge University Press, Cambridge, England, pp. 113–129.

Pearsall, D. M., 1995b, Domestication and Agriculture in the New World Tropics, in: *Last Hunters-First Farmers: New Perspectives on the Prehistoric Transition to Agriculture* (T. D. Price and A. B. Gebauer, eds.), School of American Research, Santa Fe, New Mexico, pp. 157–192.

Pearsall, D. M., 1999, Agricultural Evolution and the Emergence of Formative Societies in Ecuador, in: *Pacific Latin America in Prehistory: The Evolution of Archaic and Formative Cultures* (M. Blake, ed.), Washington University Press, Pullman, pp. 161–170.

Pearsall, D. M., 2000, *Paleoethnobotany: A Handbook of Procedures*, 2nd ed., Academic Press, San Diego, California.

Pearsall, D. M., 2003, Plant Food Resources of the Ecuadorian Formative: An Overview and Comparison to the Central Andes, in: *Archaeology of Formative Ecuador: A Symposium at Dumbarton Oaks* (S. Raymond and R. Burger, eds.), Dumbarton Oaks, Washington, DC, pp. 213–257.

Pearsall, D. M., 2004, *Plants and People in Ancient Ecuador: The Ethnobotany of the Jama River Valley*, Wadsworth/Thomson Learning, Belmont, California.

Pearsall, D. M., and Piperno, D. R. (eds.), 1993, Current Research in Phytolith Analysis: Applications in Archaeology and Paleoecology, *MASCA Research Papers in Science and Archaeology* 10, University Museum, University of Pennsylvania, Philadelphia.

Pearsall, D. M., and Zeidler, J. A., 1994, Regional Environment, Cultural Chronology, and Prehistoric Subsistence in Northern Manabí, in: *Regional Archaeology in Northern Manabí, Ecuador,* Volume 1. *Environment, Cultural Chronology, and Prehistoric Subsistence in the Jama River Valley* (J. A. Zeidler and D. M. Pearsall, eds.), *University of Pittsburgh Memoirs in Latin American Archaeology* 8:201–216.

Pinilla, A., Juan-Tresserras, J., and Machado, M. J. (eds.), 1997, *The State-of-the-Art of Phytoliths in Soils and Plants*, Centro de Ciéncias Medioambientales del Consejo Superior de Investigaciones Científicas, Madrid.

Piperno, D. R., 1984, A Comparison and Differentiation of Phytoliths from Maize and Wild Grasses: Use of Morphological Criteria, *American Antiquity* 49:361–383.

Piperno, D. R., 1988, *Phytolith Analysis: An Archaeological and Geological Perspective*, Academic Press, San Diego, California.

Piperno, D. R., 2006, *Phytoliths: A Comprehensive Guide for Archaeologists and Paleoecologists*, Altamira Press, Lanham, Maryland.

Piperno, D. R., and Pearsall, D. M., 1998, *The Origins of Agriculture in the Lowland Neotropics*, Academic Press, San Diego.

Rapp, G., Jr., and Mulholland, S. C. (eds.), 1992, *Phytolith Systematics, Emerging Issues*, Plenum Press, New York.

Roosevelt, A. C., 1991, *Moundbuilders of the Amazon: Geophysical Archaeology on Marajó Island, Brazil*, Academic Press, San Diego.

Stahl, P. W. (ed.), 1995, *Archaeology in the Lowland American Tropics: Current Analytical Methods and Recent Applications*, Cambridge University Press, Cambridge, England.

Stahl, P. W., 2000, Archaeofaunal Accumulation, Fragmented Forests, and Anthropogenic Landscape Mosaics in the Tropical Lowlands of Prehispanic Ecuador, *Latin American Antiquity* 11:241–257.

Steward, J. H., 1948, The Circum-Caribbean Tribes: An Introduction, in: *The Handbook of South American Indians,* Volume 4. *Bureau of American Ethnology Bulletin* 143, Smithsonian Institution, Washington, DC, pp. 1–41.

Steward, J. H., 1949, South American Cultures: An Interpretive Summary, in: *The Handbook of South American Indians,* Volume 5. *Bureau of American Ethnology Bulletin* 143, Smithsonian Institution, Washington, DC, pp. 669–772.

Veintimilla, C., 1998, *Analysis of Past Vegetation in the Jama River Valley, Manabí Province, Ecuador*, M.A. thesis, Department of Anthropology, University of Missouri, Columbia.

Veintimilla, C., 2000, Reconstrucción Paleo-ambiental y Evolución Agrícola en el Valle del Río Jama, Provincia de Manabí, Ecuador, *Revista del Museo Antropológico del Banco Central del Ecuador, Guayaquil* 9:135–151.

Zeidler, J. A., 1994, Archaeological Testing in the Middle Jama Valley, in: *Regional Archaeology in Northern Manabí, Ecuador,* Volume 1. *Environment, Cultural Chronology, and Prehistoric Subsistence in the Jama River Valley* (J. A. Zeidler and D. M. Pearsall, eds.), *University of Pittsburgh Memoirs in Latin American Archaeology* 8:71–98.

Zeidler, J. A., Buck, C. E., and Litton, C. D., 1998, Integration of Archaeological Phase Information and Radiocarbon Results from the Jama River Valley, Ecuador: A Bayesian Approach, *Latin American Antiquity* 9:160–179.

Zeidler, J. A., and Isaacson, J. S., 2003, Settlement Process and Historical Contingency in the Western Ecuadorian Formative, in: *Archaeology of Formative Ecuador: A Symposium at Dumbarton Oaks* (S. Raymond and R. Burger, eds.), Dumbarton Oaks, Washington, DC, pp. 69–123.

Zeidler, J. A., and Kennedy, R., 1994, Environmental Setting, in: *Regional Archaeology in Northern Manabí, Ecuador,* Volume 1. *Environment, Cultural Chronology, and Prehistoric Subsistence in the Jama River Valley* (J. A. Zeidler and D. M. Pearsall, eds.), *University of Pittsburgh Memoirs in Latin American Archaeology* 8:13–41.

Zeidler, J. A., and Pearsall, D. M. (eds.), 1994, *Regional Archaeology in Northern Manabí, Ecuador,* Volume 1. *Environment, Cultural Chronology, and Prehistoric Subsistence in the Jama River Valley, University of Pittsburgh Memoirs in Latin American Archaeology* 8:1–224.

Zimmerman, L. S., 1994, Palynological Analysis, in: *Regional Archaeology in Northern Manabí, Ecuador, Volume 1. Environment, Cultural Chronology, and Prehistoric Subsistence in the Jama River Valley* (J. A. Zeidler and D. M. Pearsall, eds.), *University of Pittsburgh Memoirs in Latin American Archaeology* 8:175–163.

Chapter 14

What Seasonal Diet at a Fort Ancient Community Reveals About Coping Mechanisms

Gail E. Wagner

This case study examines how an A.D. 1250 Fort Ancient community in the midcontinental United States coped with potential short- and long-term food shortages while relying upon a narrow range of food resources. Societies that depend on a narrow range of foods may be seen as especially vulnerable to shortages and resource fluctuations. Short-term or seasonal (predictable) pressure on resources leads people to develop coping mechanisms such as food storage (Brenton 1988). Food storage is less effective for dealing with long-term food shortages, however, since dried food or seed for planting can rarely be stored for more than a year or two (Colson 1979). Long-term (unpredictable) pressure on resources can lead to social, political, and economic changes (Johnson and Earle 1987). Thus, the composition and periodicity of a community's diet have implications beyond answering questions about nutrition and health (e.g., Messer 1988).

Archaeologists are not always successful at reconstructing seasonal patterns in diet (Begler and Keatinge 1979; Cross 1988; Dennell 1979), due in no small part to the dual difficulties of identifying unmixed deposits and quantifying the plant and animal remains recovered from them. In a few cases, good arguments for identification to season are based on macrobotanical remains (e.g., Hally 1981; King and McMillan 1975; Miller 1988;

Watson and Yarnell 1989). As is apparent in the following case study, however, macrobotanical remains are not necessarily good indicators of seasonality if the community stores food. Instead, archaeologists should rely upon combinations of zoological, palynological, and macrobotanical remains for season of deposition (Cross 1988; Monks 1981; Reinhard and Bryant 1992).

This chapter begins with a brief characterization of the overall Fort Ancient settlement-subsistence system. The SunWatch site is described, concentrating on the recovery of botanical and faunal remains from 39 underground storage/refuse pits that can be attributed to a season of deposition based on the faunal remains. The differences in diet between seasons are detailed. Given that the Fort Ancient people preferred a focal diet based on a relatively small selection of plant and animal foods, one sign of dietary stress would be a diversification of the diet. Analysis reveals that plants gathered or grown in one season were successfully preserved and stored for consumption in other seasons. Among plant remains, highest diversity and species evenness occurred in the spring, an expected season of dietary stress for agricultural societies like the Fort Ancient. Comparison is made to other Fort Ancient sites with deposits identified to season of fill, and the rate of pit filling is addressed. The social and practical aspects of storage are discussed. The chapter concludes with a summary of the ways in which Fort Ancient people managed to cope with potential short- and long-term food shortages.

FORT ANCIENT SOCIETY AND SUNWATCH SITE

In their model of Fort Ancient society in northeastern Kentucky, Pollack and Henderson (1992:282) view Fort Ancient "as a distinct cultural expression situated on the periphery of Mississippian developments." By late Fort Ancient times, communities show increasing intraregional interaction (e.g., Cook 2004). Fort Ancient societies, centered around the middle Ohio River valley, are seen as developing out of local Late Woodland societies by about A.D. 1000. By this time, local societies had adopted maize (*Zea mays*) agriculture and a suite of changes in material culture. Fort Ancient societies disintegrated around A.D. 1670 due to colonial expansion and introduction of European diseases.

The Fort Ancient peoples examined in this study are considered consummate maize agriculturists (Wagner 1987, 1989) who followed the ethnohistoric Miami-Potawatomi pattern of settlement and subsistence (Fitting and Cleland 1969). During the spring and summer, from April through September, communities gathered in villages with permanent houses. Family groups dispersed to winter hunting camps between October and April, leaving a reduced contingent behind in the village (Essenpreis 1978; Shane 1988; Wagner and Shane 1980; for an alternate interpretation for one site, see Carskadden 1992).

Botanical and faunal studies from a number of village sites indicate that Fort Ancient peoples practiced a focal economy, relying on relatively few plant and animal sources for most of their calories (Shane 1988; Wagner and Shane 1980). White-tailed deer (*Odocoileus virginianus*) or American elk (*Cervus elaphus*) in the northern Fort Ancient range, or deer, elk, and black bear (*Ursus americanus*) in the southern Fort Ancient range, supplied over 82% of the meat in the diet (Breitburg 1992:232). These dietary mainstays were supplemented with meat from smaller animals such as turkey (*Meleagris gallopavo*), raccoon (*Procyon lotor*), squirrel (*Sciurus* sp.), turtle, and fish (Breitburg 1992; Shane 1988; Wagner and Shane 1980). Maize was an important plant food. It was supplemented by hickory nut (*Carya* spp.), black walnut (*Juglans nigra*), domesticated bean (*Phaseolus vulgaris*), and domesticated and wild chenopod (*Chenopodium* spp.), as well as a variety of other wild plant resources (Wagner 1987).

The SunWatch (formerly Incinerator) site (33MY57) is located in southwestern Ohio in the floodplain of the Great Miami River, a major tributary emptying into the Ohio River approximately 83 km downstream (Figure 14-1).

This portion of North America experiences a continental climate characterized by extreme differences between winter and summer temperatures and a mean of 168 frost-free days (Ruffner 1978). Topography and soils are dominated by glacial drift dating to the last and most recent Wisconsonian glaciation (Fenneman 1938; Thornbury 1965). The site is situated within a floodplain that is 1 km wide. It is approximately 220 m above sea level on a low outwash terrace 200 m west of the present course of the river. Although the native vegetation is part of the beech-maple forest region within the deciduous forest formation (Braun 1950), outliers of mesic prairies once existed along the Great Miami River (Sears 1926). The animals identified at the site include a number of prairie species (Shane 1988; Wagner and Shane 1980).

With over 60% of the village uncovered by systematic excavation (Nass 1989), the SunWatch site is the best-known Fort Ancient village. This site is particularly well suited for studies of seasonal occupation because it was inhabited for a brief period of time, as short as 15–20 years (Turnbow 1989) or as long as 40 years (Cook 2004:183), leaving many features relatively intact. Extensive archaeological recovery efforts combined with excellent bone and charred macrobotanical preservation provide an extraordinary opportunity for in-depth studies of Fort Ancient subsistence.

SunWatch is a typical middle Fort Ancient Anderson phase village, occupied for only a few decades around A.D. 1250 (Turnbow 1989) and representing a

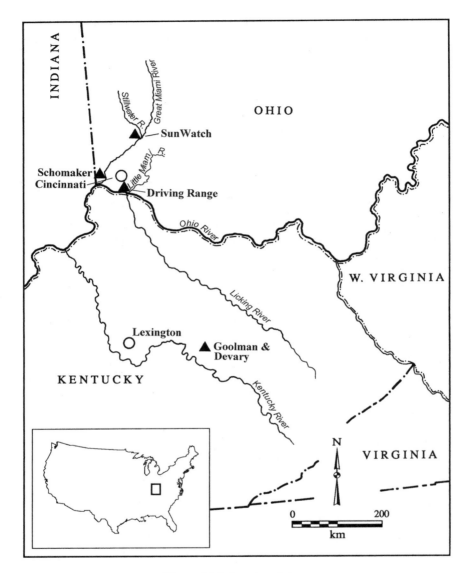

Figure 14-1. Location of sites.

spring-summer permanent settlement in the ethnohistoric model (Figure 14-1). SunWatch was an oval-shaped palisaded village arranged in concentric residential, work, mortuary, and public zones (Figure 14-2). The western wedge appears to have served as the ceremonial and public center of the village, as shown by the size of the structures, mortuary behavior (Cook 2004), the number and contents of the storage/refuse pits (Cook 2004; Nass 1987; Shane 1988), and the

Figure 14-2. SunWatch site (33MY57) village.

distributions of stone tools/flintknapping debris (Robertson 1984, 1988) and pottery (Heilman 1988).

The bone and charred plant remains were recovered from underground storage/refuse pits located for the most part between the structures and the cemetery (Figure 14-2). Among the different types of pits found, bell-shaped (N = 147) and deep flat-bottomed (N = 53) pits (Nass 1987) were probably used originally for storage, as evidenced by their shape, size, and occasional remnants of grass lining preserved by burning. Preserved fragments of charred grass lining as well as stacked maize cobs at other Fort Ancient village sites attest to the widespread use of this storage technique (Wagner 1987; see also Bendremer et al. 1991). A number of written accounts for eastern North America describe similar underground storage pits (e.g., Wilson 1917). Subsequently, these pits were receptacles for refuse such as midden and hearth sweepings. The plant remains examined here are from secondary use of these pits for refuse.

Of the storage/refuse pits, 39 can be attributed to a season of deposition based on the recovery through dry, 1/4-inch screening of 103 young deer mandibles (Shane and Wagner 1993). Deer age at death was determined by the stages of dental eruption, bone resorption at the alveolus of deciduous teeth, and tooth wear relative to tooth eruption on mandibles from deer less than 2 years of age (Severinghaus 1949; Shane 1988:203–205; Shane and Wagner 1993). The season of kill is estimated by adding the age at death of the deer to an average birth date of June 1 (Shane and Wagner 1993). Of the 39 pits, 6 contain deer mandibles indicating deposition in spring (April–June); 17 in summer (late May–September); and 16 in winter (October–April) (Shane and Wagner 1993). Because few pits contain faunal seasonal indicators spanning more than one season (Shane and Wagner 1993) and many pits show ceramic refits that tie together multiple levels of fill (Heilman 1988), these pits likely were filled with refuse during the season of kill indicated by the deer mandibles. Identification of the charred macrobotanical remains from seasonal pits allows a partial reconstruction of the diet during the season the pit was filled.

Flotation samples were taken from some of the 39 pits, but only a small number of these samples have been analyzed. In all, 11 samples totaling 62.5 liters of fill containing 58.96 g of charred plant remains were analyzed from 3 spring pits; 13 samples totaling 32.0 liters of fill containing 49.85 g of charred plant remains were analyzed from 6 summer pits; and 13 samples totaling 33.0 liters of fill containing 57.44 g of charred plant remains were analyzed from 7 winter pits. Flotation was by hand, using 0.8 mm mesh to capture heavy fractions and a 0.4 mm scoop to capture light fractions (Wagner 1979). All charred plant remains 2.0 mm in size and larger were identified, counted, and weighed. All remains under 2.0 mm in size were scanned for seeds and for the presence or absence of plant taxa (Table 14-1). Up to 20 pieces of wood were identified from each sample (Table 14-2).

Table 14-1. Summary of Carbonized Plant Remains Recovered from Seasonal Pits[a]

Taxa	Spring	Summer	Winter
Domesticates			
Maize cupule[b] (*Zea mays*)	53	77	89
Maize kernel	77	126	169
Squash rind (*Cucurbita pepo*)	2	–	3
Common bean (*Phaseolus vulgaris*)	–	4	2
Chenopod (*Chenopodium berlandieri*)	–	1	1
Tobacco (*Nicotiana* spp.)	–	1	–
Nuts			
Hickory (*Carya* spp.)	156	172	768
Black walnut (*Juglans nigra*)	2	30	6
Hickory/black walnut	–	2	5
Hazelnut (*Corylus* spp.)	8	2	–
Acorn (*Quercus* spp.)	2	–	1
Acorn/hazelnut	–	1	–
Wild/weedy seeds	91	59	68
Wood remains			
Wood	4429	4536	3464
Bark	43	17	58
Fungal bodies	51	92	46
Miscellaneous			
Grass stem (Poaceae)	1	2	5
Root	–	4	–
Unidentified	71	91	81
Flotation samples (*N*)	11	13	13
Features (*N*)	3	6	7
Soil floated (liters)	62.5	32.0	33.0

[a] The count was of all carbonized plant remains 2.0 mm in size and larger (except all sizes of squash rind, bean, and small seeds). A maize shank fragment was recovered from one winter pit.
[b] A cupule is a maize cob fragment.

Table 14-2. Summary of Wood Identified from Seasonal Pits, Weight (g)

Taxa	Spring	Summer	Winter
Red oak group (*Quercus* spp.)	0.237	0.600	0.130
White oak group (*Quercus* spp.)	4.950	4.306	2.594
Unidentified oak	0.080	0.321	0.563
Total oak	5.267	5.227	3.287
Hickory (*Carya* spp.)	2.099	1.093	4.108
Black walnut (*Juglans nigra*)	(0.130)	0.898	0.116
Ash (*Fraxinus* spp.)	0.185	0.516	0.358
Black locust (*Robinia pseudoacacia*)	0.415	0.097	0.140
Elm (*Ulmus* spp.)	0.509	0.100	0.250
Hackberry (*Celtis* spp.)	0.256	–	–
Red mulberry (*Morus rubra*)	–	0.434	0.110
Sycamore (*Platanus occidentalis*)	0.344	0.510	0.230
Willow family (Salicaceae)	0.898	–	0.152
Basswood (*Tilia americana*)	–	–	(0.025)

(Continued)

Table 14-2. (*Continued*)

Taxa	Spring	Summer	Winter
Maple (*Acer* spp.)	0.309	–	0.190
Hornbeam (*Carpinus caroliniana*)	0.560	–	–
Unidentified ring porous hardwood	0.312	0.300	0.250
Unidentified diffuse porous hardwood	0.967	0.216	0.474
Unidentified wood	0.038	0.150	0.433
Total identified wood	12.289	9.541	10.123
Total wood ≥ 2.0 mm	39.693	44.070	54.108

RESULTS

Shane's analysis of mandibles indicates that deer bones were deposited in pits during all seasons, but significantly more deer remains (chi-square $= 10.34$, $p < 0.02$) were deposited between April and September than during the winter months, when most of the Fort Ancient population left for winter hunting camps (Shane 1988; Shane and Wagner 1993). As predicted from the ethnohistoric model, summer pits also have significantly higher ($p < 0.01$) numbers of elements per deer, whereas winter pits contain only selected cuts of meat (primarily legs) that presumably were returned to the village to provision the remnant population (Shane 1988:203–205; Shane and Wagner 1993; Orrin Shane 1994, personal communication). Migratory waterfowl, which fly through this area in March-April and October-November, are poorly represented. Shane (1988:186–187) suggests that they may be found at the winter hunting camps.

Pits from all seasons of the year contained nearly the same basic potential plant foods, as shown by the ubiquity of major recoverable remains such as maize, hickory, black walnut, sumac (*Rhus* sp.), and purslane (*Portulaca* sp.). Likewise, the ranking by abundance of different small seeds shows little seasonal change: The two most abundant small seeds invariably are sumac and purslane, and the next four most common include bramble (*Rubus* sp.), panic grass (*Panicum* sp.), nightshade (*Solanum* sp.), and weedy chenopod (Table 14-3).

Table 14-3. Abundance of Seeds from Seasonal Pits[a]

Taxa	Spring	Summer	Winter
Domesticates			
Maize (*Zea mays*)	77	126	169
Common bean (*Phaseolus vulgaris*)	–	4	2
Tobacco (*Nicotiana* spp.)	–	1	–
Chenopod (*Chenopodium berlandieri*)	–	1	1
Wild and weedy plants			
Sumac (*Rhus* spp.)	50	24	18
Purslane (*Portulaca* spp.)	13	8	22
Bramble (*Rubus* spp.)	5	7	2
Panic grass (*Panicum* spp.)	4	3	3
Nightshade (*Solanum* spp.)	3	3	8

(*Continued*)

Table 14-3. (Continued)

Taxa	Spring	Summer	Winter
Chenopod (*Chenopodium* spp.)	2	4	2
Smartweed (*Polygonum* spp.)	2	3	4
Groundcherry (*Physalis* spp.)	1	–	–
Unidentified grasses (Poaceae)	4	3	1
Little barley (*Hordeum pusillum*)	–	–	1
Maygrass[b] (*Phalaris caroliniana*)	(1)	–	–
Foxtail grass (*Setaria* spp.)	1	–	–
Unidentified legume (Fabaceae)	–	–	3
Tick-trefoil (*Desmodium* spp.)	–	1	–
Grape (*Vitis* spp.)	–	1	–
Hackberry (*Celtis* spp.)	1	–	2[c]
Hawthorn (*Crataegus* spp.)	1	–	–
Vervain (*Verbena* spp.)	1	–	(1)
Sedge (*Scirpus* spp.)	–	2	–
Wood sorrel (*Oxalis* spp.)	1	–	–
Spurge (*Euphorbia* spp.)	1	–	1
Unidentified seeds	18	9	13
Unidentifiable seeds	30	16	25
Total identified wild/weedy seeds	91	59	68
Flotation samples (*N*)	11	13	13
Features (*N*)	3	6	7
Soil floated (liters)	62.5	32.0	33.0

[a] The count was of all seeds from all size fractions. Values in parentheses are uncertain identifications.
[b] May be a cultigen.
[c] One storage/trash pit (F22/76) contains a cache of 261 hackberry seeds that may be uncarbonized and are of questionable age.

The only outstanding difference in plant food remains among pits from different seasons is the peak in the spring pits of fruit seeds from fruits that are all out of season. In terms of volume of dirt examined through flotation, the sample from spring pits is nearly twice that of the other two seasons. Spring pits show the least intensity of burning activity, as demonstrated by the total weight of charred plant remains per volume of fill (Figure 14-3) (Miller 1988). Nearly all these plant remains are wood. Nevertheless, spring pits also contain maize, the greatest variety of nuts (hickory, black walnut, hazelnut [*Corylus* sp.], acorn [*Quercus* spp.]), and the greatest variety of small seeds of all the seasonally specific pits (Tables 14-1 and 14-3). As might be expected, given that the refuse in these pits reflects activities at the time of deposition, winter pits evidence the highest intensity of burning (Figure 14-3). Although most of the winter pit charcoal is wood, there is also a high proportion of hickory nutshell. Both hickory nutshell and maize cobs may have been used as fuel. Whereas oak (*Quercus* spp.) and hickory together constitute 60–73% of identified wood from pits of all seasons, in spring and summer pits oak is far more common by weight than hickory, and in winter pits hickory is more common than oak (Table 14-2).

Given that the Fort Ancient people appear to have preferred a diet based on a relatively small selection of plant and animal foods, at times of seasonal plenty

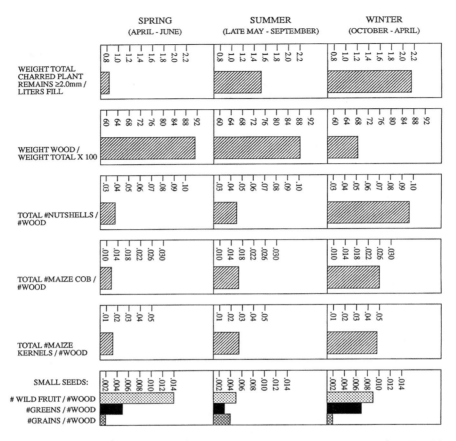

Figure 14-3. Ratio of plant remains by season. In the last row, the numbers of small wild fruit, greens, and grain seeds are compared as ratios to wood.

we might expect to find that preference marked by a low diversity of foods, with reliance on a few dominant species. As a result, evenness of food items would be low. Conversely, at times of seasonal stress people may have been forced to broaden the diversity of the foods they were willing to eat. Diversity and evenness of food items would be higher.

We can use Simpson's index (Magurran 1988:39–40) to measure the proportional abundance of the most common edible plant foods by season:

$$D = \frac{\Sigma\ (n_i(n_j - 1)}{N(N - 1)}$$

where n_i equals the number of individuals in the ith species, and N equals the total number of individuals. Simpson's index measures the proportional abundance of the commonest edible plant species looking conservatively at counts of seeds (Table 14-3) from only the most likely food sources (maize, bean, chenopod, sumac, purslane, bramble, groundcherry [*Physalis* sp.], little barley [*Hordeum pusillum*], maygrass [*Phalaris caroliniana*], grape [*Vitis* spp.], hackberry [*Celtis* spp.], hawthorn [*Crataegus* spp.], and wood sorrel [*Oxalis* spp.]). When expressed as 1-*D*, spring pits contain the highest diversity of seeds from potential plant foods (1-*D* = 2.708), summer pits less (1-*D* = 1.854), and winter pits the lowest diversity (1-*D* = 1.637).

These findings are supported by the Berger-Parker index, which "expresses the proportional importance of the most abundant species" (Magurran 1988:41):

$$d = N_{max} / N$$

where N_{max} equals the number of individuals in the most abundant species. The Berger-Parker index measures, then, the evenness of species. The Berger-Parker index is expressed as 1/*d*, so that a higher result equates to "an increase in diversity and a reduction in dominance" (Magurran 1988:41). In the refuse from each of the seasons, maize kernels are the most abundant seed species. Using the Berger-Parker index, winter pits show the highest dominance of maize and the lowest evenness of species (1/*d* = 1.2958). Maize is not quite as dominant in summer pits (1/*d* = 1.3888). Maize is least dominant in spring pits (1/*d* = 1.974), where all the species can be viewed as more even in importance.

One problem common to using either of these diversity indices is that counts of seeds are not comparable between species; that is, each species produces differing numbers of seeds per fruit, and each fruit type is of different utility to people. Additionally, counts of seeds (Table 14-3) do not distinguish between whole seeds and seed fragments, so that a count of three maize kernels, for example, may be of three fragments that were once a single kernel.

DISCUSSION

Although diet differed from season to season, the charred macrobotanical remains from seasonally-filled storage/refuse pits do not strongly indicate that any plant foods were excluded from the diet because of seasonal unavailability. Drying and storage technology for plant foods harvested in different seasons was sufficient to supply a variety of plant foods year-round to the people who remained in the main village. Fruits certainly were eaten fresh, but also were dried, because their seeds were still available for incorporation into refuse deposited out of season, in the spring. The only evidence for plant food deprivation during any season is finding a higher diversity of less desired plant foods in the spring refuse deposits. Additionally, spring pits contain the lowest amounts of plant foods and wood charcoal, perhaps linked to an increased reliance on dried, processed foods.

Seasonal Diet at Other Fort Ancient Sites

The diet of those who spent the winter at hunting camps differed from the diet of those who stayed behind at the main village. Whereas entire deer were processed at the village during the spring and summer, only selected cuts of meat were brought back during the winter to provision those who remained behind and did not disperse to the winter camps. By omission, then, one could expect to find the missing fresh cuts of deer meat at the winter hunting camps.

Although the locations of some winter hunting camps are known, few have been excavated and even fewer subjected to botanical and faunal analyses. The argument that most of these camps represent winter occupations rests mainly upon the fit between site size and location relative to the ethnohistoric Miami-Potawatomi settlement model.

The winter hunting camps with the best-documented botanical and faunal material are the Goolman (15CK146) and Devary (15CK147) sites in east central Kentucky (Figure 14-1). These sites are adjacent to one another and were occupied several centuries later than the SunWatch site (Turnbow et al. 1983; Wymer 1982). Deer antlers and dentition from Goolman firmly place occupation of this site between November and February (Turnbow and Jobe 1984). The winter occupation at Devary has been inferred from site size and location.

Maize cob and kernel fragments were recovered from both sites, albeit in low ubiquity (16–18% of the samples compared to 100% ubiquity in the seasonally-specific features reported here for the SunWatch site). Hickory nutshell is abundant and fairly ubiquitous at both winter camps. Like the SunWatch site, the small seed assemblage represents ruderal plants from open, disturbed areas, yet the types of seeds are quite different. Identified seeds that are abundant at these two winter camps include vervain (*Verbena urticifolia*), American pennyroyal (*Hedeoma pulegioides*), plantain (*Plantago* sp.), grasses (Poaceae), groundcherry, and knotweed (*Polygonum* sp.). Only a few sumac and chenopod seeds were recovered (Wymer 1982).

The presence of both maize kernels and cobs indicates that maize was carried to winter hunting camps, as were hickory and black walnut. Other weedy edible plants associated with open, disturbed areas occupied by humans were also present at these sites. Maize may have been less abundant than at the permanent village, but it was a part of the winter camp diet.

The Driving Range site (33HA586), a late Fort Ancient seasonal village in southwestern Ohio (Figure 14-1), appears to have been occupied repeatedly during the fall to winter (Purtill 1999). Admittedly, the site is multicomponent and the two young deer mandibles that indicate season of occupation were recovered from deposits only inferred to be Fort Ancient. Unlike the isolated winter hunting camps of Goolman and Devary, the Driving Range site both overlooks a large floodplain and is near permanent villages. It has both storage/refuse pits and burials. Unlike permanent villages such as the SunWatch site and like the winter hunting camps of Goolman and Devary, maize kernels occur in only 8% of the analyzed flotation samples. Other plant remains recovered include hickory, black walnut, and one possibly domesticated chenopod (*Chenopodium berlandieri*) seed. The bones of the most common and ubiquitous animal species, white-tailed deer, were processed for marrow extraction. As Purtill (1999) points out, only a limited number of activities were undertaken and a low variety of favored foods were eaten at the fall/winter Driving Range seasonal village.

Limited excavation at a middle Fort Ancient permanent village, the Schomaker site (33HA400), included identification through faunal analysis of seasonally stratified deposits in two refuse-filled storage pits, Features 3 and 5 (Cowan et al. 1990) (Figure 14-1). The oval-shaped village is located on a terrace above the west bank of the Great Miami River. Although the most abundant and ubiquitous animal bone is from white-tailed deer, deer constitutes less than 61% by weight of the bone sample in the summer deposits, but nearly 89% in the late winter to early spring samples. Faunal analysis suggests depletion of foods in stored pits by early spring, allowing commencement of refuse disposal in the emptied pits. Spring deposits are marked by a wide variety of animal prey, and are especially noteworthy for high amounts of fish (Cowan et al. 1990:33).

Analysis of flotation light fractions reveals that maize is present in 100% of the samples (Dunavan 1989). Nuts (mostly hickory, but also hazelnut and black walnut) are ubiquitous in very low quantities in the light fractions. The two most common small seeds are sumac and both wild and domesticated chenopod. Both are found in late winter and spring-summer deposits in Feature 5, along with one seed each of elderberry (*Sambucus canadensis*), grape, and hawthorn. The early spring to midsummer deposit in Feature 5 contains sumac, grape, chenopod, maygrass, and hawthorn. Only one of these food items (maygrass) definitely would have ripened during the period of deposition: the presence of the others may indicate that they had been stored. No seeds were recovered from the early spring deposits in Feature 3, but the summer deposits in this feature include seeds from grape and bramble. Like at SunWatch, the early spring deposit in Feature 3 shows the least intensity of burning activity (0.67 g wood charcoal per liter of dirt floated). The summer deposits from the same feature show the highest intensity of burning (2.84 g/l) (Dunavan 1989).

Rate of Pit Filling

Although the pits used in this case study likely were filled with refuse from a number of different activities, they are all activities that occurred during a single season. If dietary components vary from season to season, the subsistence-related debris in the seasonal pits should reflect those differences. In one hypothetical situation, members of a community could depend entirely on the foods available that season, without provision for other times of the year. In the simplest construction of this situation, one might expect to find maize and discarded nutshells only in summer to early winter pits, whereas spring pits might include greens, spring grass and other seeds, and spawning species of fish, but no maize or nutshell.

Fragments of the same ceramic vessel may be found in the bottom and the top of any one SunWatch pit, hinting that open pits were filled quickly with refuse (Heilman 1988). A few of the pits contain two or three deer mandibles all indicating spring kills or all indicating winter kills but in multiple levels of the pit, and some contain ceramic refits that tie together different levels. On the basis of cross-feature ceramic refits (Cook 2004:73–74; Heilman 1988), the occupants of more than one house may have contributed to the refuse in any one pit.

Although at this stage in our investigations we rely upon the generalization that pits at the SunWatch site were filled quickly, it must be admitted that each seasonal pit should be examined individually for evidence of artifact refitting throughout the fill. As others point out, different fills within any one pit likely represent different activities (e.g., Wilson 1985) or possibly different seasons, as at the Schomaker site. Of the 39 SunWatch site pits with faunal seasonal indicators, three have deer mandibles from two different seasons, but the mandibles from one season occur in the uppermost levels (Orrin Shane 1994, personal communication) and those from the other are found throughout the pit. In addition, there is a ceramic cross-fit between one summer and one winter pit. Depending upon the levels in which sherds and bones originate, most of the fill in these pits may still be attributed to activities in one season. On the other hand, two of the summer pits are linked with a ceramic cross-fit, which strengthens their contemporaneity.

The rate of pit fill depends upon the types of refuse and the number of people producing it, making it difficult to estimate rates of refuse-pit filling (Callender 1976; Hayden and Cannon 1983). Different types of refuse are discarded at different rates, varying from objects discarded once in a generation (e.g., a metate) to objects discarded each day (e.g., food refuse) (Hayden and Cannon 1983). In addition, refuse may be sorted by size and type and discarded in different locations (Hayden and Cannon 1983). Distributional data from the SunWatch site indicate that where broken ceramic vessel fragments (Heilman 1988) or stone debitage/tool fragments (Robertson 1984) may be discarded maps the social landscape of reciprocal relationships within the village (Cook 2004).

Storage

As noted by Smyth (1989:93), "the structure of political power within a society should be related to the organization of storage systems." At the SunWatch site, an unknown amount of storage could have been in above-ground facilities. For example, food and other goods may be stored in lofts, rafters, or thatch of houses; in textile, bark, wooden, gourd, hide, ceramic, or other containers; or in elevated cribs (e.g., Bursey 2001). Because of the nature of the evidence, we are limited to discussion of storage in underground pits.

Supposed storage pits at the SunWatch site range in capacity from 27 to 1,956 liters (Nass 1987), but pit size is limited by physical constraints. If you wish to camouflage a pit or seal the top of it to control the temperature, humidity, and atmosphere, a small opening is easier to cover or seal than a large one. In this portion of the Great Miami River Valley, the depth of a pit is limited by the presence of glacial till at approximately 1.1 m below the ground surface. Although glacial till is difficult to excavate and manifests a sometimes precarious profile, it also ensures excellent drainage and as such contributes to the viability of this particular form of storage.

Storage at the SunWatch site appears to have been controlled by kin relationships in corporate groups rather than by a central authority. No firm evidence exists for control of storage by any one segment of the village. Although frequencies and sizes of pits differ from one excavated segment of the village to another, no one segment is outstandingly different (Nass 1987:158; for contrary views see Cook 2004; Nass and Church 1992). There is evidence that cooperation, perhaps within corporate groups, occurs within village segments: ceramic styles are consistent within segments but vary between segments, and patterns of ceramic disposal mark both separation of and ties between the various segments and corporate groups (Cook 2004; Heilman 1988). Broken stone tools link various sections of the village (Robertson 1984), a pattern that might occur when meat is given to or exchanged with someone else. For example, the tip of the point may remain in the meat while the base remains with the hunter's arrow. Taken together, the ceramic sherd disposal pattern and the pattern of stone debitage and tool fragments indicate cooperation and reciprocity within the village. It is not unreasonable to extend reciprocity in stored food minimally to within corporate groups. Thus, storage by individual households would buffer seasonal household consumption variance (per Winterhalder et al. 1999), and reciprocity in stored food between households could further serve to minimize individual household shortfalls.

In adjacent Monongahela territory to the east and northeast (centered in southwestern Pennsylvania), underground storage pits exterior to structures are found only at early sites (A.D. 1050–1250). Through time, storage becomes more visible in above-ground structures attached first to individual houses and eventually, by Late Monongahela (circa A.D. 1580), to community-controlled structures (Hart 1995). Changes in stored food occur also among Fort Ancient communities. At the short-lived SunWatch site, pit volume increases through time (Cook 2004:170). At late Fort Ancient sites, houses become larger and storage pits are more frequently hidden inside structures, away from public or non-corporate-group view (Cook 2004:48).

Why Use Underground Pits?

Why use underground storage pits? Two different views emerge from the archaeological and ethnographic literature. Storage pits are an effective way to store grain. An underground pit provides controlled temperature, light, humidity, and atmosphere, and may keep out many vermin (see Smyth 1989). Controlled experiments in England with barley (Bowen and Wood 1967; Reynolds 1967, 1969, 1974) indicate that underground storage pits can be lined or unlined, may be opened and resealed, and may be reused for at least two years and perhaps for an unlimited amount of time. In large part, the pit's effectiveness rests upon keeping out moisture and keeping in the atmosphere of carbon dioxide provided by the initially rotting grain located on the outside edges of the grain mass. Reynolds (1974) suggests that underground pits are an especially effective way to store seed meant for planting, since germination rates of grain from within the grain mass remain high with this sort of treatment.

DeBoer (1988:1) views storage pits not as a subsistence strategy, but as a "social stratagem in which surplus is protected through concealment." Concealment can be a necessity for a seasonally abandoned settlement (DeBoer 1988; Smyth 1989; Ward 1985), it can be a form of resistance to a new sociopolitical order (DeBoer 1988), or it can be seen as a coping mechanism during a time of famine (Colson 1979).

The storage pits at the SunWatch site likely were used both as an effective way to store maize and to conceal stored goods during seasonal abandonment of portions of the site. There must have been some element of worry about outside raiding, given that the village was surrounded by a wooden stockade. The village was most vulnerable to raiding when most of the occupants were away at the winter hunting camps. Deep pits suitable for storage are not usually found at sites interpreted as winter hunting camps (Turnbow et al. 1983).

Concealment was directed against outsiders rather than other members of the community, since nearly all storage pits were located outside the structures and between the structures and the public area of the plaza. If concealment were a reaction to famine or to sociopolitical forces within the village, pits would have been dug inside the structures and hidden from other members of the community, as they were at late Fort Ancient sites. The human skeletal remains from the SunWatch site do not support the idea of famine (Giesen 1992).

There is no evidence that concealment was due to resistance to new sociopolitical forces. The trajectory of Fort Ancient settlement-subsistence patterns and sociopolitical organization shows gradual change through time (Pollack and Henderson 1992), but not until late Fort Ancient is there much evidence for increasing intraregional interaction. At that time, storage pits are increasingly moved to the inside of multi-family structures (Cook 2004:48).

CONCLUSIONS

The data in this study show how the Fort Ancient community at the SunWatch site avoided short-term and long-term food shortages. Fort Ancient society in general and this community specifically practiced a focal economy concentrating their food procurement strategies on maize, deer, and hickory nuts. Without appropriate coping strategies, this reliance on few resources would have left people liable to famine in bad seasons or years.

This society buffered shortages by use of a storage technology that enabled them to save both food and seed for planting. Their storage systems served not only to safely keep food and seed, but also to conceal them during seasonal near-abandonment of the village when families dispersed to their winter hunting camps. The variety of plant remains from spring pits, the season most likely to engender predictable food shortages in a heavily agricultural society, suggests that rather than being reduced to starvation or to relying on less-preferred foods, the SunWatch villagers successfully maintained year-round access to favored or dependable foods. Successful, planned storage of less-preferred foods, however, allowed the villagers to supplement their diet in the spring, when stored quantities of preferred foods were at their lowest.

The variety of animal and plant foods recovered demonstrates knowledge of a broad array of resources, yet the quantities deposited indicate that the full array of known resources was not relied upon. The food sources chosen for use likely reflect convenience (e.g., Gremillion 2004) or taste preferences or both. Reliance on only a few resources within a broad-based pattern of resource familiarity may be interpreted as a strategy for coping with long-term food shortages. A society that maintains a knowledge of a wide variety of food resources is prepared to more fully adopt that broader base should the need arise.

The short-term occupation of this village may also be interpreted as a way to cope with resource shortages. Apparently Fort Ancient villagers could periodically move a village to new land rather than continue to deplete the local environment (e.g., Nass 1988). For example, as mentioned earlier, the SunWatch site village is thought to have been occupied for only 40 years, or less (Cook 2004; Turnbow 1989).

Despite the adoption of a focal, and therefore potentially unstable or vulnerable, food economy centered on deer, maize, and hickory nuts by A.D. 1000, some Fort Ancient peoples were well able to cope with both short- and long-term shortages. Coping mechanisms included food preservation and storage, retention of a knowledge system about a broad array of potential foods, cooperation, and the ability to periodically relocate populations.

ACKNOWLEDGMENTS

This case study depends on data shared by a number of SunWatch analysts, and in particular Orrin C. Shane, who analyzed the faunal remains. All work at the site was encouraged and facilitated by James M. Heilman of the Dayton Museum of Natural History. This study benefitted from conversations with these two researchers and with others, but the views presented herein are solely those of the author. Figure 14-2 was provided by the Dayton Museum of Natural History, and additional work on the figures was done by Gisela Weis Gresham.

REFERENCES

Begler, E. B., and Keatinge, R. W., 1979, Theoretical Goals and Methodological Realities: Problems in the Reconstruction of Prehistoric Subsistence Economies, *World Archaeology* 11:208–226.

Bendremer, J. C. M., Kellogg, E. A., and Largy, T. B., 1991, A Grass-Lined Maize Storage Pit and Early Maize Horticulture in Central Connecticut, *North American Archaeologist* 12:325–349.

Bowen, H. C., and Wood, P. D., 1967, Experimental Storage of Corn Underground and Its Implications for Iron Age Settlements, *Bulletin of the Institute of Archaeology* (London) 7:1–14.

Braun, E. L., 1950, *Deciduous Forests of Eastern North America*, Blakiston, Philadelphia, Pennsylvania.

Breitburg, E., 1992, Vertebrate Faunal Remains, in: *Fort Ancient Cultural Dynamics in the Middle Ohio Valley* (A. G. Henderson, ed.), *Monographs in World Archaeology* 8, Prehistory Press, Madison, Wisconsin, pp. 209–241.

Brenton, B. P., 1988, The Seasonality of Storage, in: *Coping with Seasonal Constraints* (R. Huss-Ashmore, J. J. Curry, and R. K. Hitchcock, eds.), *MASCA Research Papers in Science and Archaeology* 5, University Museum, University of Pennsylvania, pp. 45–54.

Bursey, J. A., 2001, Storage Behavior in the Northeast: A Review of the Evidence, *North American Archaeologist* 22:179–199.

Callender, D. W., 1976, Reliving the Past: Experimental Archaeology in Pennsylvania, *Archaeology* 29:173–177.

Carskadden, J., 1992, Possible Summer and Winter Houses at the Fort Ancient Philo II Site, *West Virginia Archeologist* 44:26–39.

Colson, E., 1979, In Good Years and In Bad: Food Strategies of Self-Reliant Societies, *Journal of Anthropological Research* 35:18–29.

Cook, R. E., 2004, *Upper Mississippian Village Structure and Formation: Spatial Analysis of SunWatch, a Fort Ancient Site in Southwest Ohio*, Ph.D. dissertation, Department of Anthropology, Michigan State University, East Lansing.

Cowan, C. W., Dunavan, S., Nass, J. P., Jr., and Scott, S., 1990, The Schomaker Site, a Middle Period Fort Ancient Town on the Great Miami River, Hamilton County, Ohio, *West Virginia Archeologist* 42:11–35.

Cross, J. R., 1988, Expanding the Scope of Seasonality Research in Archaeology, in: *Coping with Seasonal Constraints* (R. Huss-Ashmore, J. J. Curry, and R. K. Hitchcock, eds.), *MASCA Research Papers in Science and Archaeology* 5, University Museum, University of Pennsylvania, pp. 55–63.

DeBoer, W. R., 1988, Subterranean Storage and the Organization of Surplus: The View from Eastern North America, *Southeastern Archaeology* 7:1–20.

Dennell, R. W., 1979, Prehistoric Diet and Nutrition: Some Food for Thought, *World Archaeology* 2:121–135.

Dunavan, S. L., 1989, Botanical Remains from the Schomaker Site (33HA400), *Museum of Anthropology, University of Michigan, Ethnobotanical Laboratory Report* 587, Ann Arbor.

Essenpreis, P. S., 1978, Fort Ancient Settlement: Differential Response at a Mississippian–Late Woodland Interface, in: *Mississippian Settlement Patterns* (B. D. Smith, ed.), Academic Press, New York, pp. 141–167.

Fenneman, N. M., 1938, *Physiography of Eastern United States*, McGraw-Hill, New York.

Fitting, J. E., and Cleland, C. E., 1969, Late Prehistoric Settlement Patterns in the Upper Great Lakes, *Ethnohistory* 16:289–302.

Giesen, M. J., 1992, *Late Prehistoric Populations in the Ohio Area: Biological Affinities and Stress Indicators*, Ph.D. dissertation, Department of Anthropology, Ohio State University, Columbus.

Gremillion, K. J., 2004, Seed Processing and the Origins of Food Production in Eastern North America, *American Antiquity* 69:215–233.

Hally, D. J., 1981, Plant Preservation and the Content of Paleobotanical Samples: A Case Study, *American Antiquity* 46:723–742.

Hart, J. P., 1995, Storage and Monongahela Subsistence-Settlement Change, *Archaeology of Eastern North America* 23:41–56.

Hayden, B., and Cannon, A., 1983, Where the Garbage Goes: Refuse Disposal in the Maya Highlands, *Journal of Anthropological Archaeology* 2:117–163.

Heilman, J. M., 1988, Ceramics as Indicators of Social Organization, in: *A History of 17 Years of Excavation and Reconstruction-A Chronicle of 12th Century Human Values and the Built Environment*, Volume 1 (J. M. Heilman, M. C. Lileas, and C. A. Turnbow, eds.), Dayton Museum of Natural History, Dayton, Ohio, pp. 242–260.

Johnson, A. W., and Earle, T., 1987, *The Evolution of Human Societies: From Foraging Groups to Agrarian State*, Stanford University Press, Stanford, California.

King, F. B., and McMillan, R. B., 1975, Plant Remains from a Woodland Storage Pit, Boney Spring, Missouri, *Plains Anthropologist* 20:111–115.

Magurran, A. E., 1988, *Ecological Diversity and Its Measurement*, Princeton University Press, Princeton, New Jersey.

Messer, E., 1988, Seasonal Hunger and Coping Strategies: An Anthropological Discussion, in: *Coping with Seasonal Constraints* (R. Huss-Ashmore, J. J. Curry, and R. K. Hitchcock, eds.), *MASCA Research Papers in Science and Archaeology 5*, University Museum, University of Pennsylvania, pp. 131–141.

Miller, N., 1988, Ratios in Paleoethnobotanical Analysis, in: *Current Paleoethnobotany: Analytical Methods and Cultural Interpretations of Archaeological Plant Remains* (C. A. Hastorf and V. S. Popper, eds.), University of Chicago Press, Chicago, Illinois, pp. 72–85.

Monks, G., 1981, Seasonality Studies, in: *Advances in Archaeological Method and Theory*, Volume 4 (M. B. Schiffer, ed.), Academic Press, San Diego, California, pp. 177–240.

Nass, J. P., Jr., 1987, *Use-Wear Analysis and Household Archaeology: A Study of the Activity Structure of the Incinerator Site, An Anderson Phase Fort Ancient Community in Southwestern Ohio*, Ph.D. dissertation, Department of Anthropology, Ohio State University, Columbus.

Nass, J. P., Jr., 1988, Fort Ancient Agricultural Systems and Settlement: A View from Southwestern Ohio, *North American Archaeologist* 9:319–347.

Nass, J. P., Jr., 1989, Household Archaeology and Functional Analysis as Procedure for Studying Fort Ancient Communities in the Ohio Valley, *Pennsylvania Archaeologist* 59:1–13.

Nass, J. P., Jr., and Church, F., 1992, Central Ohio during the Late Prehistoric: Subsistence/Settlement System Responses to Risk, paper presented at the 57th Annual Meeting of the Society for American Archaeology, Pittsburgh, Pennsylvania.

Pollack, D., and Henderson, A. G., 1992, Toward a Model of Fort Ancient Society, in: *Fort Ancient Cultural Dynamics in the Middle Ohio Valley* (A. G. Henderson, ed.), *Monographs in World Archaeology* 8, Prehistory Press, Madison, Wisconsin, pp. 281–294.

Purtill, M. P., 1999, Evidence for a Late Fort Ancient Fall/Winter Occupation in Southwestern Ohio, *North American Archaeologist* 20:105–133.

Reinhard, K. J., and Bryant, V. M., Jr., 1992, Coprolite Analysis: A Biological Perspective on Archaeology, in: *Archaeological Method and Theory*, Volume 4 (M. B. Schiffer, ed.), University of Arizona Press, Tucson, pp. 245–288.

Reynolds, P. J., 1967, Experiment in Iron Age Agriculture, *Transactions of the Bristol and Gloucester Archaeological Society* 86:60–73.

Reynolds, P. J., 1969, Experiment in Iron Age Agriculture: Part Two, *Transactions of the Bristol and Gloucester Archaeological Society* 88:29–33.

Reynolds, P. J., 1974, Experimental Iron Age Storage Pits: An Interim Report, *Proceedings of the Prehistoric Society* 40:118–131.

Robertson, J. A., 1984, Chipped Stone and Functional Interpretations: A Fort Ancient Example, *Midcontinental Journal of Archaeology* 9:251–267.

Robertson, J. A., 1988, The Classification and Analysis of Chipped Stone Tools, in: *A History of 17 Years of Excavation and Reconstruction-A Chronicle of 12th Century Human Values and the Built Environment*, Volume 1 (J. M. Heilman, M. C. Lileas, and C. A. Turnbow, eds.), Dayton Museum of Natural History, Dayton, Ohio, pp. 214–241.

Ruffner, J. A. (comp.), 1978, *Climates of the States: National Oceanic and Atmospheric Administration Narrative Summaries, Tables, and Maps for Each State*, 2 Volumes, Gale Research, Detroit, Michigan.

Sears, P. B., 1926, The Natural Vegetation of Ohio II, The Prairies, *Ohio Journal of Science* 26:128–146.

Severinghaus, C. W., 1949, Tooth Development and Wear as a Criterion of Age in White-tailed Deer, *Journal of Wildlife Management* 13:196–216.

Shane, O. C., III, 1988, An Interim Summary Report of the Vertebrate Faunal Remains from the Incinerator Site (33MY57), Montgomery County, Ohio, in: *A History of 17 Years of Excavation and Reconstruction-A Chronicle of 12th Century Human Values and the Built Environment*, Volume 1 (J. M. Heilman, M. C. Lileas, and C. A. Turnbow, eds.), Dayton Museum of Natural History, Dayton, Ohio, pp. 157–213.

Shane, O. C., III, and Wagner, G. E., 1993, Seasonality and Diet at a Fort Ancient Site, paper presented at the 16th Annual Meeting of the Society for Ethnobiology, Boston, Massachusetts.

Smyth, M. P., 1989, Domestic Storage Behavior in Mesoamerica: An Ethnoarchaeological Approach, in: *Archaeological Method and Theory*, Volume 1 (M. B. Schiffer, ed.), University of Arizona Press, Tucson, pp. 89–138.

Thornbury, W. D., 1965, *Regional Geomorphology of the United States*, John Wiley, New York.

Turnbow, C., 1989, Radiocarbon Determinations of the SunWatch Site (33My57), Montgomery County, Ohio, manuscript on file, Department of Anthropology, Dayton Museum of Natural History, Dayton, Ohio.

Turnbow, C. A., and Jobe, C. E., 1984, The Goolman Site: A Late Fort Ancient Winter Encampment in Clark County, Kentucky, in: *Late Prehistoric Research in Kentucky* (D. Pollack, C. D. Hockensmith, and T. N. Sanders, eds.), Kentucky Heritage Council, Frankfort, pp. 25–48.

Turnbow, C., Jobe, C., and O'Malley, N., 1983, *Archaeological Excavations of the Goolman, Devary, and Stone Sites in Clark County, Kentucky, Archaeological Report 78*, Program for Cultural Resource Assessment, Department of Anthropology, University of Kentucky, Lexington.

Wagner, G. E., 1979, The Dayton Museum of Natural History Flotation Procedure Manual, revised edition, manuscript on file, Dayton Society of Natural History, SunWatch Village, Dayton, Ohio.

Wagner, G. E., 1987, *Uses of Plants by the Fort Ancient Indians*, Ph.D. dissertation, Department of Anthropology, Washington University, St. Louis, Missouri.

Wagner, G. E., 1989, The Corn and Cultivated Beans of the Fort Ancient Indians, *Missouri Archaeologist* 47:107–135.

Wagner, G. E., and Shane, O. C., III, 1980, Fort Ancient Subsistence in Southwestern Ohio, paper presented at the 45th Annual Meeting of the Society for American Archaeology, Philadelphia, Pennsylvania.

Ward, H. T., 1985, Social Implications of Storage and Disposal Patterns, in: *Structure and Process in Southeastern Archaeology* (R. S. Dickens, Jr., and H. T. Ward, eds.), University of Alabama Press, Tuscaloosa, pp. 82–101.

Watson, P. J., and Yarnell, R. A., 1989, Lost John's Last Meal, *The Missouri Archaeologist* 47:241–255.

Wilson, G. L., 1917, *Agriculture of the Hidatsa Indians: An Indian Interpretation, University of Minnesota Studies in the Social Sciences* 9, Minneapolis.

Wilson, J. H., Jr., 1985, Feature Zones and Feature Fill: More Than Trash, in: *Structure and Process in Southeastern Archaeology* (R. S. Dickens, Jr., and H. T. Ward, eds.), University of Alabama Press, Tuscaloosa, pp. 60–81.

Winterhalder, B., Lu, F., and Tucker, B., 1999, Risk-Sensitive Adaptive Tactics: Models and Evidence from Subsistence Studies in Biology and Anthropology, *Journal of Archaeological Research* 7:301–348.

Wymer, D. A., 1982, Paleoethnobotanical Material from the Stone (15CK89), Goolman (15CK146), and Devary (15CK147) Sites, Clark County, Kentucky, Ethnobotanical Laboratory Report 4, manuscript on file, Department of Anthropology, Ohio State University, Columbus.

Chapter **15**

Game Procurement among Temperate Horticulturists: The Case for Garden Hunting by the Dolores Anasazi

SARAH W. NEUSIUS

This case study examines some of the evidence for game procurement among the Dolores Anasazi who resided in southwestern Colorado (Figure 15-1) between approximately A.D. 600 and A.D. 1000 (Kane 1986). A large archaeological database for these Anasazi was generated under the auspices of the Dolores Archaeological Program (DAP), a Bureau of Reclamation cultural resource management program undertaken by the University of Colorado and various subcontractors between 1978 and 1985 (Lipe 1998; Robinson et al. 1986). On the basis of interpretations of these data, the Dolores Anasazi had a mixed subsistence base that included the cultivation of maize (*Zea mays*), beans (*Phaseolus* spp.), and gourd/squash (*Cucurbita* spp.), use of a variety of wild plants, and procurement of several types of game (Matthews 1986; Neusius 1986; Petersen 1986a). Study of the archaeological assemblages associated with these people provides an opportunity to explore the relationship between animal and plant use in a temperate horticultural system based on shifting cultivation (Orcutt 1985; Petersen 1987).

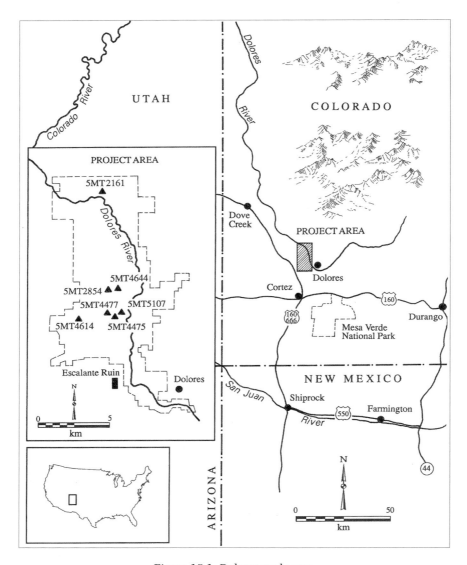

Figure 15-1. Dolores study area.

Anthropologists long have been interested in categorizing patterns of human subsistence and in relating subsistence patterns to cultural change. In this regard, horticulturists are an important focus of interest. These are small-scale food producers who farm without draft animals, the plow, or irrigation, and usually practice some form of shifting cultivation (Netting 1977:57–82). Archaeology in general, and environmental archaeology in particular, has a great deal to contribute to the anthropological understanding of horticultural systems. Not only does the archaeological record give us more cases in which to examine horticulture, but also, and more importantly, it provides cases from a broader geographic range. Although the vast majority of horticultural systems known from ethnography are from tropical settings, horticulturists formerly were found in temperate environments as well. The North American Southwest is one region in which temperate-zone horticulturists existed for many centuries.

From the perspective of environmental archaeology, one obviously significant question is: To what extent were the horticultural systems of the temperate Southwest like or unlike tropical horticultural systems? Perhaps reconstruction of these North American systems will require us to modify or rethink our characterizations of horticulturists in general. This case study contributes to the investigation of these issues by focusing on the zooarchaeological record for the Dolores Anasazi. The specific question asked is whether there is evidence in the Dolores faunal assemblages for garden hunting, a type of game procurement strategy common among horticulturists elsewhere (Linares 1976).

GARDEN HUNTING AMONG HORTICULTURISTS

In order to understand the significance of this question to the general problem of the nature and variability of horticultural systems, it is necessary to adopt an ecological perspective on horticulture. From this perspective, the critical issue is not so much what foods people are eating or the technology and knowledge required to produce food but the fact that food production involves modifying the natural ecosystems in which people live. In this view, horticulturists establish anthropogenic ecosystems that include people, plants, and animals in a set of complex interactions (Ford 1984, 2000). The existence of new habitats produces new selective pressures with long-term evolutionary consequences for people, plants, and nonhuman animals (Rindos 1984). Although the impact of pre-Columbian North Americans on the landscape is traditionally ignored, considerable evidence of the creative role of native practices exists (e.g., Doolittle 2000). Among the complex interactions between native people and local habitats, horticultural practices were significant agents of modification in many North American areas.

It is important to recognize that animals are as much a part of these interactions as plants. Domestic animals are easily viewed as part of agricultural systems, but human actions also alter the distribution and abundance of nondomestic animals (Grayson 2001). Horticultural systems create a mosaic of ecological patches, which, in turn, increase the heterogeneity of resources overall and alter the density and distribution of specific animal species (Hammett 2000). Various species of insects as well as many vertebrates are favored by horticultural practices establishing the conditions for coevolution. In addition, the animals attracted to fields and gardens are readily available sources of high-quality protein that people logically might exploit. The practice of acquiring game in and around fields is reported for various tropical settings (Conklin 1975; Linares 1976; Peterson 1977) where anthropologists have called it *garden hunting*.

Some evidence for this practice is found in the North American ethnohistorical and ethnographic records (Neusius 1990; Neusius and Walker 2002, 2003). For example, Beaglehole (1936) reports the Hopi set traps at the edges of fields in order to prevent crop depredation, and Hill (1938) explains the many ways the Navajo killed animal pests in their fields. Hammett (1997) has amassed evidence for land use patterns suggesting that anthropogenic environments helped maintain game resources among indigenous North Americans. For example, she notes that in the Eastern Woodlands, domestic structures with small dooryard gardens typically were associated with nearby fields of crops, old fields lying fallow and at farther remove, a wooded zone that served as an orchard, wood lot, and hunting park. However, archaeological investigations of garden hunting in temperate as well as tropical settings can add to understanding of horticultural adaptations.

Three assumptions are inherent in the garden hunting model when applied archaeologically. First, the fields and gardens of horticultural peoples as well as field borders and fallow fields are characterized by high density and diversity of animal species, especially in comparison with the undisturbed habitats that surround them. Clearing fields allows both cultigens and weedy pioneer plants to grow, thereby potentially increasing the plant food available for herbivorous species. Horticulturists often allow certain plants that grow in disturbed habitats (ruderal plants) to remain among cultigens because their food value increases the attractiveness of such settings to some animals. Insects also are attracted to fields, as are insect-eating birds, rodents, and carnivores that prey on rodents. Animals may find adequate cover at the borders of fields or even in trees and stumps not removed by burning. The disturbed soils of fields and gardens may be ideal for burrowing species. Although the specific impact of shifting cultivation may be variable (Leonard 1989:115–116), archaeologists have envisioned one common result to be the concentration of resources in habitat patches corresponding to active and abandoned fields and gardens (Ford 1984; Semé 1984). A related argument that does not focus on garden hunting per se is Dean's (2005) suggestion that high representation of certain rodent and bird species favored by farming practices might be used as a proxy for the development of agricultural intensification and sedentism in the Hohokam region.

Garden hunting models also assume that exploitation of animal species that frequent fields and gardens has two distinct benefits for horticulturists. First, of course, these animals provide high-quality protein with which to supplement the diet. Although a diet based on maize, beans, and squash has much nutritional value, the addition of some meat certainly is advantageous. Second, in exploiting many of the animals found in and around their fields and gardens, horticulturists effectively reduce competition for their crops. Conklin (1975:102–103) notes the following reasons for garden hunting among the Hanunóo: "Partly for swidden protection, partly for food, and partly for sport, men and boys also set traps . . . for pestiferous wild mammals and birds, and hunt others with spears . . . and other means . . . ."

Finally, garden hunting has scheduling implications. This kind of game procurement does not necessarily require special preparation and may be embedded in other horticultural activities, particularly because non-selective garden hunting is a low-cost and low-risk means of acquiring meat. Thus, small animals, which are less likely to be preferred game in other hunting strategies, are expected prey for garden hunters. Fields and gardens attract these species, making them reliable and easily obtained sources of protein. Linares (1976) envisioned garden hunters as selecting key species with large body sizes, but Neusius (1984) stressed the probability of nonselective acquisition of meat on the grounds that game procurement was largely incidental to other activities such as planting, weeding, and harvesting. The ethnographic record for the greater Southwest supports the existence of both selective and non-selective garden hunting (Neusius 1990).

Thus, the argument that garden hunting is a common game procurement strategy among horticulturists is not purely ecological but also behavioral. If it were simply ecological, the assertion would be that clearing, planting, and fallowing fields creates a more abundant and diverse complement of animals that people exploit. Human resource exploitation, however, seldom mirrors what is present in an environment. A common response to resource richness is specialization on a few resources rather than generalization. Yet, as formulated here, garden hunting involves the procurement of less attractive species and generalization on a wide variety of species. This strategy is adopted because cultivating plants reduces the time and labor available for hunting at the same time that it alters the natural distribution and abundance of some animal species. Some of the animal species affected might be preferred by hunter-gatherers because of their size and abundance, but others become preferred as a result of peoples' de facto encouragement of environments in which the latter species can flourish and because people no longer can invest as much energy in game procurement. The garden hunting model proposes both that anthropogenic habitats have particular resource structures and that the horticulturists who create such habitats are faced with new behavioral constraints that affect their selection of resources.

ZOOARCHAEOLOGICAL EVIDENCE FOR GARDEN HUNTING

Although the preceding discussion establishes why garden hunting can be predicted for horticulturists like the Dolores Anasazi, it does not address how garden hunting can be recognized archaeologically. Doing so is a difficult methodological problem that has not been resolved. Certain characteristics can be expected, however, in faunal assemblages resulting from garden hunting as opposed to those resulting from other game procurement strategies.

Archaeologists examine three characteristics in testing for garden hunting. The simplest characteristic is the relative proportion of species that are likely targets of garden hunting. These are species with known habitat preferences and behavior that make them likely to frequent fields and gardens. When the remains of such animals occur more commonly in a faunal assemblage than the remains of species not expected to frequent gardens, the existence of garden hunting can be inferred (e.g., Emslie 1981a, b; Neusius 1988a). The problem with utilizing this test alone is that the animals which may have frequented garden habitats probably also frequented nonanthropogenic habitats. Thus, they were not necessarily procured through garden hunting (see Leonard 1989:116).

A second approach is to compare the taxonomic composition of the faunal assemblage to natural species distribution and abundance (e.g., Linares 1976) or to presumed preferences for species based on other factors. Significant deviation from the expected taxonomic composition toward species presumed to prefer garden habitats may be taken as support for the hypothesis of garden hunting. Such comparisons, however, will be only as good as the environmental reconstructions or predicted preferences on which they are based.

A third test for garden hunting is the measurement of diversity in the faunal assemblage (e.g., Neusius and Gould 1988; Semé 1984). The assumption is that because garden hunting is a nonselective strategy, the faunal assemblages associated with it should be highly diverse. However, the use of several procurement strategies at once might result in a diverse faunal assemblage. In addition, garden hunting may sometimes focus on only a few species (Linares 1976). Of course, care must be taken in interpreting diversity within faunal assemblages as some remains could have become incorporated by nonhuman agents (Schmitt and Lupo 1995). Faunal assemblage diversity alone is therefore a less than perfect indicator of garden hunting.

None of the tests for garden hunting proposed by zooarchaeologists is individually conclusive. Examining a variety of possible indicators, however, provides a basis for greater confidence in the interpretation of garden hunting because it is less likely that multiple indicators will be uniformly misleading. Thus, in this study, all three of these tests are performed and the implications are considered.

TESTING FOR GARDEN HUNTING IN THE DOLORES RECORD

Between approximately A.D. 600 and 1000, the Dolores River valley of southwestern Colorado, particularly that portion immediately downstream from the town of Dolores, Colorado (Figure 15-1), was inhabited by Anasazi peoples who apparently moved into and out of the valley from the south and west. The population peak for the Dolores Anasazi occurred during the middle and end of the 9th century A.D. (Kane 1986; Schlanger 1986, 1988). Around A.D. 900, however, a major depopulation or abandonment took place, so that by A.D. 980 very few Anasazi remained in the area. Though use of the Dolores valley before and after these dates can be documented, the dates delimit the major period of human occupation of the valley prior to historic times. The Dolores Anasazi sequence generally corresponds to Late Basketmaker III through Early Pueblo II within the Pecos classification (Kidder 1927), spanning the pithouse to pueblo transition in the northern Southwest. The archaeological record indicates that these people were horticulturists who used wild plants; ruderal plants in their fields; and wild game, especially rabbits and hares (Lagomorpha), rodents (Rodentia), and deer (*Odocoileus* spp.)

One of the most important factors in the colonization and later abandonment of the Dolores valley appears to be changes in climate resulting in an altitudinal shifting of the dry-farming belt (Petersen 1986b). This belt is the area in which crops have an adequate growing season and do not require irrigation. In southwestern Colorado, the elevation of this belt shifts in response to climate. When climate is relatively warm and dry in lower elevations, the belt rises to moister, higher elevations. When climate is relatively cool and moist, the dry-farming belt drops to a lower elevation where the growing season is longer. The Anasazi settlement pattern in the Dolores valley can be seen as a response to fluctuations in length of growing season and in moisture availability (Lipe and Kane 1986; Schlanger 1988). However, it seems probable that rapid population growth and settlement aggregation, and perhaps, attendant social changes, that occurred during the 9th century added to the stress of these environmental fluctuations making the abandonment of the Dolores valley that took place early in the 10th century inevitable (Kohler 1992).

Obviously, climatic changes affected the natural distribution of plants and animals during the Dolores Anasazi's occupation of the valley. Vegetational reconstructions for the study area indicate that the primary result of climatic change was altitudinal shifting and changes in the areal extent of three principal habitat types: sagebrush, pinyon-juniper woodland, and mountain brush (Petersen 1987). Along the Dolores River itself, a variety of aquatic habitats and riparian forests were found. At the highest elevations, aspen (*Populus tremuloides*), ponderosa pine (*Pinus ponderosa*), and Douglas fir (*Pseudotsuga menziesii*) forests were present. A wide variety of wild plants and animals in these habitats provided abundant resources with which to supplement crop foods.

The colonization and settlement of the Dolores valley, specifically the shifting cultivation of Dolores Anasazi horticulturists, undoubtedly disturbed the natural distributions and dynamic equilibrium of these communities. Dolores practices probably involved clearing land through burning, cultivating some fields/gardens, and abandoning fields for extended periods after cultivation had depleted soil nutrients. Petersen et al. (1987) and Kohler and Matthews (1988) review some evidence for these practices in archaeological, geological, and archaeobotanical data. As discussed above, the effect of shifting cultivation on animal distributions may be envisioned as the creation of resource patches in which animals are concentrated (Ford 1984). Such patches overlay the natural habitats where land is arable.

The large archaeological data base generated by the Dolores Archaeological Program (Robinson et al. 1986) seems an obvious place to look for evidence of garden hunting in the temperate Southwest. In fact, garden hunting is addressed in a number of studies of Dolores Anasazi faunal remains and resource use (e.g., Neusius 1984, 1988a, b, 1991; Neusius and Gould 1988; Neusius and Matthews 1985; Neusius and Phagan 1985; Petersen 1986a). These studies focused, however, on the temporal changes in garden hunting expected to occur in response to population increases and agricultural intensification; they did not first establish the existence of garden hunting. Thus, when results were inconclusive, it was unclear which assumptions required reformulation (Neusius 1988b).

This case study addresses this ambiguity by examining the evidence for garden hunting itself. Its starting point is the idea that garden hunting is a normal means of game procurement in all horticultural systems regardless of their location. If this is the case, garden hunting by the Dolores Anasazi is expected, and evidence should be found in related faunal assemblages.

In this study, 28,463 faunal specimens divided into ten faunal assemblages collected from seven habitation sites at Dolores form the database (Figure 15-1 and Table 15-1). These assemblages are further subdivided into specimens that could be identified to genus or species (identifiable assemblage) and those which could not. Individual site assemblages were chosen because of the concern that aggregation of site data as in Neusius (1988a) could introduce hidden biases. Temporal subdivisions were recognized only in the large McPhee Village (5MT4475) assemblage. For purposes of this study, it is assumed that the assemblages in Table 15-1 are characteristic of habitation sites in the Dolores Anasazi sequence. A more exhaustive exploration of the data is not attempted here.

All data used in this study are based on the Number of Individual Specimens (NISP) rather than on Minimum Number of Individuals (MNI) or any of a number of other possible zooarchaeological units (see Lyman 1994). NISP is used as the unit of measurement because of the complexity of aggregation issues inherent in MNI calculation in these data, because of the small sample sizes for these assemblages, and because MNI is a function of NISP (Grayson 1984).

The first characteristic of the Dolores faunal assemblages is the proportion of each assemblage that consists of animals likely to be in gardens and fields. Determining this characteristic means considering which of the taxa represented have habitat preferences and behavioral characteristics that would make fields and gardens especially attractive to them. As is common in faunal

Table 15-1. Dolores Faunal Assemblages Used in This Study

Site	Dates	NISP[a] Total	NISP[a] Identifiable
Prince Hamlet 5MT2161	A.D. 780–900	1401	361
Aldea Sierritas 5MT2854	A.D. 720–800	3086	972
McPhee Village 5MT4475			
Element[b] 1	A.D. 760–850	1209	232
Element 2	A.D. 850–900	7228	1049
Element 3	A.D. 920–940	2891	817
Element 4	A.D. 940–980	4040	901
Masa Negra Pueblo 5MT4477	A.D. 860–980	2736	629
Prairie Dog Hamlet 5MT4614	A.D. 700–780	2459	806
Windy Wheat Hamlet 5MT4644	A.D. 720–820	1291	354
Pueblo de las Golondrinas 5MT5107	A.D. 760–900	2122	465
Total		28,463	6586

[a] Total NISP is the number of bone specimens; Identifiable NISP is the number of bone specimens that can be assigned to a genus or species.
[b] In the Dolores Archaeological Project, an element is equivalent to a temporal component or occupation. Robinson et al. (1986) provide a revised assessment of elements at McPhee Village that postdates the faunal analysis on which this study is based.

assemblages, the majority of the faunal remains are not sufficiently diagnostic to be assessed in this manner. For example, specimens identified as artiodactyl (which includes elk [*Cervus elaphus*], mule deer [*Odocoileus hemionus*], bighorn sheep [*Ovis canadensis*], and antelope [*Antilocapra americana*]) cannot be considered as coming from a taxon that is likely either to be found or not to be found in fields and gardens. Since no conclusion can be reached about specimens identified to class, order, or family, only the identifiable assemblage is analyzed here. This limitation is unfortunate because it further reduces the size of the assemblages (Table 15-1) and because largely unknown biases in preservation, completeness, or morphology may be affecting which specimens are diagnostic to genus or species and become part of the identifiable assemblage. Without further study, it only can be assumed that such factors are randomly distributed between garden and nongarden taxa. Since this classification is not based on either size or taxonomy, there is some reason for confidence in such an assumption.

Table 15-2 lists those taxa from the identifiable assemblage that are considered likely to be found in fields and gardens. The assumptions upon which this classification are based are discussed in Neusius (1984, 1986, 1988a; Neusius and Gould 1988). The mule deer requires further discussion, however. Behavioral accounts of mule deer indicate that they often are attracted to crops. There is some evidence, however, that mule deer herds in the Dolores vicinity spent the warmest months of the year at higher elevations, far removed from Anasazi fields and gardens (Burkhardt et al. 1980). Thus, the proportions of garden taxa given in Table 15-3 include both a calculation with mule deer and, in parentheses, a calculation from which mule deer are excluded.

Table 15-2. Garden Taxa in the Identifiable Faunal Assemblages[a]

Cottontail rabbits[b] (*Sylvilagus* spp.)	Cooper's hawk (*Accipiter cooperii*)
Jackrabbits (*Lepus* spp.)	Red-tailed hawk (*Buteo jamaicensis*)
White-tailed jackrabbit (*L. townsendii*)	Swainson's hawk (*Buteo swainsoni*)
Black-tailed jackrabbit (*L. californicus*)	Golden eagle (*Aquila chrysaetos*)
Yellow-bellied marmot (*Marmota flaviventris*)	Bald eagle (*Haliaeetus leucocephalus*)
Spotted ground squirrel (*Spermophilus spilosoma*)	American kestrel (*Falco sparverius*)
Golden-mantled ground squirrel (*Spermophilus lateralis*)	Sage grouse (*Centrocercus urophasianus*)
Rock squirrel (*Spermophilus variegatus*)	Turkey (*Meleagris gallopavo*)
Gunnison's prairie dog (*Cynomys gunnisoni*)	Sandhill crane (*Grus canadensis*)
Pocket gophers (*Thomomys* spp.)	Mourning dove (*Zenaida macroura*)
Valley pocket gopher (*Thomomys bottae*)	Screech owl (*Otus asio*)
Apache pocket mouse (*Perognathus apache*)	Great horned owl (*Bubo virginianus*)
Northern grasshopper mouse (*Onychomys leucogaster*)	Western kingbird (*Tyrannus verticalis*)
Mexican vole (*Microtus mexicanus*)	Common raven (*Corvus corax*)
Muskrat (*Ondatra zibethicus*)	Common crow (*Corvus brachyrhynchos*)
	Black-billed magpie (*Pica pica*)
Red fox (*Vulpes vulpes*)	Robin (*Turdus migratorius*)
Gray fox (*Urocyon cinereoargenteus*)	Meadowlarks (*Sturnella* spp.)
Mule deer[c] (*Odocoileus hemionus*)	Yellow-headed blackbird (*Xanthocephalus xanthocephalus*)
Canada goose (*Branta canadensis*)	
Mallard (*Anas platyrhynchos*)	

[a] Included in garden taxa are species and genera that are likely to frequent fields and gardens (Neusius 1984, 1986, 1988a; Neusius and Gould 1988).
[b] All cottontail bones, including Nuttall's cottontail (*S. nuttalli*) and desert cottontail (*S. audubonii*), were lumped together.
[c] See the discussion of mule deer in the text.

Table 15-3. Indicators of Garden Hunting in the Dolores Faunal Assemblages

Site	Identifiable NISP	Garden[a]	Absent[b]	r_s[c]	J[c]
Prince Hamlet	361	73.9% (40.7%)	40.0%	.510	.624
Aldea Sierritas	972	88.4% (87.3%)	60.0%	.476	.555
McPhee Village					
Element 1	232	89.6% (82.7%)	52.0%	.505	.702
Element 2	1049	83.2% (75.6%)	28.0%	.314	.609
Element 3	817	86.3% (77.2%)	32.0%	.579	.604
Element 4	901	80.2% (69.0%)	20.0%	.439	.668
Masa Negra Pueblo	629	90.9% (75.6%)	32.0%	.313	.644
Prairie Dog Hamlet	806	90.1% (85.0%)	36.0%	.420	.582
Windy Wheat Hamlet	354	90.9% (82.4%)	68.0%	.382	.671
Pueblo de las Golondrinas	465	82.0% (63.9%)	36.0%	.338	.720

[a] Garden taxa are those listed in Table 15-2. The percentages in parentheses are the percentages of these taxa if mule deer are excluded.
[b] The Absent percentages are the proportions of the preferred mammalian species listed in Table 15-4 that are not present in the faunal assemblage.
[c] The possible range of both Spearman's rho (r_s) and J is from 0.00 to 1.00.

Regardless of whether or not this adjustment is made, garden taxa appear to predominate in these assemblages (Table 15-3). The range is between 74% and 91% of NISP if mule deer is included, and between 41% and 87% if mule deer are excluded. Only in one assemblage, that from Prince Hamlet (5MT2161), are garden taxa less than 60% of NISP, and then only if mule deer fragments are excluded. Prince Hamlet's location at the northern end of the study area away from the other sites in this study (Figure 15-1) may be a factor. The implication of these data is that fields and gardens were important sources of game. Because many of these species might be procured in other habitats, such an interpretation is more probable if the faunal assemblages deviate in composition from predictions based on natural distributions.

Data were collected by the Dolores Archaeological Program on the natural distribution and abundance of mammals in the study area (Neusius 1986), and predictions of those mammals most likely to be exploited were made on the basis of these data. These predictions assumed that, if all else were equal, the preferred game species would be those with high species biomass. Species biomass is calculated by multiplying average body weight per individual by density in the project area (Neusius 1986). Of course, the garden hunting model assumes that all else is not equal. Table 15-4 lists the top 25 mammals based on relative biomass. It is instructive to compare this list with that of animals likely to frequent fields and gardens (Table 15-2). Moreover, many of the mammals considered likely to be preferred are lacking altogether from the faunal assemblages examined here. Depending on the assemblage, from 20% to 68% of these mammalian species are simply not represented (Absent % in Table 15-3).

Table 15-4. Most Preferred 25 Mammalian Resources in the Dolores Area[a]

1. Black bear (*Ursus americanus*)	15. Raccoon (*Procyon lotor*)
2. Mule deer (*Odocoileus hemionus*)	16. Gray fox (*Urocyon cinereoargenteus*)
3. American elk (*Cervus elaphus*)	17. Gray wolf (*Canis lupus*)
4. Pronghorn antelope (*Antilocapra americana*)	18. Red fox (*Vulpes vulpes*)
5. Black-tailed jackrabbit (*Lepus californicus*)	19. Yellow-bellied marmot (*Marmota flaviventris*)
6. Beaver (*Castor canadensis*)	20. White-tailed jackrabbit (*Lepus townsendii*)
7. Bighorn sheep (*Ovis canadensis*)	21. Golden-mantled ground squirrel (*Spermophilus lateralis*)
8. Desert cottontail (*Sylvilagus audubonii*)	22. Snowshoe hare (*Lepus americanus*)
9. Coyote (*Canis latrans*)	23. Striped skunk (*Mephitis mephitis*)
10. Muskrat (*Ondatra zibethicus*)	24. Nuttall's cottontail (*Sylvilagus nuttalli*)
11. Badger (*Taxidea taxus*)	25. Marten (*Martes americana*)
12. Bobcat (*Lynx rufus*)	
13. Porcupine (*Erethizon dorsatum*)	
14. Mountain lion (*Puma concolor*)	

[a] Species are listed in order of preference based on biomass estimates adjusted for seasonal fluctuations in availability (Neusius 1986:204–212).

When the apparent rank order importance of different mammalian species as suggested by the composition of these faunal assemblages is compared with the predicted rank order of preferences for mammals, the correspondence is not strong. Spearman's rho, a commonly used rank-order correlation coefficient (Thomas 1976:395–406), varies between 0.313 and 0.579 in these assemblages (rho can range from 0.00 to 1.00). Tests for statistical significance indicate that the null hypothesis of no correlation between rank orders can be rejected only for McPhee Village, Element 3. In all other cases, the correlation between the rank order in the assemblages and that predicted by biomass estimates is not high enough to preclude chance association ($\alpha = 0.01$). These data support the idea that biomass is not the primary determinant of mammalian exploitation. Since the proportion of taxa likely to be found in fields and gardens is high, the argument for garden hunting is supported, with the possible exception of McPhee Village, Element 3.

Assuming that garden hunting at Dolores was an embedded, nonselective game procurement strategy, a wide variety of garden taxa is expected. This prediction can be evaluated by examining taxonomic diversity in the faunal assemblages. Archaeological studies employ a number of diversity indices (Leonard and Jones 1989). In this study, evenness, or J, is used. This statistic can be derived from the Shannon-Wiener index of diversity (H) as follows:

$$J = \frac{H}{H_{max}}$$

where H_{max} is the log of the number of categories or taxa, H is $-\Sigma p_i \log p_i$, and p_i is the proportion of the ith category. This measure of evenness was chosen because it varies between 0.00 and 1.00, making it easy to understand. As shown in Table 15-3, J varies from 0.555 to 0.720 in these faunal assemblages. It is difficult to interpret these figures without comparative data. Evenness certainly is not low, but does it deviate significantly from what would be expected in a nongarden strategy? Although subjectively the answer to this question is yes, there is no clear way to test this impression within the Dolores database itself. Kintigh (1984, 1989) provides a method of assessing the significance of deviations in evenness among assemblages that takes into account varying sample sizes. This method is not pursued here because the issue really is not whether one or the other of these assemblages is significantly more diverse than the others. Instead, the issue is whether diversity is high enough to suggest garden hunting in all these assemblages. Development of a theoretical distribution for J based on predictions for alternative game procurement strategies is beyond the scope of this study, though it would be one way of evaluating the significance of the evenness figures listed in Table 15-3.

DISCUSSION

Considered together, the results of these three tests support the hypothesis that the Dolores Anazasi did garden hunt. The percentage of taxa in the identifiable assemblages that could represent garden hunting is quite high. These proportions suggest that the garden habitat was a preferred source of game. Moreover, the apparent preferences in game procurement generally do not reflect those expected if biomass were the main determinant of resource selection. In other words, efficiency in maximizing the amount of game or meat procured does not seem to have structured resource selection unless anthropogenic habitats are taken into account. In addition, although it is difficult to assess the actual significance of the diversity estimates obtained, the diversity of taxa in these assemblages is consistent with the expectation of a nonselective game procurement strategy such as garden hunting.

A comparison of NISP for the identifiable assemblages and each of the measures of garden hunting given in Table 15-3 suggests that sample size is not a prime determinant of the variability that exists in these measures. The proportion of preferred mammalian species that are not present may be inversely correlated with sample size. The Aldea Sierritas (5MT2854) assemblage, however, is not consistent with this pattern. With a few exceptions as noted previously, these indicators support the expectations based on garden hunting. Thus, in the absence of an alternative, equally plausible explanation for these findings, the garden hunting strategy is confirmed for Dolores Anasazi horticulturists.

This conclusion is far from unassailable, but it does allow a research program based on the garden hunting model to be developed for investigation of Dolores Anasazi faunal use. All the tests used in this study rely only on measures of taxonomic composition. Other aspects of the faunal record should be evaluated for additional evidence of garden hunting. For example, skeletal element distributions might be informative. Resource selection probably is not the only way in which garden hunting differs from other exploitation strategies. Processing practices for animals procured in fields and gardens might differ from those for animals procured some distance from villages during long-distance hunts. Moreover, the consumption of meat and the disposal of bones and other waste acquired through garden hunting might be structured by different constraints on food sharing than the consumption and disposal of game acquired through cooperative hunts. Ubiquity measures, which examine the presence of garden species across excavated contexts, might be informative and isotopic analyses of deer bones might indicate whether the deer taken actually were eating crops. It seems clear that other tests for garden hunting could be explored.

Nevertheless, it is reasonable to assert that garden hunting was an important game procurement strategy for the Dolores Anasazi. Making this assertion suggests new areas of investigation. For example, does the reliance on garden hunting vary among site types? Faunal assemblages from habitation sites might have a different composition from those from field houses and other special-purpose sites. Both Prince Hamlet and Element 3 at McPhee Village should be studied more closely for variations in the degree of reliance on garden hunting. It is possible that there is evidence for seasonal variation in game procurement strategies. Winter procurement of large game is proposed elsewhere for the Dolores Anasazi (Neusius 1988a, 1991; Neusius and Phagan 1985). The question of temporal variation in reliance on garden hunting must be revisited. As noted, attempts to recognize such variation within the Dolores record (Neusius 1984, 1988a; Neusius and Gould 1988) have produced inconclusive results, but more thorough analyses may provide new perspectives. Through the investigation of questions such as these, much greater understanding of the significance of the Dolores Anasazi faunal data will be gained.

More importantly, the Dolores case should encourage environmental archaeologists to evaluate the evidence for garden hunting in other contexts. Leonard (1989:182) has observed that although archaeologists interested in the Southwest have accomplished much through the development of normative generalizations about subsistence and other cultural attributes, the investigation of intersite variation and interregional contrasts is important. Even though intersite variation is not the focus of this case study, such variability is suggested as a problem for further work, and the Dolores case should be compared with data from other areas. It would be inappropriate to use the Dolores case to generalize about game procurement among the Anasazi or among all southwestern horticulturists. Instead, the tests used here, as well as others yet to be devised, should be employed in investigating the specifics of game exploitation elsewhere. Is the evidence similar or is it not? If not, what are the probable reasons for observed differences? In this way, the garden hunting model can provide a structure for exploring variation in animal procurement among horticulturists and its causes.

CONCLUSION

Through the application of the garden hunting model, environmental archaeologists can move beyond questions about what resources were used by the inhabitants of particular sites toward a reconstruction of the exploitation strategies. Over time, a body of evidence for variation in strategies can be generated that will lead to greater understanding of temporal and spatial variation in southwestern horticultural adaptations and their relationship to similar horticultural adaptations in other parts of the world. This body of evidence contributes to a more representative characterization of horticultural systems than anthropologists now have. It is mainly as a contribution to this larger process that this case study, and indeed any characterization of Dolores Anasazi faunal exploitation, is significant.

ACKNOWLEDGMENTS

The data utilized in this study were generated under the auspices of the Dolores Archaeological Program, which was funded by a contract between the University of Colorado and the United States Bureau of Reclamation, Upper Colorado Region (Contract No. 8–07–40-S0562). Some of the faunal data included here were generated under the direction of Steven D. Emslie prior to 1981. The author was in charge of zooarchaeological analyses for this project between November 1981 and August 1984, and gratefully acknowledges having been granted continuing access to the database since that time. This faunal data base is stored at the Anasazi Heritage Center, a Bureau of Land Management curation facility and museum located in Dolores, Colorado. Phil Neusius's comments on the first draft of this chapter also are much appreciated even when the author stubbornly has persisted in her folly.

REFERENCES

Beaglehole, E., 1936, Hopi Hunting and Hunting Ritual, *Yale University Publications in Anthropology* 4: 3–26.

Burkhardt, W. T., Graham, V. K., and Hanna, D. D., 1980, Wildlife Assessment for Dolores River Project Proposed Recreation Sites, submitted to U.S. Fish and Wildlife Service, copies available from Department of Natural Resources, Colorado Division of Wildlife, Denver.

Conklin, H. C., 1975, *Hanunóo Agriculture: A Report on an Integral System of Shifting Cultivation in the Phillipines*, Elliot's Books, Northfield, Connecticut.

Dean, R. M., 2005, Site-Use Intensity, Cultural Modification of the Environment, and the Development of Agricultural Communities in Southern Arizona, *American Antiquity* 70(3):403–431.

Doolittle, W. E., 2000, *Cultivated Landscapes of Native North America*, Oxford University Press, Oxford.

Emslie, S. D., 1981a, Birds and Prehistoric Agriculture: The New Mexican Pueblos, *Human Ecology* 9(3):305–329.

Emslie, S. D., 1981b, Prehistoric Agricultural Ecosystems: Avifauna from Pottery Mound, New Mexico, *American Antiquity* 46(4):853–861.

Ford, R. I., 1984, Ecological Consequences of Early Agriculture in the Southwest, in: *Papers on the Archaeology of Black Mesa, Arizona*, Volume II (S. Plog and S. Powell, eds.), Southern Illinois University Press, Carbondale, pp. 127–138.

Ford, R. I., 2000, Human Disturbance and Biodiversity: A Case Study from Northern New Mexico, in: *Biodiversity and Native America* (P. E. Minnis and W. J. Elisens, eds.), University of Oklahoma Press, Norman, pp. 207–222.

Grayson, D. K., 1984, *Quantitative Zooarchaeology*, Academic Press, Orlando, Florida.

Grayson, D. K., 2001, The Archaeological Record of Human Impacts on Animal Populations, *Journal of World Prehistory* 15(1):1–68.

Hammett, J. E., 1997, Interregional Patterns of Land Use and Plant Management in Native North America, in: *People, Plants and Landscapes: Studies in Paleoethnobotany* (K. J. Gremillion, ed.), University of Alabama Press, Tuscaloosa.

Hammett, J. E., 2000, Ethnohistory of Aboriginal Landscapes in the Southeastern United States, in: *Biodiversity and Native America* (P. E. Minnis and W. J. Elisens, eds.), University of Oklahoma Press, Norman, pp. 248–299.

Hill, W. W., 1938, Agricultural and Hunting Methods of Navajo Indians, *Yale University Publications in Anthropology* 18: 1–194.

Kane, A. E., 1986, Prehistory of the Dolores River Valley, in: *Final Synthetic Report* (D. A. Breternitz, C. K. Robinson and G. T. Gross, comps.), Dolores Archaeological Program Series, Engineering and Research Center, Bureau of Reclamation, U.S. Department of the Interior, Denver, Colorado, pp. 353–435.

Kidder, A. V., 1927, The Southwestern Archaeological Conference, *Science* 66:489–491.

Kintigh, K. W., 1984, Measuring Archaeological Diversity by Comparison with Simulated Assemblages, *American Antiquity* 49(1):44–54.

Kintigh, K. W., 1989, Sample Size, Significance and Measures of Diversity, in: *Quantifying Diversity in Archaeology* (R. D. Leonard and G. T. Jones, eds.), Cambridge University Press, Cambridge, England, pp. 25–36.

Kohler, T. A., 1992, Fieldhouses, Villages and the Tragedy of the Commons in the Early Northern Anasazi Southwest, *American Antiquity* 57(4):617–635.

Kohler, T. A., and Matthews, M. H., 1988, Long-term Anasazi Land Use and Forest Reduction: A Case Study from Southwest Colorado, *American Antiquity* 53(3):537–564.

Leonard, R. D., 1989, *Anasazi Faunal Exploitation: Prehistoric Subsistence on Northern Black Mesa, Arizona, Center for Archaeological Investigations Occasional Paper* 13, Southern Illinois University, Carbondale.

Leonard, R. D., and Jones, G. T. (eds.), 1989, *Quantifying Diversity in Archaeology*, Cambridge University Press, Cambridge, England.

Linares, O. F., 1976, "Garden Hunting" in the American Tropics, *Human Ecology* 4(4):331–349.

Lipe, W. D., 1998, View From the Lake: Legacies of the Dolores Archaeological Program, SW Colorado, paper presented at the 63rd Annual Meeting of the Society for American Archaeology, Seattle, Washington.

Lipe, W. D., and Kane, A. E., 1986, Evaluations of the Models with Dolores Area Data, in: *Final Synthetic Report* (D. A. Breternitz, C. K. Robinson and G. T. Gross, comps.), Dolores Archaeological Program Series, Engineering and Research Center, Bureau of Reclamation, U.S. Department of the Interior, Denver, Colorado, pp. 703–707.

Lyman, R. L., 1994, Quantitative Units and Terminology in Zooarchaeology, *American Antiquity* 59(1):36–71.

Matthews, M. H., 1986, The Dolores Archaeological Program Macrobotanical Data Base: Resource Availability and Mix, in: *Final Synthetic Report* (D. A. Breternitz, C. K. Robinson and G. T. Gross, comps.), Dolores Archaeological Program Series, U.S. Department of the Interior, Bureau of Reclamation, Engineering and Research Center, Denver, Colorado, pp. 151–184.

Netting, R. McC., 1977, *Cultural Ecology*, Cummings Publishing, Menlo Park, California.

Neusius, S. W., 1984, Garden Hunting and Anasazi Game Procurement: Perspectives from Dolores, paper presented at the 49th Annual Meeting of the Society for American Archaeology, Portland, Oregon.

Neusius, S. W., 1986, The Dolores Archaeological Program Faunal Data Base: Resource Availability and Resource Mix, in: *Final Synthetic Report* (D. A. Breternitz, C. K. Robinson, and G. T. Gross, comps.), Dolores Archaeological Program Series, Engineering and Research Center, Bureau of Reclamation, U.S. Department of the Interior, Denver, Colorado, pp. 199–303.

Neusius, S. W., 1988a, Faunal Exploitation during the McPhee Phase: Evidence from the McPhee Community Cluster, in: *Anasazi Communities at Dolores: McPhee Village* (A. E. Kane and C. K. Robinson, comps.), Dolores Archaeological Program Series, Engineering and Research Center, Bureau of Reclamation, U.S. Department of the Interior, Denver, Colorado, pp. 1208–1291.

Neusius, S. W., 1988b, Population Growth, Climatic Change and Hunting Strategies among the Dolores Anasazi, paper presented at the 53rd Annual Meeting of the Society for American Archaeology, Phoenix, Arizona.

Neusius, S. W., 1990, Hunting Strategies and Horticulturalists: The Anasazi and Iroquoian Cases, paper presented at the 6th Conference of the International Council of Archaeozoology, Washington, DC.

Neusius, S. W., 1991, Seasonality and Game Procurement among the Dolores Anasazi, paper presented at the 56th Annual Meeting of the Society for American Archaeology, New Orleans, Louisiana.

Neusius, S. W., and Gould, M., 1988, Faunal Remains: Implications for Dolores Anasazi Adaptations, in: *Anasazi Communities at Dolores: Grass Mesa Village* (W. D. Lipe, J. N. Morris, and T. A. Kohler, comps.), Dolores Archaeological Program Series, Engineering and Research Center, Bureau of Reclamation, U.S. Department of the Interior, Denver, Colorado, pp. 1049–1135.

Neusius, S. W., and Matthews, M. H., 1985, Resource Mix: Changes in the Relative Importance of Plant and Animal Foods of the Dolores Anasazi, paper presented at the 50th Annual Meeting of the Society for American Archaeology, Denver, Colorado.

Neusius, S. W., and Phagan, C. J., 1985, Patterns of Large and Small Game Procurement among the Dolores Anasazi, A.D. 600–900, *Dolores Archaeological Program Technical Reports DAP-221*, submitted to the Bureau of Reclamation, Upper Colorado Region, Salt Lake City, Contract 8–07–40-S0562.

Neusius, S. W., and Walker, R. B., 2002, Garden Hunting: Part of Northeastern Horticultural Systems or Not? paper presented in the Symposium "The History and Impacts of Maize-Beans-Squash Agriculture," The Northeast Natural History Conference VII, Albany, New York.

Neusius, S. W., and Walker, R. B., 2003, Garden Hunting in the Native Northeast: Ethnohistorical and Archaeological Evidence, paper presented at the 68th Annual Meeting of the Society for American Archaeology, Milwaukee, Wisconsin.

Orcutt, J., 1985, Modeling Prehistoric Agricultural Ecology in the Dolores Area, *Dolores Archaeological Program Technical Reports DAP-227*, submitted to the Bureau of Reclamation, Upper Colorado Region, Salt Lake City, Contract No. 8–07–40-S0562.

Petersen, K. L., 1986a, Resource Studies, in: *Final Synthetic Report* (D. A. Breternitz, C. K. Robinson and G. T. Gross, comps.), Dolores Archaeological Program Series, Engineering and Research Center, Bureau of Reclamation, U.S. Department of the Interior, Denver, pp. 3–50.

Petersen, K. L., 1986b, Climatic Reconstruction for the Dolores Project, in: *Final Synthetic Report* (D. A. Breternitz, C. K. Robinson and G. T. Gross, comps.), Dolores Archaeological Program Series, Engineering and Research Center, Bureau of Reclamation, U.S. Department of the Interior, Denver, pp. 311–325.

Petersen, K. L., 1987, Vegetation Reconstruction, in: *Supporting Studies: Settlement and Environment* (K. L. Petersen and J. D. Orcutt, comps.) Dolores Archaeological Program Series, Engineering and Research Center, Bureau of Reclamation, U.S. Department of the Interior, Denver, pp. 147–184.

Petersen, K. L., Clay, V. L., Matthews, M. H., and Neusius, S. W., 1987, Implications of Anasazi Impact on the Landscape, in: *Supporting Studies: Settlement and Environment* (K. L. Petersen and J. D. Orcutt, comps.), Dolores Archaeological Program Series, Engineering and Research Center, Bureau of Reclamation, U.S. Department of the Interior, Denver, pp. 147–184.

Peterson, J., 1977, Ecotones and Exchange in Northeast Luzon, in: *Economic Exchange and Social Interaction in Southeast Asia* (K. L. Hutterer, ed.), *Michigan Papers on Southeast Asia* 13, University of Michigan, Ann Arbor.

Rindos, D., 1984, *The Origins of Agriculture: An Evolutionary Perspective*, Academic Press, San Diego.

Robinson, C. K., Gross, G. T., and Breternitz, D. A., 1986, Overview of the Dolores Archaeological Program, in: *Final Synthetic Report* (D. A. Breternitz, C. K. Robinson and G. T. Gross, comps.), Dolores Archaeological Program Series, Engineering and Research Center, Bureau of Reclamation, U.S. Department of the Interior, Denver, pp. 3–50.

Schlanger, S. H., 1986, Population Studies, in: *Final Synthetic Report* (D. A. Breternitz, C. K. Robinson, and G. T. Gross, comps.), Dolores Archaeological Program Series, Engineering and Research Center, Bureau of Reclamation, U.S. Department of the Interior, Denver, Colorado, pp. 493–524.

Schlanger, S. H., 1988, Patterns of Population Movement and Long-Term Population Growth in Southwestern Colorado, *American Antiquity* 53(4):773–793.

Schmitt, D. N., and Lupo, K. D., 1995, On Mammalian Taphonomy, Taxonomic Diversity, and Measuring Subsistence Data in Zooarchaeology, *American Antiquity* 60(3):496–514.

Semé, M., 1984, The Effects of Agricultural Fields on Faunal Assemblage Variation, in: *Papers on the Archaeology of Black Mesa, Arizona*, Volume II (S. Plog and S. Powell, eds.), Southern Illinois University Press, Carbondale, pp. 139–157.

Thomas, D. H., 1976, *Figuring Anthropology*, Holt, Rinehart and Winston, New York.

Chapter **16**

The Emergence of Maize Farming in Northwest Mexico

ROBERT J. HARD, KAREN R. ADAMS, JOHN R. RONEY, KARI
M. SCHMIDT, AND GAYLE J. FRITZ

The emergence of farming economies is among the most significant developments in human history. In many places the shift from reliance on wild
resources to reliance on domesticated plants and animals coincides with
sedentary communities the evolution of complex societies, population growth,
elaboration in technology, and art. Since typical farming requires more work
and usually results in a decline in health compared with most hunting and
gathering strategies, anthropologists are interested in understanding why
farming was adopted in so many parts of the world (Barlow 2002; Cohen 1987;
Hames 1992; Roosevelt 1984).

In the Americas, the spread of maize (*Zea mays* ssp. *mays*) cultivation from
its origins in Mesoamerica profoundly affected human populations as they
shifted to a farming way of life. This case study focuses on the earliest known
maize cultivation efforts in northwest Chihuahua, Mexico but it has relevance
to a much wider area. The adoption of maize in the southwest United States
(Arizona, New Mexico, and portions of Colorado and Utah) is well studied,
but the topic has been little explored in northwest Mexico (states of
Chihuahua, Sonora, and parts of Durango and Sinaloa), despite the fact that
these areas form one culture area, generally known as the "Greater Southwest"
(Figure 16-1). Northwest Mexico is particularly important as it lies between
Mesoamerica, where maize originated, and the southwest United States, where
it became fundamental to ancient economies. We assess the integration of

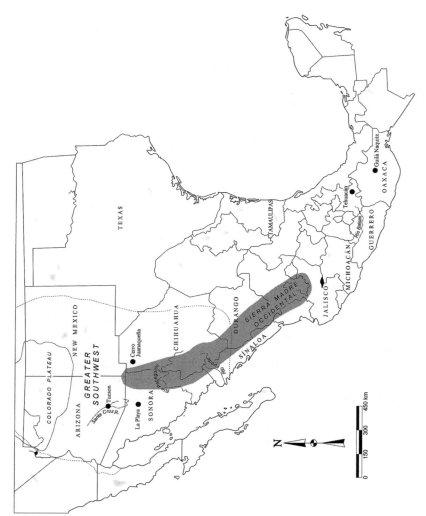

Figure 16-1. The Greater Southwest culture area, shown by the dotted line, includes northwest Mexico and southwest United States (modified from MacWilliams et al. 2008)

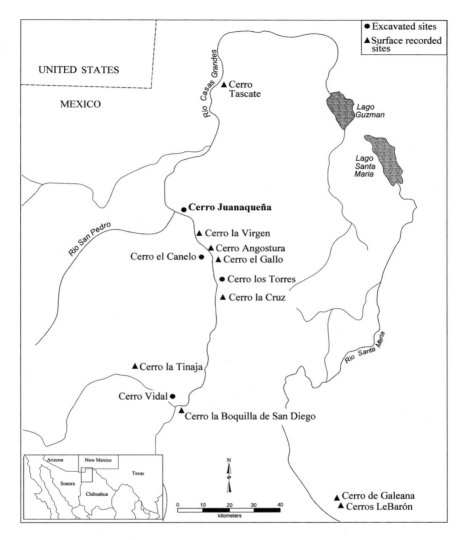

Figure 16-2. Map of Cerro Juanaqueña and other cerros de trincheras sites in northwest Chihuahua, Mexico.

farming into the local hunting and gathering economies at a series of terraced hillside sites (cerros de trincheras), including the large site of Cerro Juanaqueña (Figure 16-2). Our research reveals that maize became an important component of the diet at Cerro Juanaqueña by 1350–1300 B.C. This early and unusual farming adaptation has relevance for understanding the timing and movements of maize and agricultural adaptations from Mesoamerica into northwest Mexico and the southwest United States.

Our multi-disciplinary team of researchers worked together over a number of years to develop lines of evidence for the emergence of a maize-based farming strategy. We review the theoretical context and major findings of the project including chronology, paleoecology, site architecture, and the roles of the cerros de trincheras sites as defensive and domestic settlements. We then present the records of plants and animals in these terraced hillside communities, and examine food-processing implements to understand the role domestic plants played in cerros de trincheras economies. Finally, we evaluate the ecological and cultural contexts prevailing when maize was integrated into the subsistence base.

THEORETICAL FRAMEWORK

Understanding the spread of maize agriculture from Mesoamerica to northwest Mexico and then into the southwest United States is fundamental to addressing the impetus for the adoption of farming and its consequences. We focus here on maize, as it is typically the most frequently recovered domesticate, the best studied, and was a critical subsistence resource to ancient populations.

Theoretical discussions of the transition to farming focus on two key processes: the introduction of agriculture and the formation of an agricultural economy. Our concern here is the latter. Smith (2001b) delineates three adaptive patterns relevant to understanding the formation of an agricultural economy: (1) low-level food production without domesticates; (2) low-level food production with domesticates; and (3) agriculture. According to Smith (2001b:16–17), the transition between low-level food production with domesticates and agriculture occurs when domesticates make up 30–50% of the diet, depending upon local conditions. Agricultural adaptations are accompanied by other cultural changes including shifts in settlement, seasonal movement, social organization, and material culture.

Domesticated maize and some species of beans (*Phaseolus* spp.) and gourd/squash (*Cucurbita* spp.) originated in tropical Mesoamerica and then spread to other parts of the Americas (Benz and Long 2000; Benz and Staller 2006; Mangelsdorf 1986; Matsuoka et al. 2002). DNA comparisons between maize and its wild ancestral plant, teosinte (*Zea mays* ssp. *parviglumis*), suggest that the divergence between teosinte and maize has a 95% confidence level of occurring in the range of 3739–11,143 B.C., if DNA mutation rates are correctly inferred (Benz 2006; Matsuoka et al. 2002). This variety of teosinte is currently found along the Rio Balsas in the Mexican states of Michoacán and Guerrero where maize may have evolved (Benz 2006). However, direct evidence of this scenario is lacking because primitive maize of that age has not been found in the region.

The earliest primitive maize, from a cave in Oaxaca known as Guilá Naquitz, is directly radiocarbon dated to 4300 B.C. (Piperno and Flannery 2001). From there, maize spread northward to Puebla's Tehuacán Valley by about 3500 B.C. (Long et al. 1989:1039), to Tamaulipas by 2400 B.C. (Smith 1997), and then into Arizona and New Mexico by 2100 B.C. (Huber 2005;

Huckell et al. 1999; Long et al. 1989:1039; Shackley 2005; Smith 1997, 2001a:1326, 2005; see Fowler, this volume). The Early Agricultural period (2100 B.C. to A.D. 100) is the interval during which domesticated maize reached northwest Mexico and the southwest United States.

In northwest Mexico and the southwest U.S., maize was integrated into a culture where pottery use was limited and the major subsistence strategy was hunting and gathering. Quite possibly some groups practiced manipulation of wild plants, that is low-level food production without domestic plants (Hard et al. 2006; Huber 2005; Huckell 1995, 1996). This could include such strategies as burning and other forms of landscape disturbance, transplanting wild plants, and casual broadcasting of seeds (see Adams 2004; Smith 2001b). Direct data on wild plant manipulations from pre-maize contexts (prior to circa 1500–2000 B.C.) in the Greater Southwest are rare. As we describe below, amaranth (*Amaranthus* spp.) was being cultivated, if not already domesticated, alongside maize at Cerro Juanaqueña at circa 1300 B.C. In later time periods, there is evidence that a number of wild species were being manipulated (e.g., Bohrer 1991; Minnis 1992:123). Such practices may have a long time depth.

Based on recent evidence from central Arizona and west-central New Mexico, maize entered the Southwest by 4100 years ago (Huber 2005; Shackley 2005). By 3000–3500 years ago (e.g., Hard and Roney 1998; Wills 1985), maize was widespread. Although variability characterizes the process of maize introduction and use, minor reliance on maize, that is low-level food production with domesticates, is documented for much of the Southwest (Adams 1994; Benz and Long 2000:36, 47; Diehl 1997; Hard et al. 1996; Haury 1962:118).

At one time it was assumed that in the southwest United States low-level maize cultivation lasted for several thousand years before agricultural settlements appeared circa A.D. 100–700 or later (Haury 1962). Discoveries in southern Arizona, on the Colorado Plateau, in northern Sonora and in northwest Chihuahua currently challenge this "slow transition" (Figure 16-1). It now appears that the transition from low-level food production with domesticates to an agricultural adaptation occurred rapidly in a few locales, including in our study area in northwest Chihuahua. In southern Arizona, particularly in the Tucson Basin, Late Archaic period sites (circa 1200 B.C.–A.D. 100) consist of numerous pithouses and storage features associated with the emergence of elaborate material culture, irrigation, limited use of ceramics, and an adaptation in which both maize and wild plants were important (Gregory 1999; Gregory and Diehl 2002; Mabry 1998, 2002; Mabry et al. 1997; Schurr and Gregory 2002; Sliva 2005). At the multi-component site of La Playa in northern Sonora, occupants were involved in significant levels of farming as well as foraging (Carpenter et al. 2003:14–17; Carpenter et al. 2005; Montero et al. 2004). Basketmaker II occupations on the Colorado Plateau were substantially dependent on maize no later than circa 100 B.C. (Coltrain et al. 2006; Hard et al. 1996; Matson 1991; Matson and Chisholm 1991). Cerros de trincheras sites in northwest Chihuahua provide additional evidence for settled communities relying on maize much earlier than previously thought.

CERROS DE TRINCHERAS SITES

Cerros de trincheras are hilltop sites with numerous stone terraces and other stone constructions in northwest Chihuahua, Mexico (Figure 16-2). Most are located on isolated volcanic hills adjacent to wide flood plains along a 70 km stretch of the Rio Casas Grandes; a few are found in the Rio Santa María river valley to the east (Roney and Hard 2002). Although this study focuses on Cerro Juanaqueña, data from three other intermediate-sized cerros de trincheras (Cerro el Canelo, Cerro los Torres, and Cerro Vidal), and nine small hilltop sites in the region show similar patterns (Roney and Hard 2004).

A series of AMS dates on maize or other short-lived plant species indicates the occupants of Cerro Juanaqueña relied on farming for a significant part of their economy from 1350–1300 B.C. until about 1100 B.C. (Adams and Hanselka 2005; Hard and Roney 2005a; Roney and Hard 2002). Cerro el Canelo and Cerro los Torres yielded AMS dates that indicate the occupations of these sites were contemporaneous with Cerro Juanaqueña. A later, limited, reoccupation of two terraces at Cerro Juanaqueña was AMS dated to between 400–300 B.C. and about A.D. 1. These two later occupations were on isolated terraces on the lower slopes, locations distinct from the earlier, primary occupation on the upper terraces. The second occupation at Cerro Juanaqueña was similar to the first (primary) occupation. During this same time, circa 400 B.C.–A.D. 1, Cerro los Torres also was occupied, apparently for the first time (Roney and Hard 2002).

The Rio Casas Grandes flows north from the Sierra Madre Occidental through the high basin and range country of northwest Chihuahua. Here, the plains and semidesert grasslands form a fluctuating continuum that ranges from grasslands to shrublands, with a particularly high species diversity that includes numerous perennials and annuals (Bowers and McLaughlin 1982). Summer temperatures are hot (July mean 24.8 °C) and winters are mild (January mean 7 °C). The average annual precipitation is 33.4 cm, two-thirds of which falls between July and October. Paleoclimatic studies suggest a warm and dry middle Holocene (circa 7000 B.C. to 3000 B.C.) followed by: (1) widespread increased moisture from circa 3000 B.C. to 1000 B.C.; (2) a return to somewhat arid conditions after 1000 B.C.; and (3) a return to mesic conditions at circa A.D. 500 (Hard and Roney 2005a, b; Nordt 2003; Ortega-Ramírez et al. 1998).

Based on these paleoclimatic data, it appears that people occupied Cerro Juanaqueña during a time that was ideal for riverine farming. During the two millennia or so preceding the occupation, high magnitude flooding along the Rio Casas Grandes deposited arable silty sands (Ely 1997; Nordt 2003). After 1500 B.C., destructive flooding abated and rainfall seems to have been conducive to farming in what was by that time a wide floodplain composed of deep soils. Farmers also benefitted from moisture accessible via a high water table, perennial river flow, and occasional low velocity overbank flooding that replenished the soil with nutrients and minimized dense competition from native vegetation. In addition, Cerro Juanaqueña is situated within a large flood basin formed by volcanic bedrock that constricts the flood plain some distance downstream and promotes formation of deep soils upstream. Good maize yields may have required minimal planting effort in such a setting.

Cerro Juanaqueña is the largest of the cerros de trincheras sites studied. The site consists of 100 rock rings (alignments of rocks in circular to oval patterns that may be foundations for impermanent structures), 550 terraces, and 8 km of stone walls. It sits 120 m high on an unremarkable hill on the western edge of a small, unnamed basaltic mountain range. The terraces are distributed in two groups, the main group consists of over 300 terraces within a 6 ha area on the top and upper slopes of the hill. This area was thoroughly modified through clearing stones and constructing the terraces; artifact density is typically high. The second group consists primarily of terraces on the lower west flank of the hill about 20 to 40 m above the floodplain. The estimated population at Cerro Juanaqueña ranged from 100 to 300 people between 1350–1300 B.C. and 1100 B.C. (Hard and Roney 2005a; Roney and Hard 2002).

Several lines of evidence suggest the terraces were not built for agricultural purposes. If the 3.6 ha of total surface area behind all 550 terraces were planted in maize, the harvest would feed only six people for a year, assuming rainfall typical of higher elevations in the Sierra Madre (Hard et al. 1999). Further, the terraces average only about 51.5^2 m in surface area, and lack attributes typical of agricultural terraces, such as water control devices or forward slopes. With their costly construction and low potential yield, the terraces seem an impractical farming solution when compared to the production potential of the adjacent floodplain. Instead, we believe most terraces were constructed as house and activity platforms. Our excavations revealed that some terraces contained midden deposits with ample burned and unburned animal remains, wood charcoal and other plant remains, and lithic debris, all typical residue of habitations (Hard and Roney 1998; Roney and Hard 2002). Finally, 31,000 m^3 of stone and sediment were incorporated into the terrace constructions at Cerro Juanaqueña, representing about 30 person years of labor, an effort similar to that required to build a 600-room stone pueblo, and far more than what is expected for this time period in the southwest United States or northwest Mexico (Hard et al. 1999).

We argue elsewhere that these hilltop settlements served a defensive function, based on multiple lines of evidence (Hard and Roney 2004, 2005a, 2007). The sites meet many of LeBlanc's (1999) criteria for defensive sites, including use of defensible land forms; evidence of planning and a defensive layout, such as the perimeter berms; large settlements, such as Cerro Juanaqueña; coordinated rapid construction as opposed to uncoordinated accretionary growth; and intersite visibility. From the top of most of the 12 cerros de trincheras sites, others can be seen, thereby facilitating line of sight signaling (Swanson 2003).

Thus we suggest that Cerro Juanaqueña, and many of the other cerro de trincheras sites in northwest Chihuahua, were constructed in response to raiding and warfare (Hard and Roney 2005a, 2007). High populations and warfare, together make it likely that mobility was constrained. In this competitive environment, the residents of Cerro Juanaqueña maintained a defensive posture while participating in a mixed farming, hunting, gathering, and fishing economy (Hard et al. 2006). In the sections that follow, we define and discuss some of the contributions of zooarchaeological and archaeobotanical analyses, and then focus on the Cerro Juanaqueña biological record and other supporting data to evaluate the role of agriculture in northern Mexico for the two centuries that began around 1350–1300 B.C.

We argue that maize dependence was high, and that settlement construction effort, food processing intensity, and population levels are similar to that of an agricultural strategy rather than a low-level food production strategy, although elements of hunting and gathering do persist. In contrast, during this period, in most of the Greater Southwest a hunter-gatherer adaptation using minor amounts of maize was the norm (low-level farming with domesticates). Therefore, Cerro Juanaqueña and related cerros de trincheras represent the formation of an agricultural strategy that precedes by 1500–2000 years widespread agricultural dependence in the Greater Southwest. This early agricultural dependence is also known from southern Arizona and northern Sonora. We conclude by offering a tentative explanation of the conditions that encouraged early, substantial dependence on farming in northwest Chihuahua.

METHODS

Animal remains were recovered from the Cerro Juanaqueña by two different processes. First, excavation focused on individual terrace and rock ring features and virtually all fill from the terraces was passed through 1/8-inch screens from which all vertebrate specimens were collected. Second, to collect animal remains smaller than 1/8-inch, we picked through the sediment or heavy fraction of the flotation samples. The fill processed by flotation represents only a small fraction of the excavated fill compared to the screened samples which represent all excavated fill. Thus, the materials collected from the screens and flotation are two separate samples that must be considered independently. Both the number of identified specimens (NISP) and the minimum number of individuals (MNI) were derived for the assemblage, here we emphasize NISP. Materials were identified using the reference collection at the University of New Mexico. These and other methods are described in more detail elsewhere (Schmidt 2005).

Archaeobotanical samples were collected as flotation samples and macrofossil samples. Archaeobotanical methods are described in more detail elsewhere (Adams and Hanselka 2005; Bohrer and Adams 1977; Fritz 2005).

Identification was done with the aid of reference materials in the possession of Adams, specimens in the University of Arizona Herbarium, and published texts (Martin and Barkley 1961). In general, deeper excavation units provided better preservation of charred plant remains than did units closer to the modern ground surface.

Amaranth became a special target of investigations when it became clear that seeds lumped into a "cheno-am" (*Chenopodium-Amaranthus*) category had the highest ubiquity (% of samples in which the taxon occurs) of any potential plant food other than maize. Using methods described elsewhere (Fritz 2005), a number of cheno-am seeds could be identified to the genus *Amaranthus* and others to the genus *Chenopodium*. Most of the amaranths were possibly domesticated, whereas all chenopods exhibited wild-type morphology (Fritz 2005).

The role of amaranths in early farming systems is poorly understood because their tiny seeds are very difficult to distinguish from chenopod seeds in a charred and fragmented state, and because of the challenges of separating domesticated seeds from wild or weedy ones. Two species of cultigen amaranth (*Amaranthus cruentus* and *A. hypochondriacus*) were grown by indigenous farmers of northwest Mexico and the southwest U. S. in the 1800s and early 1900s (Nabhan 1979; Sauer 1950, 1967, 1969, 1993), and at least one of these, *A. hypochondriacus*, has been recovered from pre-A.D. 1500 southwest United States archaeological contexts (Bohrer 1962). Cultigen amaranth is included in some discussions of ancient crops (Miksicik 1987, Gasser and Kwiatkowski 1991), but it does not figure prominently in surveys of early agriculture in the southwest United States.

RESULTS

The Cerro Juanaqueña collection contained 33,165 vertebrate remains (Table 16-1; Schmidt 2005). Of this total, 54% of the specimens were recovered from excavations where 1/8-inch mesh screens were used. This discussion focuses on the excavation samples. Black-tailed jack rabbits (*Lepus californicus*), desert cottontail rabbits (*Sylvilagus audubonii*), and indeterminate rabbits (Leporidae) contribute 76% of the NISP. Jack rabbits dominate the assemblage at Cerro Juanaqueña, followed by cottontail rabbits. Despite the high rabbit NISP, MNI estimates at the site level are only 25 jack rabbit individuals and 5 cottontail individuals, values which appear to grossly under-represent rabbit use.

Fish and rodent specimens were primarily recovered in the heavy fraction samples. The fish were Cyprinidae (chubs, redheads, and dace) and Catostomidae (suckers) with estimated standard lengths of less than 10 cm. We extrapolated the number of fish specimens recovered per unit volume of flotation samples to predict the number of fish specimens that would have been recovered if all of the

Table 16-1. Animal Remains from Cerro Juanaqueña (Excavation and Flotation) by NISP and Percent, Summarized by Class or Order

Class or order	From excavation		From flotation		Total NISP
	NISP[a]	% of identified assemblage	NISP	% of identified assemblage	
Fishes (Actinopterygii)	6	0.2	88	26.8	94
Reptiles (Reptilia)	307	12.0	20	6.1	327
Amphibians (Amphibia)	3	0.1	4	1.2	7
Birds (Aves)	23	1.0	1	0.3	24
Rodents (Rodentia)	140	5.6	183	55.9	323
Rabbits and Hares (Lagomorpha)	1929	75.6	29	8.8	1958
Carnivores (Carnivora)	11	0.4	–	–	11
Artiodactyls (Artiodactyla)	131	5.1	3	0.9	134
Total Identified NISP	2550	–	328	–	2878
Total Unidentified NISP	15,241	–	15,046	–	30,287
Total NISP	17,791	–	15,374	–	33,165

[a] NISP = Number of identified specimens.

excavated sediments had been floated. This exercise suggests that fish would have been the third most abundant taxon, ranking after rabbits and indeterminate rodents. Unidentified rodent remains also form a large portion of the flotation portion, yet are a small part of the screened sample. Based on these extrapolations, it is likely that small fishes were key subsistence resources. Many of the rodents may be intrusive, but probably some were consumed.

Plant use at Cerro Juanaqueña included both domestic and wild resources (Table 16-2). The presence of maize in 51% of the flotation samples suggests people frequently consumed it. The maize ears contain 8–12 rows of kernels and appear to be a landrace of popcorn, with small, rounded, hard flint kernels and a small portion of floury endosperm. People may have grown a domesticated amaranth, which is discussed below. A diversity of wild plants also provided food. Chenopod and amaranth are annuals that thrive in disturbed locations such as fields and floodplains. Cerro Juanaqueña occupants gathered other wild plants from a diversity of habitats.

People appear to have lived at Cerro Juanaqueña during all seasons of the year. They harvested chia (*Salvia* spp. type) seeds in the springtime, collected the seeds of weedy plants whose growth was spurred by the summer rains, and gathered ripe lovegrass (*Eragrostis intermedia* type) grains in the fall. They probably were in the area during the early spring preparing their fields for planting and then later in the fall to harvest and dry their crops. They also may have occupied their dwellings during the winter, the one season of the year for which plant evidence is often mute (Adams and Bohrer 1998).

Table 16-2. Presence of Charred Reproductive and Non-reproductive Parts Recovered in Flotation Samples (Including Sterile Samples) from Cerro Juanaqueña[a]

Scientific name	Part represented	Total	%
Reproductive part, domestic	cob fragments and segments,		
Maize (*Zea mays*)	cupules, kernel, kernel fragment	80[b]	50.9
Reproductive part, wild			
cheno-am (*Chenopodium-Amaranthus*)	seed	41	26.1
Horse purslane (*Trianthema* type[c])	seed, seed fragment	6	3.8
Unidentified taxon	seed, seed fragment	5	3.2
Chia (*Salvia* type)	seed	4	2.5
Bulrush (*Scirpus* type)	achene	4	2.5
Lovegrass (*Eragrostis intermedia* type)	caryopsis (grain)	3	1.9
Grass (Poaceae type)	caryopsis (grain)	3	1.9
Milkvetch (*Astragalus nuttalliana* type)	seed	2	1.3
Legume (Fabaceae type)	seed	2	1.3
Barrel cactus (*Ferocactus* type)	seed	1	0.6
Legume (Fabaceae type)	cotyledon fragment	1	0.6
Mesquite (*Prosopis* type)	cotyledon fragment	1	0.6
Lemonade berry (*Rhus aromatica* type)	seed	1	0.6
Globemallow (*Sphaeralcea* type)	seed	1	0.6
Wild gourd/squash (*Cucurbita digitata* type)	seed interior	_[b]	-
Walnut (*Juglans* type)	nutshell fragment	_[b]	-
Non-reproductive parts			
Mesquite (*Prosopis* type)	wood	33[b]	36.3
Ocotillo (*Fouquieria* type)	wood	12[b]	13.2
Saltbush (*Atriplex* type)	wood	11[b]	12.0
Unidentified taxon	wood	9	9.9
Grass (Poaceae type)	stem fragment	6[b]	6.6
Walnut (*Juglans* type)	wood	4[b]	4.4
Creosote bush (*Larrea* type)	wood	4	4.4
Legume (Fabaceae type)	wood	4[b]	4.4
Cottonwood/willow (*Populus/Salix* type)	wood	4[b]	4.4
Monocot (Monocotyledon type)	tissue fragment	3[b]	3.3
Pine (*Pinus* type)	wood	2	2.2
Sunflower (Compositae type)	wood	1[b]	1.1
Ash (*Fraxinus* type)	wood	1	1.1
Juniper (*Juniperus* type)	wood	1[b]	1.1
Monocot (Monocotyledon type)	stem fragment	1	1.1
Unidentified taxon	tissue fragment	1	1.1

[a] Charred reproductive parts $N = 157$; charred non-reproductive parts $N = 91$.

[b] Also or only present in larger macrofossil samples. See text for a discussion of amaranth (*Amaranthus* spp.) seeds.

[c] Use of "type" indicates the specimen(s) closely resembles the plant taxon/taxa named, but due to poor condition and incomplete knowledge of local flora, may also compare well to other related or unrelated taxa.

Similarity between the ancient and modern environment is suggested because a modern list of plants growing in the vicinity of Cerro Juanaqueña contains many of the ancient taxa recovered from the site. However, use of pine (*Pinus* spp.) and juniper (*Juniperus* spp.) wood in the past cannot be explained by the present landscape, and may represent the presence of conifer woodlands closer to the site, gathering of driftwood carried by the Rio Casas Grandes, or travel to distant areas for these woods

The specialized amaranth study recognized 38 *Amaranthus* spp. seeds, 24 *Chenopodium* spp. seeds, and 115 seeds that are classified as cheno-ams due to loss of seed coat or otherwise fragmentary condition (Fritz 2005). Electron microscopy revealed that amaranth seed coat thickness from Cerro Juanaqueña and Cerro los Torres falls within the range of cultigen populations. For the 14 measured amaranth seeds, testa thickness averages 8.9 microns (range is 5–13 microns; S.D. is 2 microns). This contrasts with a range of 16–34 microns for wild amaranth seeds from five species measured by McClung de Tapia et al. (1996) and Fritz (2005). A reasonable explanation is that domesticated amaranth was grown in northern Chihuahua during the Early Agricultural period, although, because of the imperfect condition of the amaranth seeds, alternative explanations cannot be ruled out.

Documenting cultigen amaranth at 1350–1300 B.C. in northwest Chihuahua has consequences for understanding the diversity and potential resilience of early farming systems. The presence of such a highly drought-resistant crop (Cole 1979) reduced risk as well as increased overall food productivity. Amaranth seeds also are high in lysine and tryptophan, two amino acids that are rare in maize (National Research Council 1985). Using amaranth seeds could have improved the health and overall security of maize growers living along the Rio Casas Grandes during the Early Agricultural period.

The ground stone assemblage indicates that processing maize and seeds were major activities in domestic settings. The deeply worn, trough-like metates and over-sized one-hand manos suggest a level of maize processing not seen on typical hunter-gatherer sites in this area. Most of the nearly 700 whole and fragmentary basin metates recovered from Cerro Juanaqueña exhibit heavy wear; many are worn out. In the southwest United States, typical Archaic period basin-shaped metates have shallow, saucer-like grinding surfaces formed by both rotary (circular) and reciprocal (back and forth) grinding with a mano. In contrast, trough-shaped metates typical of later farming settlements are worn with a reciprocal motion made with a larger, rectangular mano. This tool form and the mechanics of its use are effective for grinding maize kernels (Adams 2002; Morris 1990). In the southwest United States, the shift from basin to trough metates accompanies increased maize usage. The atypical Cerro Juanaqueña basin metates have wear surfaces that are deeply worn (average depth 7.8 cm) into elongated ovals. This wear pattern indicates a reciprocal motion appropriate for grinding substantial quantities of maize kernels, rather than small seeds, and is more like that found on trough metates. In effect, the deep basin metate used in a reciprocal fashion can be considered a transitional form between basin metates and trough metates.

The Cerro Juanaqueña manos have a mean grinding surface area of 122.2 cm² compared with typical hunter-gatherer manos of the same age from elsewhere in the southwest United States, which tend to have a mean grinding surface area of about 75 cm² (Hard et al. 1996; Morris 1990). The larger manos of Cerro Juanaqueña are consistent with studies that suggest the larger mano grinding surface area correlates with increased dependence on maize (Hard and Roney 2004; Hard et al. 1996; Horsfall 1987; Morris 1990).

DISCUSSION

The botanical, architectural, ground stone, and other data indicate that around 1350–1100 B.C. an average of 200 people lived on Cerro Juanaqueña during most of the year. They farmed the nearby floodplain, hunted, gathered, and fished. They also lived on a steep, terraced hill. Previously it was thought that population aggregations, substantial use of maize, and investment in construction were not underway until the early centuries A.D. in this region and that these changes were preceded by many centuries during which people slowly converted to farming. What this and other studies show is that this scenario is not universally valid: the transition to farming proceeded at varying rates under various conditions in northwest Mexico and the southwest United States (Hard et al. 2006; Huckell 1995).

We can readily identify a few of the cultural and ecological factors that contributed to this early agricultural adaptation in northwest Chihuahua. During a period of increased rainfall, the substantial populations in the area were attracted to stable flood basins with deep soils and abundant moisture that reduced the effort required to prepare fields for planting. This particularly productive farming environment became an attractive niche, especially in a setting where increasing human population levels and warfare restricted group mobility and required the increased productivity that agriculture provides.

Cerro Juanaqueña, and the other cerros de trincheras, are steep terraced hills which were occupied for no more than two centuries. Despite the intensive use of the valley by farmers in later centuries, Cerro Juanaqueña was not subsequently used as a residential site. This suggests that occupation of Cerro Juanaqueña during the Early Agricultural period was not based on the desire to sustain an agrarian adaptation, but that other factors promoted the presence of significant numbers of people in a region and that mobility for hunting and gathering had become limited. Although the inventory of wild plant and animal foods exploited is rich, virtually all taxa utilized, except juniper and pinyon pine (*Pinus edulis*), are found near Cerro Juanaqueña. These two exceptions may be driftwood picked up from the edges of the Rio Casas Grandes. Further, the plants and animals used are dominated by life forms offering relatively low returns for the labor invested compared to other taxa. Small-seeded forbs and grasses tend to have low returns relative to shrubs and trees; rabbits have much lower returns than artiodactyls (Barlow 2002; Hard and Roney 2005b). Even the fish were small. In fact, maize has a similar or greater return than many small seeds, particularly in a productive floodplain where field preparation costs may be low (Barlow 2002; Hard and Roney 2005b). The process of intensification, that is the extraction of greater amounts of resources from available areas, was well underway, and likely pushed along by population density and group conflict. Yet the occupants adapted with an effective mixed subsistence strategy that took advantage of the desert grasslands setting and local floodplains whose fertility was enhanced by ecological changes.

CONCLUSION

We identify a number of factors encouraging an unusually early agricultural adaptation in northwest Mexico. We do not know if the conditions that appear to have stimulated the formation of this early and quite large agricultural settlement were unique to the Rio Casas Grandes valley, or if they were duplicated in other settings. Together with the other early agricultural adaptations in southern Arizona and northern Sonora, these records form an important data set that can be used to model the conditions that bring about the transition from low-level food production with domesticates to an agricultural adaptation, a shift that is fundamental to a general understanding of the spread of agriculture. More excavations from time periods preceding and following the occupation of the cerros de trincheras, coupled with large regional surveys to better estimate population densities, would allow a more detailed examination of some of the driving forces behind agricultural adaptations relevant to regions all over the world.

ACKNOWLEDGMENTS

The authors appreciate the assistance and support of many individuals and institutions during the multi-year excavation project reported here. We acknowledge our home institutions: The University of Texas at San Antonio (Hard), Crow Canyon Archaeological Center (Adams), The Bureau of Land Management, Albuquerque, New Mexico (Roney), the University of New Mexico (Schmidt), and Washington University, St. Louis, Missouri (Fritz). We especially appreciate the assistance of the field team members who worked virtually every year, including Kevin Hanselka, Bruce Moses, Jennifer Nisengard, Gerry Raymond, Kari Schmidt, José Zapata, and Bridget Zavala. Jennifer Nisengard painstakingly picked through the flotation heavy fraction samples. SEM work was conducted in the Biology Department of Washington

University in St. Louis using a Hitachi S-540 instrument by Michael Veith, Microscopy Technician, along with Carol Heffern and Kimberly Schaefer, two student laboratory assistants in the Paleoethnobotany Laboratory of Washington University, all helped with the study of the amaranth seeds. Permission to excavate in Mexico was granted by INAH officers based in Mexico City and Chihuahua City. Funding for this project came from a number of sources, including the National Science Foundation (BCS-0219185, SBR-9809839, SBR-9708610), the National Endowment for the Humanities (FA-37215–02), the National Geographic Society (6749–00), and University of Texas at San Antonio faculty research grants, a faculty development leave, and support from UTSA's Center for Archaeological Research and Department of Anthropology. Art MacWilliams prepared the base map for Figure 16-1.

REFERENCES

Adams, J., 2002, *Ground Stone Analysis, A Technological Approach*, University of Utah Press, Salt Lake City.

Adams, K. R., 1994, A Regional Synthesis of *Zea mays* in the Prehistoric American Southwest, in: *Corn and Culture in the Prehistoric New World* (S. Johannessen and C. A. Hastorf, eds.), Westview Press, Boulder, pp. 273–302.

Adams, K. R., 2004, Anthropogenic Ecology of the North American Southwest, in: *People and Plants in Ancient Western North America* (P. E. Minnis, ed.), Smithsonian Books, Washington, DC, pp. 167–204.

Adams, K. R., and Bohrer, V. L., 1998, Archaeobotanical Indicators of Seasonality: Examples from Arid Southwestern United States, in: *Seasonality and Sedentism, Archaeological Perspectives from Old and New World Sites* (T. R. Rocek and O. Bar-Yosef, eds.), Peabody Museum of Archaeology and Ethnology Bulletin 6, Harvard University, Cambridge, Massachusetts, pp. 129–141.

Adams, K. R., and Hanselka, J. K., 2005, Plant Use in the Late Archaic Period, in: *Early Farming and Warfare in Northwest Mexico* (R. J. Hard and J. R. Roney, eds.), University of Utah Press, Salt Lake City, manuscript in preparation.

Barlow, K. R., 2002, Predicting Maize Agriculture Among the Fremont: An Economic Comparison of Farming and Foraging in the American Southwest, *American Antiquity* 67(1):65–88.

Benz, B. F., 2006, Maize in the Americas, in: *Histories of Maize: Multidisciplinary Approaches to the Prehistory, Linguistics, Biogeography, Domestication, and Evolution of Maize* (J. Staller, R. Tykot and B. Benz, eds.), Academic Press, Burlington, Massachusetts, pp. 9–20.

Benz, B. F., and Long, A., 2000, Prehistoric Maize Evolution in the Tehuacan Valley, *Current Anthropology* 41(3):459–465.

Benz, B. F., and Staller, J. E., 2006, The Antiquity, Biogeography, and Culture History of Maize in the Americas, in: *Histories of Maize: Multidisciplinary Approaches to the Prehistory, Linguistics, Biogeography, Domestication, and Evolution of Maize* (J. Staller, R. Tykot, and B. Benz, eds.), Academic Press, Burlington, Massachusetts, pp. 665–673.

Bohrer, V. L., 1962, Nature and Interpretation of Ethnobotanical Materials from Tonto National Monument, in: *Archaeological Studies of Tonto National Monument, Arizona* (C. R. Steen, L. M. Pierson, V. L. Bohrer, and K. P. Kent, eds.), Southwestern Monuments Association Technical Series Volume 2, Globe, Arizona., pp. 75–114.

Bohrer, V. L., 1991, Recently Recognized Cultivated and Encouraged Plants Among the Hohokam, *Kiva* 56:227–235.

Bohrer, V. L., and Adams, K. R., 1977, Ethnobotanical Techniques and Approaches at Salmon Ruin, New Mexico, *Eastern New Mexico University Contributions in Anthropology* 8(1):1–220.

Bowers, J. E., and McLaughlin, S. P., 1982, Plant Species Diversity in Arizona, *Madroño* 29:227–233

Carpenter, J. P., Sanchez, G., and Villalpando C. M. E., 2003, Sonora Precerámica: del Arcaico y del Surgimiento de Aldeas Agrícolas, *Arqueología* 29:5–29.

Carpenter, J. P., Sanchez, G., and Villalpando C. M. E., 2005, The Late Archaic/Early Agricultural Period in Sonora, Mexico, in: *The Late Archaic Across the Borderlands: From Foraging to Farming* (B. J. Vierra, ed.), University of Texas Press, Austin, pp. 13–40.

Cohen, M. N., 1987, The Significance of Long-Term Changes in Human Diet and Food Economy, in: *Food and Evolution: Toward a Theory of Human Food Habits* (M. Harris and E. B. Ross, eds), Temple University Press, Philadelphia, Pennsylvania, pp. 261–283.

Cole, J., 1979, *Amaranth: From the Past for the Future*, Rodale Press, Emmaus, Pennsylvania.

Coltrain, J. B., Janetski, J. C., and Carlyle, S. W., 2006, The Stable and Radio-Isotope Chemistry of Eastern Basketmaker and Pueblo Groups in the Four Corners Region of the American Southwest: Implications for Anasazi Diets, Origins, and Abandonments in Southwestern Colorado, in: *Histories of Maize: Multidisciplinary Approaches to the Prehistory, Linguistics, Biogeography, Domestication, and Evolution of Maize* (J. Staller, R. Tykot and B. Benz, eds.), Academic Press, Burlington, Massachusetts, pp. 275–287.

Diehl, M. W., 1997, Rational Behavior, the Adoption of Agriculture, and the Organization of Subsistence During the Late Archaic Period in the Greater Tucson Basin, in: *Rediscovering Darwin* (C. M. Barton and G. A. Clark, eds.), *Archaeological Papers of the American Anthropological Association* 7:251–265.

Ely, L. L., 1997, Response of Extreme Floods in the Southwestern United States to Climatic Variations in the Late Holocene, *Geomorphology* 19:175–202.

Fritz, G. J., 2005, Amaranth and Chenopod Seeds from Early Agricultural Cerros de Trincheras Sites, in: *Early Farming and Warfare in Northwest Mexico* (R. J. Hard and J. R. Roney, eds.), University of Utah Press, Salt Lake City, manuscript in preparation.

Gasser, R. E., and Kwiatkowski, S. M., 1991, Food for Thought: Recognizing Patterns in Hohokam Subsistence, in: *Exploring the Hohokam: Prehistoric Desert Peoples of the American Southwest* (G. J. Gumerman, ed.), University of New Mexico Press, Albuquerque, pp. 417–459.

Gregory, D. A., 1999, Data Integration and Synthesis, in: *Excavations in the Santa Cruz River Floodplain: The Middle Archaic Component at Los Pozos* (D. A. Gregory, ed.), Anthropological Papers 20, Center for Desert Archaeology, Tucson, Arizona, pp. 85–123.

Gregory, D. A., and Diehl, M. W., 2002, Duration, Continuity, and Intensity of Occupation at a Late Cienega Phase Settlement in the Santa Cruz River Floodplain, in: *Traditions, Transitions, and Technologies: Themes in Southwestern Archaeology, Proceedings of the 2000 Southwest Symposium* (S. H. Schlanger, ed.), University Press of Colorado, Boulder, pp. 200–223.

Hames, R., 1992, Time Allocation, in: *Evolutionary Ecology and Human Behavior* (E. Alden Smith and B Winterhalder, eds.), Aldine de Gruyter, New York, pp. 203–235.

Hard, R. J., Mauldin, R. P., and Raymond, G. R., 1996, Mano Size, Stable Carbon Isotope Ratios, and Macrobotanical Remains as Multiple Lines of Evidence of Maize Dependence in the American Southwest, *Journal Archaeological Method and Theory* 3:253–318.

Hard, R. J., MacWilliams, A. C., Roney, J. R., Adams, K. R., and Merrill, W. L., 2006, Early Agriculture in Chihuahua, Mexico, in: *Histories of Maize: Multidisciplinary Approaches to the Prehistory, Linguistics, Biogeography, Domestication, and Evolution of Maize* (J. Staller, R. Tykot and B. Benz, eds.), Academic Press, Burlington, Massachusetts, pp. 471–485.

Hard, R. J., and Roney, J. R., 1998, A Massive Terraced Village Complex in Chihuahua, Mexico, 3000 Years Before Present, *Science* 279 (5357):1661–1664.

Hard, R. J., and Roney, J. R., 2004, Late Archaic Period Hilltop Settlements in Northwestern Chihuahua, Mexico, in: *Identity, Feasting, and the Archaeology of the Greater Southwest* (B. Mills, ed.), University of Colorado Press, Boulder, pp. 276–294.

Hard, R. J., and Roney, J. R. (eds.), 2005a, *Early Farming and Warfare in Northwest Mexico*, University of Utah Press, Salt Lake City, manuscript in preparation.

Hard, R. J., and Roney, J. R., 2005b, The Transition to Farming on the Rio Casas Grandes and in the Southern Jornada Mogollon Region in the North American Southwest, in: *Current Research on the Late Archaic Across the Borderlands: From Foraging to Farming* (B. Vierra, ed.), University of Texas Press, Austin, pp. 141–186.

Hard, R. J., and Roney, J. R., 2007, Cerros de Trincheras in Northwestern Chihuahua: The Arguments for Defense, in: *Enduring Borderlands Traditions: Trincheras Sites in Time, Space and Society* (S. K. Fish, P. R. Fish, and E. Villalpando, eds.), University of Arizona Press, Tucson, in press.

Hard, R. J., Zapata, J. E., Moses, B. K., and Roney, J. R., 1999, Terrace Construction in Northern Chihuahua, Mexico: 1150 B.C. and Modern Experiments, *Journal of Field Archaeology* 26(2):129–146.

Haury, E. W., 1962, The Greater American Southwest, in: *Courses Toward Urban Life* (R. J. Braidwood and G. R. Willey, eds.), Viking Fund Publications in Anthropology 32, Wenner-Gren Foundation for Anthropological Research, New York, pp. 106–131.

Horsfall, G. A., 1987, A Design Theory Perspective on Variability in Grinding Stones, in: *Lithic Studies Among the Contemporary Highland Maya* (B. Hayden, ed.), University of Arizona Press, Tucson, pp. 333–375.

Huber, E. K., 2005, Early Maize at the Old Corn Site (LA 137258), in: *Fence Lake Project: Archaeological Data Recovery in the New Mexico Transportation Corridor and First Five-Year Permit Area, Fence Lake Coal Mine Project, Catron County, New Mexico, Synthetic Studies and Summary* (E. K. Huber and C. R. Van West, eds.), Technical Series 84, Volume 4, Statistical Research, Tucson, Arizona, pp. 36.1–36.33.

Huckell, B. B., 1995, *Of Marshes and Maize, Preceramic Agricultural Settlements in the Cienega Valley, Southeastern Arizona*, Anthropological Papers 59, University of Arizona Press, Tucson.

Huckell, B. B., 1996, The Archaic Prehistory of the North American Southwest, *Journal of World Prehistory* 10(3):305–373.

Huckell, B. B., Huckell, L. W., and Shackley, M. S., 1999, McEuen Cave, *Archaeology Southwest* 13(1):12.

LeBlanc, S. A., 1999, *Prehistoric Warfare in the American Southwest*, The University of Utah Press, Salt Lake City.

Long, A., Benz, B. F., Donahue, D. J., Jull, A. J. T., and Toolin, L. J., 1989, First Direct AMS Dates on Early Maize from Tehuacán, Mexico, *Radiocarbon* 3:1035–1040.

Mabry, J. B., 1998, Conclusions, in: *Archaeological Investigations of Early Village Sites in the Middle Santa Cruz Valley: Analysis and Synthesis, Part II* (J. B. Mabry, ed.), Anthropological Papers 19, Center for Desert Archaeology, Tucson, Arizona, pp. 757–791.

Mabry, J. B., 2002, The Role of Irrigation in the Transition to Agriculture and Sedentism in the Southwest: A Risk Management Model, in: *Traditions, Transitions, and Technologies: Themes in Southwestern Archaeology*, Proceedings of the 2000 Southwest Symposium (S. H. Schlanger, ed.), University of Colorado Press, Boulder, pp. 178–199.

Mabry, J. B., Swartz, D. L., Wöcherl, H., Clark, J. J., Archer, G. H., and Lindeman, M. W., 1997, *Archaeological Investigations of Early Village Sites in the Middle Santa Cruz Valley: Description of the Santa Cruz Bend, Square Hearth, Stone Pipe, and Canal Sites*. Anthropological Papers 18, Center for Desert Archaeology, Tucson, Arizona.

MacWilliams, A. C., Hard, R. J., Roney, J. R., Adams, K. A., Merrill, W. L., 2008, The Setting of Early Agriculture in Southern Chihuahua, Mexico, in: *Archaeology Without Borders: Contact, Commerce and Change in the U.S. Southwest and Northwestern Mexico* (L. Webster and M. McBrinn, eds.). University Press of Colorado, Boulder, in press

Mangelsdorf, P. C., 1986, The Origin of Corn, *Scientific American* 255(2):80–86.

Martin, A. C., and Barkley, W. D., 1961, *Seed Identification Manual*, University of California Press, Berkeley.

Matson, R. G., 1991, *The Origins of Southwestern Agriculture*, University of Arizona Press, Tucson.

Matson, R. G., and Chisholm, B., 1991, Basketmaker II Subsistence: Carbon Isotopes and Other Dietary Indicators from Cedar Mesa, Utah, *American Antiquity* 56(3):444–459.

Matsuoka, Y., Vigouroux, Y., Goodman, M. M., Sanchez G. J., Buckler, E., and Doebley, J., 2002, A Single Domestication for Maize Shown by Multilocus Microsatellite Genotyping, *Proceedings of the National Academy of Science* 99(9):6080–6084.

McClung de Tapia, E., González Vázquez, J., Zurita Noguera, J., and Morales, E. I., 1996, *La Domesticación Prehispánica de Amaranthus spp.* Universidad Nacional Autónoma de México, Instituto de Investigaciones Antropológicas, Cuadernos de Trabajo IIa, Mexico, DF.

Miksicek, C. H., 1987, Late Sedentary-Early Classic Period Hohokam Agriculture: Plant Remains from the Marana Community Complex, in: *Studies in the Hohokam Community of Marana* (G. E. Rice, ed.), Office of Cultural Resource Management, Department of Anthropology, Arizona State University, Tempe, pp. 197–216.

Minnis, P. E., 1992, Earliest Plant Cultivation in the Desert Borderlands of North America: An International Perspective, in: *The Origins of Agriculture* (C. W. Cowan and P. J. Watson, eds.), Smithsonian Institution Press, Washington, DC, pp. 121–141.

Montero, C., Carpenter, J., Barnes, E., Rohn, A., and Watson, J., 2004, Early Agricultural Period Burials at La Playa (SON F. 10:03), Sonora, Mexico: Further Preliminary Results, paper presented at the 69th Annual Meeting of the Society for American Archaeology, Montreal.

Morris, D. H., 1990, Changes in Groundstone Following the Introduction of Maize in the American Southwest, *Journal of Anthropological Research* 46:177–194.

Nabhan, G. P., 1979, Grain Amaranth and Other Rare Crops of Northern Mexico, *Dry Country News* 3:20–22.

National Research Council, 1985, *Amaranth: Modern Prospects for an Ancient Crop.* Report of an Ad Hoc Panel of the Advisory Committee on Technology Innovation Board on Science and Technology for International Development, Office of International Affairs, Rodale Press, Emmaus, Pennsylvania.

Nordt, L., 2003, Late Quaternary Fluvial Landscape Evolution in Desert Grasslands of Northern Chihuahua, Mexico, *Geological Society of America Bulletin* 115:596–606.

Ortega-Ramírez, J. R., Valiente-Banuet, A., Urrutia-Fucugauchi, J., Mortera-Gutiérrez, C. A., and Alvarado-Valdez, G., 1998, Paleoclimatic Changes during the Late Pleistocene-Holocene in the Laguna Babícora, near the Chihuahuan Desert, México, *Canadian Journal of Earth Science* 35:1168–1179.

Piperno, D. R., and Flannery, K. V., 2001, The Earliest Archaeological Maize (*Zea mays* L.) from Highland Mexico: New Accelerator Mass Spectrometry Dates and Their Implications, *Proceedings of the National Academy of Science* 98(4):2101–2103.

Roney, J. R., and Hard, R. J., 2002, Early Agriculture in Northwestern Chihuahua, in: *Traditions, Transitions, and Technologies: Themes in Southwestern Archaeology, Proceedings of the 2000 Southwest Symposium* (S. Schlanger, ed.), University of Colorado Press, Boulder, pp. 163–180.

Roney, J. R., and Hard, R. J., 2004, A Review of Cerros de Trincheras in Northwestern Chihuahua, in: *Surveying the Archaeology of Northwest Mexico* (G. E. Newell and E. Gallaga, eds.), University of Utah Press, Salt Lake City, pp. 127–148.

Roosevelt, A. C., 1984, Population, Health, and the Evolution of Subsistence: Conclusions from the Conference, in: *Paleopathology at the Origins of Agriculture* (M. N. Cohen and G. J. Armelagos, eds.), Academic Press, Orlando, pp. 572–574.

Sauer, J. D., 1950, The Grain Amaranths: A Survey of Their History and Classification, *Annals of the Missouri Botanical Garden* 37:561–632.

Sauer, J. D., 1967, The Grain Amaranths and their Relatives: A Revised Taxonomic and Geographic Survey, *Annals of the Missouri Botanical Garden* 54(2):103–137.

Sauer, J. D., 1969, Identity of Archaeological Grain Amaranths from the Valley of Tehuacan, Puebla, Mexico, *American Antiquity* 34:80–81.

Sauer, J. D., 1993, *Historical Geography of Crop Plants: A Select Roster*, CRC Press, Boca Raton, Florida.

Schmidt, K. M., 2005, Faunal Remains, in: *Early Farming and Warfare in Northwest Mexico* (R. J. Hard and J. R. Roney, eds.), University of Utah Press, Salt Lake City, manuscript in preparation.

Schurr, M. R., and Gregory, D. A., 2002, Fluoride Dating of Faunal Materials by Ion-Selective Electrode: High Resolution Relative Dating at an Early Agricultural Period Site in the Tucson Basin, *American Antiquity* 67(2):281–299.

Shackley, S., 2005, Chronometry and Geochemistry at McEuen Cave: The Radiocarbon and Obsidian Geochemical Data, paper presented at the 70th Annual Meeting of the Society for American Archaeology, Salt Lake City, Utah.

Sliva, R. J. (ed.), 2005, *Material Culture and Lifeways of Early Agricultural Communities in Southern Arizona, Center for Desert Archaeology,* Anthropological Papers 35, Tucson, Arizona.

Smith, B. D., 1997, Reconsidering the Ocampo Caves and the Era of Incipient Cultivation in Mesoamerica, *Latin American Antiquity* 8:342–383.

Smith, B. D., 2001a, Documenting Plant Domestication: The Consilience of Biological and Archaeological Approaches, *Proceedings of the National Academy of Sciences* 98(4):1324–1326.

Smith, B. D., 2001b, Low-level Food Production, *Journal of Archaeological Research* 9(1):1–43.

Smith, B. D., 2005, Reassessing Coxcatlan Cave and the Early History of Domesticated Plants in Mexico, *Proceedings of the National Academy of Sciences* 102(27):9438–9445.

Swanson, S. J., 2003, Documenting Prehistoric Communication Networks: A Case Study in the Paquimé Polity, *American Antiquity* 68:753–769.

Wills, C. H., 1985, *Early Prehistoric Agriculture in the American Southwest*, School of American Research Press, Santa Fe, New Mexico.

Part IV

Social and Economic Strategies

Chapter *17*

Meat Consumption and Bone Use in a Mississippian Village

MELINDA A. ZEDER AND SUSAN R. ARTER

A central feature of Mississippian chiefdoms in eastern North America is a marked sharpening in lines of social demarcation. The increased complexity of Mississippian societies is manifest in the differentiation of settlement types into centers, villages, and farmsteads; in the large-scale conscription of labor to construct massive earthen ceremonial mounds; in markedly different mortuary practices for commoners and elites; and in elaborate ceramic vessels and decorative items accessible to a privileged few. In recent years there has been growing recognition that social relations in early ranked societies are also reflected in foodways (Colton 1992). As a result, "biological artifacts," traditionally associated with reconstructing ancient environments and past diets, are increasingly used to build a better understanding of relations between commoners and elites, as well as among homesteads, villages, and centers (Jackson and Scott 1995; Michals 1992a; Powell 1988; Schoeninger and Peebles 1981; Welch and Scarry 1995). Such studies look for status-based distinctions in access to food resources, as well as for tribute systems in which food was one of the items extracted from satellite villages by an increasingly powerful elite located in large ceremonial centers. Another promising avenue of inquiry is the study of foodways *within* communities to detect differential access to resources and patterns of resource sharing between households. Such access bears on the nature of social and economic relations within a single tier of society. The Powers Phase Project combines faunal and botanical studies with analysis of material culture to formulate a detailed picture of life in one village in a Mississippian chiefdom.

Powers phase (circa A.D. 1300) sites are found on the western edge of the lower Mississippi valley (Figure 17-1). The 50 or so Powers phase sites include small farmsteads, perhaps 10 larger fortified villages of 40–100 houses each, and a single fortified ceremonial center (Powers Fort) with four mounds around a central plaza. With the exception of Powers Fort, most sites were occupied for as little as 5–10 years before being burned and abandoned. Although other villages to the south and east in the Mississippi valley display a longer history of

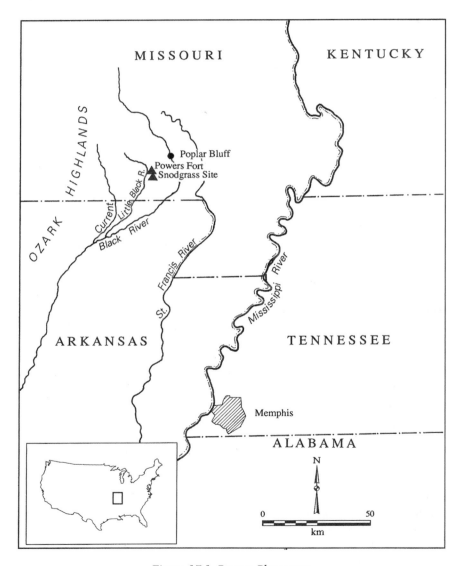

Figure 17-1. Powers Phase area.

occupation than Powers phase sites, as well as a richer array of material culture, this longevity and complexity of occupation often makes it difficult to tease out patterns of social and economic interaction defining daily life in these communities. In contrast, because of their short lifespan, Powers phase villages represent thin slices of time, frozen decades, that offer an unprecedented opportunity for a fine-grained study of social and economic interactions.

SNODGRASS SITE STRUCTURE AND FUNCTION

Excavation of the Snodgrass and Turner villages was conducted by the University of Michigan in the late 1960s and early 1970s. Through close to complete excavation of all features (93 structures and 97 pits) and complete recovery and piece plotting of artifacts, bones, and botanical remains (more than 20,000 objects), excavators hoped to discover patterns of intracommunity socioeconomic organization. A number of monographs based on these excavations were published in the 1970s (Black 1979; Price and Griffin 1979; Smith 1975). Without more advanced data processing systems, however, the sheer volume of material made it difficult to achieve the project goals. In 1989, the Powers Phase Project resumed under the direction of Smith with the goal of using new archaeological methods and computer technology to revisit questions that motivated the original research. The case study presented here uses vertebrate remains from one of these sites, Snodgrass, to examine intra-community social and economic interaction.

Price and Griffin (1979) describe Snodgrass as a village containing 93 structures surrounded by a large ditch and a palisade that probably formed an impressive fortification around the village (Figure 17-2). Surrounding 38 of the 93 structures were traces of a fairly flimsy wall constructed of posts and interwoven cane plastered with white swamp clay, known as "the white wall." Structures within this wall were generally larger than structures outside the wall (averaging 30 m²) and contained a greater density and diversity of artifacts. A large number of pits were associated with these interior structures. Structures outside the white wall generally averaged 17 m², contained a substantially lower density of a different array of artifact types, and had fewer refuse pits.

Price and Griffin (1979:140–143) raise a number of questions about social and economic relations at the Snodgrass site. Are there differences in the duration and intensity of occupation of these different sectors of the village? Are there socioeconomic differences between households located in these different sectors? Is it possible that some structures, especially those outside the white wall, were only seasonally occupied? What is the nature of interaction among households in the village, both between those in the inner and outer sectors of the site, and those within each sector? The Snodgrass vertebrate remains provide some clues that, when combined with the other biological and cultural data, may solve some of these puzzles.

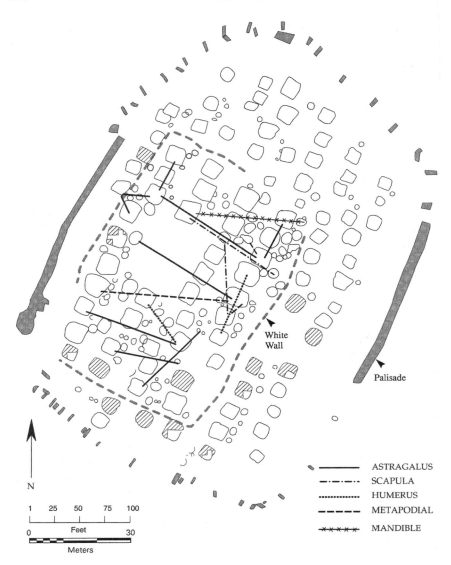

Figure 17-2. Snodgrass site with between-feature right/left matches indicated. The Square features are structures; round features, pits. The stippled areas are exterior palisade and interior "white wall"; hatching indicates the unexcavated areas.

INTENSITY AND DURATION OF OCCUPATION

Price and Griffin (1979:50, 141) concluded that since all Snodgrass structures suffered the same fiery destruction, all areas of the village were occupied when the village was burned and abandoned. But were all structures, both inside and outside the "white wall," occupied throughout the entire life of the village? The lower density of artifacts recovered from the outside structures suggests that intensity of occupation of these two sectors differed and that the outside structures were occupied either later in the life of the village, as a result of village growth, or more sporadically, as would be the case if these structures were only seasonally occupied or were used for a different variety of activities. The intensity of occupation can be addressed by looking at the density of animal bones recovered from different contexts and their degree of fragmentation.

Density of Animal Bones

Table 17-1 presents several different measures of the density of bones recovered from different contexts: the number and weight of the total sample per square meter, the number and weight of deer bones per square meter, the average Minimum Number of Individuals (MNI) of white-tailed deer (*Odocoileus virginianus*) per feature, and the average weight of bone fragments as a measure of the degree of bone fragmentation. The average MNI for deer is derived by calculating the MNI of deer in each structure or pit (based on the most common deer bone in each structure) and dividing the total of all the MNIs in a sector of the site by the number of features in that sector.

In both number and weight, the density of bones is much greater in the structures inside the white wall than in those outside the wall. Not surprisingly, deer bones (more than 80% of the sample from any sector at the site) also show this pattern. There is also a disparity between the two sectors in the number of individual deer per structure. Inside structures have, on average, a minimum of five deer per structure, while outside structures have only two. Overall bone density, in both number and weight, is greater in refuse pits than on house floors. Pits associated with structures inside the walled enclosure, however, contain a greater number and mass of bones than pits associated with outside structures, and this is true both for the total sample of bones and for the deer elements. Finally, although pits associated with inside structures have an average minimum of one deer per pit, pits associated with outside structures sometimes have no identifiable deer elements at all.

Bone Fragmentation

The lower density of bones in the outside sector might be attributable to greater postdepositional breakage and loss of bone in this sector as a result of trampling or the destruction of bones by dogs (*Canis familiaris*). If so, this would argue for a *greater* intensity of use of this area, or at least for less protection from roving scavengers. The degree of bone fragmentation is measured here by computing the average weight per fragment of bone (weight/NISP). If roughly equivalent samples are compared in terms of species composition and skeletal part distribution, the higher the average bone weight,

Table 17-1. Density and Fragmentation of Animal Remains by Context[a]

Location	(1) Area (m²)	Total (2) NISP/m²	(3) g/m²	Deer (4) NISP/m²	(5) g/m²	(6) MNI	(7) Total g/NISP	(8) Deer g/NISP
Structures								
Inside	1042	11.8	38.8 g	4.3	31.6 g	5	3.3 g	7.3 g
Outside	800	2.1	13.1 g	1.2	12.0 g	2	6.2 g	10.0 g
Total	1842	7.5	27.7 g	2.9	23.1 g	3	3.7 g	8.0 g
Pits								
Inside	186	19.7	52.3 g	6.1	46.5 g	1	2.7 g	7.6 g
Outside	45	3.6	13.8 g	1.4	12.2 g	0.5	3.8 g	8.7 g
Total	231	17.1	46.9 g	5.2	39.8 g	1	2.7 g	7.7 g

[a] Columns: (1) total area (m²) of features in different sectors of the site; (2) number of specimens (NISP) per square meter computed using the entire sample of bones; (3) weight of specimens (g) per square meter; (4, 5) number (NISP) and weight (g) of deer specimens per square meter; (6) deer minimum number of individuals (MNI); (7) total weight (g) divided by NISP; (8) total weight of deer specimens (g) divided by deer NISP.

the less fragmentation; conversely, the lower the average weight per bone fragment, the higher the degree of bone fragmentation.

Looking at fragmentation differentials between bones in these two sectors of the site, it is readily apparent that although outside structures yielded proportionately far fewer bones than inside structures, the bones they did contain were recovered in much larger fragments than bones from inside structures (Table 17-1). This pattern is particularly clear in the deer bones. This pattern is also mirrored in the pits, with deer bone in pits associated with outside structures weighing an average 8.7 g per bone and those in pits associated with inside structures averaging 7.6 g per bone. The average weight of fragments in outside structures may be elevated to some degree by the high proportion of relatively heavy antler fragments in outside structures, discussed below. The average fragment weight for different skeletal parts (Table 17-2) shows, however, that essentially all skeletal parts were found in larger fragments in outside structures than in either inside structures or pits.

Thus, although the density of bones, in both number and weight, is much less outside the walled enclosure, the individual bone fragments found there

Table 17-2. White-tailed Deer Skeletal Part Distributions

Skeletal parts	NISP (%)			Weight/NISP (g)		
	Inside	Outside	Pits	Inside	Outside	Pits
Region						
Antler	19	40	18	12.2	11.5	4.9
Skull	16	20	14	8.3	14.1	5.9
Axial	5	2	6	3.2	9.0	6.1
Limb	60	38	62	7.8	8.7	9.7
Individual elements (without antlers)						
Crania	4	5	6	4.4	18.8	4.7
Maxilla	4	2	4	4.1	5.6	4.9
Mandible	12	25	6	9.5	14.0	7.7
Vertebra	6	2	8	3.2	9.0	6.1
Scapula	9	15	5	7.6	9.9	15.7
Humerus	4	2	6	9.4	13.0	13.0
Radius	5	8	5	13.0	13.4	10.3
Ulna	4	5	2	5.6	5.8	5.8
Carpals	8	3	9	6.7	6.7	8.5
Pelvis	2	2	4	11.4	7.2	11.7
Femur	3	1	5	7.8	10.6	7.8
Tibia	6	3	8	10.8	9.0	11.6
Astragalus	6	10	4	7.9	7.2	9.2
Other tarsals	4	4	5	7.1	8.7	10.6
Metacarpal	6	3	7	8.6	17.7	9.7
Metatarsal	8	5	3	8.2	6.0	9.9
Metapodial	6	1	5	2.4	2.6	2.5
Phalanges	6	1	5	2.4	3.2	2.0
Total NISP	4221	931	1248	8.4	10.9	8.2
Total weight	35,313 g	10,140 g	10,195 g			

are substantially larger than those recovered from inside the white wall. The lower density of bones in the outside sector of the site is not, then, a function of a greater use of this area. Instead, the faunal data strongly support the impression given by artifact density that the outside structures were occupied less intensively, and possibly for a shorter period of time than the larger structures inside the white wall. Additional aspects of the Snodgrass faunal assemblage must be examined to assess whether the duration of occupation was shorter because this sector was settled some time after the initial establishment of the village, because occupation here was more sporadic, because these two areas were used in fundamentally different ways, or because of some combination of these possibilities.

DIFFERENTIALS IN DIET AND BONE USE

On the basis of structure sizes, pottery types, personal ornaments, and possible ritual or clan-marking paraphernalia, Price and Griffin (1979:141–142) suggested that there were status differences between the households located inside the white wall and those occupying the structures on the outer perimeter. A difference in stone tools and ceramics recovered from these two areas implies that there was also a difference in the activities conducted in these two sectors. Variation in both status and productive activities might be reflected either in a difference in diet or in bone tools. Possible variation in the animal-based portion of the diet and use of modified bones might be detected in two ways: by examining species distribution and diversity and by looking at skeletal part distributions.

Species Distribution and Diversity

Deer are overwhelmingly the dominant species in all contexts at the Snodgrass site (Tables 17-1 and 17-3). There is some variation, however, in nondeer species among different contexts. Nondeer species comprise 11% of the total number of identifiable bones from structures inside the wall and 8% of the assemblage outside the wall. Pit assemblages generally contained more nondeer species than did structure assemblages, perhaps because bones of small mammals, birds, and fishes were better protected in pits than on house floors. Variation in nondeer species observed between inside and outside structures is echoed, even amplified, in pit assemblages.

   The dominance of deer and the poor representation of smaller species in the Snodgrass sample, is attributable in part to the poor preservation of faunal remains in the sandy soils at the site (Smith 1975:197–199) and to its location on the western margin of the Mississippi migratory waterfowl flyway (Smith 1975:128). There is no indication, however, that the soil conditions that contribute to poor bone preservation at Powers phase sites in general affected one

Table 17-3. Species Distribution and Diversity by Location

Taxa	Structures			Pits		
	Total	Inside	Outside	Total	Inside	Outside
Unidentifiable mammals	1364	1171	193	328	300	28
Unidentifiable large mammal	6466	5964	502	2046	1944	102
Unidentifiable small mammal	199	183	16	110	88	22
Opossum (*Didelphis virginiana*)	10	10	–	3	3	–
Rabbit (*Sylvilagus* spp.)	12	9	3	15	15	–
Squirrel (*Sciurus* spp.)	9	8	1	3	3	–
Beaver (*Castor canadensis*)	5	5	–	1	1	–
Other rodents	10	8	2	13	12	1
Canid (Canidae)	20	13	7	5	5	–
Fox (*Urocyon cinereoargenteus*)	4	3	1	2	2	–
Raccoon (*Procyon lotor*)	171	153	18	60	57	3
Mustelid (Mustelidae)	5	5	–	4	4	–
Elk (*Cervus elaphus*)	4	4	–	–	–	–
Deer (*Odocoileus virginianus*)	5409	4443	966	1245	1164	81
Bison (*Bison bison*)	2	1	1	–	–	–
Bird	89	66	23	25	25	–
Turtle	274	253	21	101	96	5
Other reptiles	17	17	–	–	–	–
Fish	12	11	1	8	8	–
Totals:	14,082	12,327	1755	3969	3727	242
Total number of specimens						
NISP identifiable	6053	5009	1044	1485	1395	90
NISP unidentifiable	8029	7318	711	2484	2332	152
Deer as a percent of identifiable	89%	89%	93%	84%	83%	90%
Other species as a percent of identifiable	11%	11%	7%	16%	17%	10%
Species per feature	3	5	2	2	2	1

sector of the Snodgrass site more than another. The larger average fragment size of bones in the outside structures suggests that this pattern is also not attributable to a greater degree of bone breakage due to trampling or dogs. All deposits excavated were screened through 1/4-inch mesh screen (Smith 1975:197), so this difference cannot be attributed to recovery techniques. Moreover, because, with very few exceptions, features were completely excavated and all the bones they contained were saved and studied, the assemblages represent the total amount of recoverable bone. Therefore, concerns over variation of species diversity due to sample size differences do not apply here. Thus, variation in the representation of nondeer species between contexts inside and outside the white wall is probably a reflection of a real difference in the proportion of species utilized in these areas.

There is also variation in the diversity of species utilized (Table 17-3). Structures inside the white wall contain, on average, 5 different species per structure; some contain as many as 13 species. Structures outside the walled enclosure contain, on average, only 2 species per structure; many contain only deer. Thus, although deer are clearly the dominant species in all areas, there is some difference, albeit slight, in the importance and variety of nondeer species in different sectors of the site.

Skeletal Part Distribution

But was there variation in the way deer were used at the site? The distribution of deer skeletal elements would argue strongly that there was. Table 17-2 presents part distribution data for inside structures, outside structures, and pits (93% of which are from inside contexts). Limb elements are dominant in both inside structures and pits. Antler fragments and elements from the skull (including the mandible) place a distant, and almost equal, second and third. In contrast, antler fragments more than double their representation in the assemblages from outside structures, making up 40%. Limb elements comprise only 38% of the deer bones from outside structures, and skull elements are better represented here than they are in either the inside structures or their associated pits.

Axial elements (vertebra, ribs, sternal elements) comprise more than 20% of the bones of a deer skeleton, yet these elements are never more than 6% of bone from any Snodgrass context. To some extent, the poor representation of axial bones may be attributed to their greater fragility, and thus to postdepositional loss of bones due to trampling, destruction by dogs, or poor soil conditions. If postdepositional loss of bones were the only factor affecting the representation of axial elements, however, one would expect to find more axial bones in the outside structures where overall fragmentation of deer bones is less. Instead, only 2% of the bones from outside structures are from the axial skeleton, as opposed to 5–6% from inside structures and pits. Although postdepositional loss of bone may be a primary cause of the paucity of axial elements at the site, it would not seem to be the only factor affecting the representation of these bones. It might be tempting to posit the transhipment of meaty lumbar bones and racks of venison ribs to elites at Powers Fort, as Styles and Purdue (1986) did to explain the high proportion of axial and meaty limb bones at Napoleon Hollow, a ritual camp site on the Illinois River (see Figure 2-1 in Deagan, this volume). However, the good representation at Snodgrass of many high-meat-yield limb elements usually associated with elite consumption (Bogan 1983; Michals 1992a; Scott 1983) makes it difficult to accept this argument. Likewise, the possibility that axial elements were left at the kill site after meat was filleted from the bones is hard to support given that skull elements and lower foot bones, perhaps even more likely to be left at the kill site, are well represented in the Snodgrass assemblage.

Essentially all antlers found at the site had been shed (Table 17-4) and were likely brought to the site without an accompanying deer carcass. Antlers have a wide range of decorative and utilitarian uses and were in all probability curated items with more direct bearing on non-food uses of animals than on ancient diet. Although the substantial representation of antlers in the outside structures carries important implications about the uses of animal resources and the nature of activities practiced in different sectors of the site, it masks other variation in element distribution. For this reason, in the lower portion of Table 17-2, antler fragments are omitted and distributions are presented by individual elements rather than by broader skeletal regions. Some remarkable differences between the assemblages inside and outside the white-walled enclosure are revealed as a result.

Most striking is the uneven distribution of elements in the outside structures. The majority of elements contribute no more than 5% of nonantler deer bones from outside structures. Three elements, however, are disproportionately well represented: mandibles, scapulae, and astragali. Although mandibles and astragali are dense bones (Lyman 1984) that should withstand postdepositional forces better than, for example, the proximal humerus, the representation of these bones seems too high to be attributed to durability alone. This argument applies even more strongly to the scapula, which is certainly more friable than other bones (Lyman 1984). The abundance of scapulae in outside structures might argue for a preference for meat from the shoulder joint in this sector, although, as we will see, an alternative may be offered. A dietary explanation for the abundance of mandibles and astragali, however, is hard to support, because these elements come from low-meat-yield portions of the carcass. In the case of the astragalus, if cuts from the ankle region were for some reason preferred, why are not other bones from this joint as well represented? Moreover, in several structures from both inside and outside the white wall, mandibles, astragali, and scapulae were found in "caches" of up to 10 elements (Price and Griffin 1979:18 [citing Bruce Smith 1973, personal communication]).

Mandibles may serve a variety of functions (Brown 1964). Ethnohistoric accounts of North American Choctaw and Seneca peoples record the use of deer mandibles as graters for processing maize (*Zea mays*) (Harrington 1908 [referenced in Smith 1978:173]; Wright 1958:160–161). Indeed, some of the Snodgrass mandibles show strong polish and wear characteristic of extensive handling. Modified astragali in late Mississippian contexts may have been gaming pieces (Lewis 1988; but see Eisenberg 1989; Lewis 1990). Although only one of the Snodgrass astragali shows evidence of having been worked, others may have been used in a relatively unmodified state or may not have been handled enough to show any polish or other use wear. No scapulae show any sign of use; however, it is possible they were used as hoes or fleshers. The spatulate end of the scapula, rarely preserved in the Snodgrass sample, is more likely to show the greatest modification or use wear than the more commonly recovered articular end.

Although mandibles, scapulae, and astragali are also relatively well represented in the inside structures (27% of the nonantler fragments), they are not nearly as common in the inside structures as they are in the smaller outside structures, where they comprise 50% of the sample. Mandibles are only half as well represented in inside structures as they are in outside structures. Scapulae and astragali are quite well represented, but other meat- and non-meat-bearing limb elements are also frequently found in these inside structures. Thus, although caches of mandibles, scapulae, and astragali are found in the inside structures, their importance here is masked by the presence of a much higher proportion of elements representing butchery and consumption of animals. These three bones do not, in any way, dominate the pit assemblages. In fact, element distribution in pits is extremely even, closely mirroring the expected distributions of elements found in a complete carcass. Thus, although curated bones remained in house structures, pits primarily received butchery and food-consumption refuse.

It is possible, however, that the lopsided distribution of curated and food-refuse bones in the outside structures compared to the inside structures and the pits may not mean that activities related to the storage and use of these curated bones were more important here than was the butchery of animals and consumption of meat. If the inside structures were occupied for a longer period of time, there was more time for food debris to accumulate. This accumulation would effectively mask the numerical importance of curated bones that predominate in more briefly occupied structures outside the white wall. However, if the only variable that distinguished the outside and inside structures was duration of occupation and the activities conducted in these two areas were essentially the same, the relative density of curated and food-refuse bones in these two areas should be similar. An examination of the density of proposed curated elements (mandibles, scapula, astragali, and antlers) and of food-refuse bones (all other bones) within each of these different contexts shows that the densities are clearly not similar.

Dividing the number of curated and food-refuse bones by the area of inside structures, outside structures, and pits yields the following bone density values:

	Inside	Outside	Pits
Curated bones/m^2	1.66	0.82	1.66
Food-refuse bones/m^2	2.39	0.39	3.70

As noted earlier, the overall density of bones in all categories is lower in the outside structures than in the inside structures or the pits, but the relationship between curated bones and food-refuse bones both within contexts and between different contexts at the site is quite telling. The density of curated bones in outside structures is more than twice the density of food-refuse bones in these structures. In contrast, the density of curated bones is approximately *two thirds* the density of food-refuse bones in inside structures and less than *half* the density of food-refuse bones in pits. Comparing the relative density of bones between different contexts, the density of food-refuse bones is six times greater in the inside structures, and ten times greater in the pits, than the density of food-refuse bones in the outside structures. The density of curated bones, however, is only two times greater in both the inside structures and pits than the density of curated bones in the outside structures. The predominance of curated bones over food-refuse bones in the outside structures relative to the inside structures strongly suggests, then, that there is a real difference in the range of activities practiced in these two areas, not merely an apparent difference caused by a difference in duration of occupation.

It is harder to say whether there were any significant dietary differences that have a bearing on the status of the occupants of the two sectors. The overwhelming predominance of curated bones in the structures outside the white wall makes it difficult to tell whether the few dietary remains found in this sector are different from those found in the inside structures. For example, the lower contribution of nondeer species to the assemblage from outside structures is likely an artifact of the relatively higher proportion of curated deer bones in this area. Furthermore, the distribution of food-refuse bones shows no appreciable differences in parts of the deer consumed in these two areas. Noncurated elements from the skull (cranial fragments and maxilla) are essentially evenly represented in both sectors. In fact, meat-bearing limb bones in the outside structures (computed without the possible curated limb bones) are more than twice as common as non-meat-bearing limb bones in the outside structures; in the inside structures, meat-bearing and non-meat-bearing limb bones are essentially evenly represented. The deer skeletal part distributions, then, give no indication that any one sector of the site received preferred cuts or that butchery of animals took place in one sector of the site and meaty cuts were moved differentially into another sector of the site.

SEASONALITY OF OCCUPATION

Price and Griffin (1979:140) raise the possibility that structures located inside the white wall were occupied throughout the year by residents who controlled and farmed the arable land around the village. They theorized that structures outside the wall were occupied by families from outlying farmsteads who visited the site during certain times of the year for ceremonial purposes, for pooling resources, or for protection (see Smith 1978:147–161). Pointing to the lack of evidence for differential control over arable land and agricultural resources, Price and Griffin do not favor this hypothesis. Yet without directly assessing seasonality of occupation, they could not rule out the possibility that the outer perimeter of the Snodgrass site was only seasonally occupied. The Snodgrass remains can be used to estimate season of occupation in two ways: (1) tooth eruption and wear of deer mandibles and (2) the presence of male deer frontal bones with either shed or attached antlers (Table 17-4).

Mandibular Data

Mandibular seasonality data are based on age determinations following eruption and wear sequences in Severinghaus (1949). Age classes for the first two years of a deer's life are quite narrowly defined. When calibrated for the months in which fawns are born in a region (in this case May and June [Smith 1975:38]), deer mandibles of individuals less than two years of age can be used to determine the season in which deer were killed. Mandibular data from Snodgrass show a very strong emphasis on animals killed in the fall (Table 17-4). More than half the 60 mandibles that could be used to determine season of kill represent age classes that coincide with the fall. Yet although hunting apparently peaked during the fall, when the availability of acorns and the fall rut may have made deer more accessible to Snodgrass hunters (Smith 1974, 1975:36–39), all seasons of the year are represented by the Snodgrass mandibles. Moreover, all seasons are represented in all contexts at the site inside structures, outside structures, and pits.

Shed and Attached Antlers

It is also possible to look at seasonality by examining the proportion of male deer frontal fragments. For most male deer in temperate North America, antlers begin to grow in May and are fully hardened by September; they are dropped in mid-January to early February. Frontal bones with attached antlers, then, mean that the animal was killed between May and February, whereas frontal bones with shed antlers indicate that the animal was taken between January and May (Smith 1975:38, 1978:152–153). Frontal bones with both shed and attached antlers are present at the Snodgrass site, with 13 of a total of 17 male crania indicating kills in the months from May to February (Table 17-4). However, four frontal bones have evidence of shed antlers. All these four frontal bones were located in the outside structures.

These data seem to argue against seasonal occupation in any sector of the village, but they may not be as definitive as they appear. The presence of mandibles belonging to age classes that coincide with all four seasons of the year certainly

Table 17-4. Seasonality Data, NISP

Location	Mandibles				Male frontal bones	
	Summer	Fall	Winter	Spring	Jan.-May	May-Feb.
Inside structures						
Year 1	–	7	–	1	–	–
Year 2	4	17	1	5	–	–
Total	4	24	1	6	–	5
Outside structures						
Year 1	1	–	–	–	–	–
Year 2	2	8	1	4	–	–
Total	3	8	1	4	4	7
Pits						
Year 1	–	1	2	–	–	–
Year 2	1	2	–	3	–	–
Total	1	3	2	3	–	1
Total	8	35	4	13	4	13

strongly suggests that the village as a whole was occupied year round. Because many of these mandibles were probably curated, the presence of a mandible from an animal killed in the summer says little about the season in which the mandible was used and deposited. Obviously the presence of shed antlers says nothing about the season of occupation of the area where they are used and disposed. Frontal bones with either attached or shed antlers, however, are likely to actually represent butchery refuse from a freshly killed animal, and so can be used to more directly examine the season of occupation of the place where they were deposited. The presence of frontal bones with evidence of shed antlers allows us to bracket a relatively short time period for the kill of the animal and the occupation of the area in which the bone was deposited (late winter to early spring). Unfortunately, the presence of frontal bones with attached antlers tells us only that the animal was killed some time in the late spring, summer, fall, or early winter.

It is possible to narrow this large window somewhat by determining the stage of antler development by examining the burr of the attached antlers (Smith 1978:149). None of the attached antlers with burrs from the Snodgrass site had either open or even recently closed nutrient foramina that are present when antlers are in an earlier stage of growth (May–September). Later stages of antler growth (September–February) are indicated. This factor plus the strong mandibular evidence for an emphasis on hunting deer in the fall suggests that the majority of frontal bones with attached antlers probably belong to the later part of the eight month period that male deer carry antlers. If the outside structures were occupied in the fall and winter months, as might be predicted under a seasonal occupation scenario, we could expect to see both attached and shed antler fragments here. Thus, although there are no grounds to conclude that any sector of the Snodgrass site was seasonally occupied, neither can we categorically rule out this possibility.

MOVEMENT OF MEAT AND BONE WITHIN THE SNODGRASS SITE

But what of interaction among Snodgrass households, both within and between the two sectors? Were animal resources, either meat or curated bones, being moved among structures, and if they were what does this movement tell us about relations among households in the Snodgrass village community? A number of studies have pieced together broken bone fragments, used metric analysis, reunited single joints, or matched right and left sides of individuals to trace the movement of bones and meat cuts within sites (Enloe and David 1992; Todd and Frison 1992; Todd and Stanford 1992). Refitting bone requires complete recovery of faunal material, careful recording of the location of bones, and a shallow time depth. These conditions are easily met by the Snodgrass faunal assemblage. All four techniques were attempted with the Snodgrass assemblage, but reuniting joints and, especially, matching right and left sides of single animals proved the most profitable.

More than 33 right/left matches of bones from single individuals were made within 19 structures and 5 pits at Snodgrass; more than 54 articulations were identified from 11 structures (Zeder and Arter 1993). Within-structure matches convey information about the ancient activities conducted in the structures and the postdepositional forces that shaped bone and artifact distribution within structures. In addition to these within-structure matches, and of more immediate interest to us here, a total of 19 right/left matches between different structures and pits at the site were also identified.

As shown in Figure 17-2, no matches are found between structures or pits located outside the white wall although there are two astragalus matches between Structure 25 outside the white wall, and nearby Structures 18 and 19 within the wall. Instead, the vast majority of matches occur between structures and pits located within the white-walled enclosure. Although a few of these matches are between structures and adjoining pits, most of the matches are between distant contexts. Contemporaneity in the occupation of structures within the white wall is strongly implied. Occupation of at least one structure outside the white wall seems also to have been contemporaneous with occupation of inside structures.

These matches highlight possible patterns of interaction among inhabitants inside the white wall. Although they occur among all inside structures, pairs seem to cluster in certain areas. Specifically, there is a cluster of pairs linking structures and pits in the southern portion inside the walled sector and a separate cluster in the northern portion. Interaction between structures in both the northeastern and western rows is also suggested. Finally, there are connections between features across the central courtyard of the site. No matches, however, link the features in the far southern sector inside the white wall with those in the far northern sector. Looking at the astragalus pairs alone, a north/south dichotomy of interaction at Snodgrass becomes even clearer. If astragali were used as gaming pieces, stronger social ties among households in the northern sector of the site and among those in the southern sector might be posited, with more social distance between northern and southern sectors.

DISCUSSION

Each of the ways in which animal remains were examined in this study sheds light on different facets of life within the Snodgrass site: intensity of occupation, social and economic differences in animal resource exploitation, seasonality of occupation, and interaction between households. When all these different facets are brought together, a picture of life in this Mississippian village community comes into focus.

On the basis of the faunal remains, the dichotomy between structures located within the white wall and those situated outside does not appear to be solely attributable to length of occupation, seasonality of occupation, or status differences. Intensity of occupation certainly appears to have been less in this sector, but is this the product of a shorter duration of occupation? Given the projected short life of the village as a whole (5–10 years), there would not seem to be enough time for members of founding families within the white wall to bud off and establish their own homes in the outer perimeter of the site. If these outside structures were built by families moving into the village, why is food refuse so poorly represented here, and why is there no evidence for the movement of resources between households? Seasonality data are equivocal. Although these data by no means support the contention that any sector of the site was seasonally occupied, neither do they rule it out. A late fall to winter occupation of the outer perimeter of the site is still a possibility. Finally, if there are any status differences between residents of these two sectors of the site, they are not reflected in the animal portion of the diet. Unlike other, larger, Mississippian communities (Jackson and Scott 1995; Michals 1992b; Scott 1983), there is no evidence of status-based differences in quality of meat resources or in movement of meat from one sector of the site to another.

Although it is relatively easy to say what cannot be supported by these data, it is harder to be definitive about what can be said. The predominance of proposed curated bone and the paucity of food refuse in the outer perimeter of the site, plus the lack of any indication of bone movement between outside structures, all suggest that the apparently less intense occupation here may be attributable to a fundamental difference in the way this area was used, rather than to a briefer or more sporadic occupation. Although some domestic activity involving the preparation, consumption, and disposal of food refuse is indicated, the dominance of curated bones in the outside sector of the site suggests that a number of structures were devoted to a variety of nonresidential activities, some of which involved storage and use of bone implements.

In contrast, the larger structures inside the white wall may have served more of a combined residential and productive function. Within the walled enclosure, there appear to have been definite patterns of interaction linking certain households and certain parts of this sector of the site (involving meat sharing, tool use, and, possibly, recreation) that may have direct bearing on kinship relations in this village community.

CONCLUSION

The picture that emerges from the animal bones is by no means complete. Ongoing studies of the ceramic, lithic, and botanical materials from the site, plus continued refinement of the faunal study, will no doubt enrich, and make more complex, our understanding of the dynamics of village life at Snodgrass. But this study does highlight the potential for biological remains to go beyond environment and diet and join the mainstream of archaeological analysis for the study of complex social and economic interactions of the past.

ACKNOWLEDGMENTS

The original excavations at the Snodgrass site were conducted by the University of Michigan under the direction of James B. Griffin and James E. Price supported by funds granted by the National Science Foundation. The ongoing Powers Phase Project is based at the National Museum of Natural History, Smithsonian Institution, under the direction of Bruce D. Smith. This work was generously supported by funds from the Smithsonian's Scholarly Studies Program and by the Center for Archaeobiological Research. Faunal analysis was conducted in the National Museum's Archaeobiology Laboratory under the direction of Melinda Zeder. Susan Arter was senior analyst on the project. Elizabeth Moore, Mark Warner, Eric Fettman, and Justin Lev-Tov assisted in the analysis of bones from the Snodgrass and Turner sites. Many thanks are given Bruce Smith for his helpful comments on an earlier draft of this chapter. All errors and omissions are the sole responsibility of the authors.

REFERENCES

Black, T. K., 1979, The Biological and Social Analyses of a Mississippian Cemetery from Southeastern Missouri: The Turner Site 23BU21, University of Michigan Museum of Anthropology Papers 68, Ann Arbor.

Bogan, A., 1983, Evidence for Faunal Resource Partitioning in an Eastern North American Chiefdom, in: Animals and Archaeology. 1. Hunters and their Prey (J. Clutton-Brock and C. Grigson, eds.), British Archaeological Reports International Series (Oxford) 163:305–324.

Brown, J. A., 1964, The Identification of a Prehistoric Bone Tool from the Midwest: The Deer-jaw Sickle, American Antiquity 29:381–386.

Colton, R., 1992, Animal Resources, Social Complexity, and the Archaeological Record, paper presented at the 57th Annual Meeting of the Society for American Archaeology, Pittsburgh, Pennsylvania.

Eisenberg, L. E., 1989, On Gaming Pieces and Culture Contact, Current Anthropology 30:345.

Enloe, J. G., and David, F., 1992, Food Sharing in the Paleolithic: Carcass Refitting at Pincevent, in: Piecing Together the Past: Applications of Refitting Studies in Archaeology (J. L. Hofman and J. G. Enloe, eds.), British Archaeological Reports International Series (Oxford) 578:296–315.

Harrington, M. R., 1908, Some Seneca Corn-Foods and their Preparation, American Anthropologist 10:575–590.

Jackson, H. E., and Scott, S., 1995, The Faunal Record of the Southern Elite: The Implications of Economy, Social Relations, and Ideology, Southeastern Archaeology 14:103–119.

Lewis, R. B., 1988, Old World Dice in the Protohistoric Southern United States, Current Anthropology 31:410–413.

Lewis, R. B., 1990, On Astragalus Dice and Culture Contact: Reply to Eisenberg, Current Anthropology 31:410–413.

Lyman, R. L., 1984, Bone Density and Differential Survivorship of Fossil Classes, Journal of Anthropological Archaeology 3:259–299.

Michals, L. M., 1992a, The Nature of Faunal Exploitation in Mississippian Societies, paper presented at the 57th Annual Meeting of the Society for American Archaeology, Pittsburgh, Pennsylvania.

Michals, L. M., 1992b, Faunal Exploitation and Complex Chiefdom Development at Moundville, Alabama, Ph.D. dissertation, Department of Anthropology, University of California, Santa Barbara.

Powell, M. L., 1988, *Status and Health in Prehistory: A Case Study of the Moundville Chiefdom*, Smithsonian Institution Press, Washington, DC.

Price, J. E., and Griffin, J. B., 1979, *The Snodgrass Site of the Powers Phase of Southeast Missouri, University of Michigan Museum of Anthropology Papers* 66, Ann Arbor.

Schoeninger, M. J., and Peebles, C. S., 1981, Notes on the Relationship between Social Status and Diet at Moundville, *Southeastern Archaeological Conference Bulletin* 24:96–97.

Scott, S. L., 1983, Analysis, Synthesis, and Interpretation of Faunal Remains from the Lubbub Creek Archaeological Locality, in: *Prehistoric Agricultural Communities in West Central Alabama* (C. S. Peebles, ed.), U.S. Army Corps of Engineers, Mobile District, Mobile, Alabama, pp. 272–390.

Severinghaus, C. W., 1949, Tooth Development and Wear as a Criteria of Age in White-tailed Deer, *Journal of Wildlife Management* 13:196–216.

Smith, B. D., 1974, Predator-Prey Relationships in the Eastern Ozarks: A.D. 1300, *Human Ecology* 2:31–44.

Smith, B. D., 1975, *Middle Mississippi Exploitation of Animal Populations, University of Michigan Museum of Anthropology Papers* 57, Ann Arbor.

Smith, B. D., 1978, *Prehistoric Patterns of Human Behavior: A Case Study in the Mississippi Valley*, Academic Press, New York.

Styles, B., and Purdue, J. R., 1986, Middle Woodland Faunal Exploitation, in: *Napoleon Hollow Site in the Lower Illinois Valley* (M. D. Wiant and C. R. McGimsey, eds.), *Kampsville Archaeological Center Research Series*, Volume 6, Center for Archaeological Research, Kampsville, Illinois, pp. 513–526.

Todd, L. E., and Frison, G. C., 1992, Reassembly of Bison Skeletons from the Horner Site: A Study in Anatomical Refitting, in: *Piecing Together the Past: Applications of Refitting Studies in Archaeology* (J. L. Hofman and J. G. Enloe, eds.), *British Archaeological Reports International Series* (Oxford) 578:63–82.

Todd, L. E., and Stanford, D. J., 1992, Applications of Conjoined Bone Data to Site Structural Studies, in: *Piecing Together the Past: Applications of Refitting Studies in Archaeology* (J. L. Hofman and J. G. Enloe, eds.), *British Archaeological Reports International Series* (Oxford) 578:21–35.

Welch, P. D., and Scarry, C. M., 1995, Status-Related Variation in Foodways in the Moundville Chiefdom, *American Antiquity* 60:397–420.

Wright, M., 1958, American Indian Corn Dishes, *Chronicle of Oklahomautoi* 36:155–166.

Zeder, M. A., and Arter, S. R., 1993, Intra-Site Animal Utilization at a Middle Mississippi Site, paper presented at the 51st Annual Meeting of the Southeastern Archaeological Conference, Raleigh, North Carolina.

Chapter *18*

Who Ate What? Archaeological Food Remains and Cultural Diversity

ELIZABETH M. SCOTT

Environmental archaeologists are paying increasing attention to evidence of socioeconomic and ethnic distinctions in human societies (e.g., Crabtree 1990; Reitz and Honerkamp 1983; Reitz and Scarry 1985; Schulz and Gust 1983). Archaeologists combine ethnohistorical, archival, and archaeological records to reveal a more complete picture of life in the past than if any of those records were used alone.

Especially in culturally mixed societies, where people of several ethnic and economic groups resided together, material culture often provides few clues to the ethnicity of the site's residents. Artifacts are indispensable for, among other things, dating, interpreting the activities that took place at the site, or determining the socioeconomic position, and sometimes even the gender, of a site's occupants. The material culture available to a community however, was that of the dominant political and economic group, so ethnic differences among the users of those material items often are not apparent to archaeologists.

Food remains, however, provide not only evidence of socioeconomic position but also some of the strongest evidence for the ethnic identity of a site's occupants. Food provides a means to emphasize or deemphasize one's differences with others, i.e., a means of denoting ethnicity or religion. Culturally determined food preferences, and the degree of departure from them, provide important clues for interpreting sites.

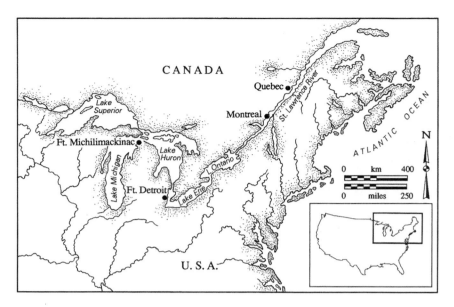

Figure 18-1. Map of the Great Lakes region of North America.

This case study presents an example from a culturally heterogeneous colonial community, an 18th-century fur-trading settlement in what is now northern Michigan (Figure 18-1). Analysis of the food remains associated with several households reveals information about residents of diverse economic, ethnic, and religious groups that might remain invisible without environmental archaeology.

METHODS

In this study, 18th-century written records were used to predict the food preferences for each class and ethnic group at the settlement. These records include travel accounts, letters, military accounts, diaries, and cookbooks. The vertebrate and botanical assemblages from each household were analyzed and compared to the predicted models for the best fit. The comparative collections of the Illinois State Museum, Springfield, were used for faunal identification; Leonard W. Blake, Washington University, St. Louis, analyzed the botanical remains, and his reports are relied upon here (Blake 1981, 1982, 1985, 1987, 1990). Optimal recovery of animal and plant remains was possible because archaeological deposits were water-screened through 1/16-inch mesh. Flotation samples were taken from the floor deposits of the one undisturbed house interior. Minimum Number of Individuals (MNI) and biomass estimates were calculated for the faunal remains from each household (Reitz and Scarry 1985:17–20; Wing and Brown 1979:127–128), although only biomass estimates are presented here. This study excludes commensal and other presumed nonfood faunal remains recovered from the households (for further discussion, see Scott 1991).

SITE DESCRIPTION

Fort Michilimackinac was located on the Straits of Mackinac connecting Lakes Michigan and Huron (Figure 18-1). The area is a mixed conifer–deciduous forest, with large deep lakes as well as smaller lakes and streams nearby. Dry, sandy soils predominate in the area, making crop cultivation difficult. The settlement was geographically isolated: very little shipping, either by canoe or sailing ship, could be carried out between November and April. The Straits of Mackinac are usually frozen over by January, and the ice often does not break up until April. Thus, for half of each year, this community was isolated and depended on local animal and plant resources for food.

Michilimackinac was established around A.D. 1715 by French missionaries, soldiers, and fur traders. Britain gained control of the settlement at the end of the French and Indian War (or Seven Years' War) in 1760, and it remained a part of Britain's colonial empire until 1781. Living in its palisaded fort and in the village outside, known as the "subarbs," was a diverse community of British, French-Canadian, German-Jewish, Native American, metis (offspring of European-Native American unions), and African-American men, women, and children, of all socioeconomic groups. Throughout the 18th century and into the 19th century, the economy in the Upper Great Lakes was based on the exchange between indigenous peoples and European colonizers of furs for European goods.

This study analyzes food remains from four households occupied in the 1760s and 1770s. They were part of one long rowhouse along the Rue de la Babillarde ("Street of the Gossips") in the southeast quarter of the fort (Halchin 1985; Heldman 1977; Heldman and Grange 1981). Written records and artifact assemblages provided some evidence about the occupants of each household, but as this study shows, much more was revealed by the food remains.

SUBSISTENCE AT MICHILIMACKINAC

When British colonists arrived at Michilimackinac in 1761, they encountered French-Canadians, metis, Ottawa, and Ojibwa using primarily local, wild resources. A few European domestic animals were raised, but their consumption was probably limited to the wealthier inhabitants (Scott 1985). The

French-Canadians and Native Americans who lived at the settlement cultivated some vegetables, but relied to a great degree on the agricultural produce of
the Ottawa women from the village of L'Arbre Croche, about 32 km south
along the Lake Michigan shore. British colonists brought with them their own
subsistence system, which they modified and adapted to the ecological and
cultural environment at Michilimackinac. Some British foods were adopted by
the French-Canadians, *metis*, and Native Americans.

The most striking feature of the subsistence system at Michilimackinac was
the interdependence among the ethnic groups at the settlement. French-
Canadian and British colonists adopted Native American methods of procur-
ing and preparing local foods, and Native Americans depended upon
Europeans for the materials they used in subsistence activities as well as for
some foodstuffs. British settlers also depended upon French-Canadians, *metis*,
and Native Americans to perform subsistence-related labor. Despite this interdependence, however, documents indicate differences in food preferences for
the various socioeconomic and ethnic groups in the community.

Ottawa and Ojibwa Diets

Ethnohistoric sources indicate only slight differences between the diets of the
Ottawa and Ojibwa who lived near Michilimackinac during this period (Feest
and Feest 1978; Gates 1965; Quaife 1921, 1922; Rogers 1978). Archaeological
evidence from 17th- and 18th-century Native American sites in the region is
available, but the sites cannot be definitely associated with one or the other of
these groups (Cardinal 1976; Colburn and Martin 1985; Martin 1981; Smith
1984). Both depended on the same wide array of animal resources, which
included wild mammals, birds, turtles, and fishes, as well as dogs (*Canis
familiaris*). European animals do occur on late 17th-century sites near the
Straits of Mackinac, but in very small numbers (Cardinal 1976; Martin 1981).
The primary differences between the two groups were in the plant foods used.
Both planted maize (*Zea mays*), but the Ottawa grew common beans
(*Phaseolus vulgaris*) and gourd/squash (*Cucurbita* sp.) as well. Both used maple
sugar (made from the sap of *Acer saccharum*) and several kinds of wild berries,
but the Ojibwa along the northern shore of Lake Superior used wild rice
(*Zizania aquatica*) and blueberries/cranberries (*Vaccinium* sp.), which were not
prevalent to the south.

French-Canadian Diet

The Swedish naturalist Peter Kalm left a remarkably vivid account of his travels to Montreal and Quebec in 1749 and 1750, including descriptions of food
(Benson 1987:381–576). French-Canadians of all socioeconomic groups in
lower Canada appear to have had access to similar kinds of vegetables, fishes,
wild birds, and game. What seems to have distinguished the diet of the "better

classes" from that of poorer soldiers and farmers was the greater reliance on meat from domestic animals and the greater variety of fruits and nuts eaten by the former. Wealthier French-Canadians relied on beef, mutton, veal, fowl, and squab (young pigeons [Columbidae]) prepared in a variety of ways, including boiling, roasting, frying, and as fricassees and ragouts (Benson 1987:473–475). The farmers and soldiers, by contrast, relied on bacon, dried meat, and only occasional beef. They kept cows (*Bos taurus*), primarily for milk, and sheep (*Ovis aries*), primarily for wool.

Descriptions of the 18th-century French-Canadian diet at Michilimackinac are much different (Gerin-Lajoie 1976; Stevens et al. 1941). They emphasize reliance on maize (usually hominy), fishes, wild birds, and wild mammals, some of which were obtained through trade with local Ottawa and Ojibwa. These records do not distinguish dietary differences between socioeconomic groups at Michilimackinac. The archaeological evidence from households of the earlier French regime (A.D. 1715–1761) at Michilimackinac largely supports the diet they describe. Although wealthier households relied more heavily on domestic animals (pig [*Sus scrofa*] being the most frequently consumed) and cultivated plants than did the relatively poorer French-Canadian households, the overwhelming picture is one of reliance on wild resources (Scott 1985).

British Diet

Two 18th-century cookbooks provide insights into food preferences in British North America. These cookbooks were intended for different audiences: *The Frugal Housewife or Complete Woman Cook*, published in 1772 by Susannah Carter (McKibbin 1976), was intended for the wealthy; *American Cookery*, published in 1796 by Simmons (1984), was intended for domestic servants or less wealthy housewives.

These recipes indicate that among British colonists in North America, as among French settlers, the wealthy enjoyed a far greater variety of dishes made from beef, pork, mutton, lamb, and veal than did those less well-off. Besides the increased variety in "meatier" cuts, the wealthier British diet was characterized by an increased use of many bony portions as well. Not only could the wealthy afford more of the meatier and tender cuts, but also they could afford the servants, cooks, equipment, and time necessary for the preparation of feet and head parts as stews, soups, hashes, pies, and fricassees.

White-tailed deer (*Odocoileus virginianus*), rabbits (*Sylvilagus* sp.), and hares (*Lepus americanus*) are the only wild mammals mentioned in the two cookbooks, along with many wild birds and fishes. The sources suggest that venison was associated with the wealthy, while rabbits and hares were consumed by both the upper and lower classes. As with domestic mammals, the wealthy enjoyed much more variety in the dishes prepared from wild birds and fishes than did those of lesser means.

British colonists at Michilimackinac, whether they emigrated from Britain or from North American British colonies, brought with them their dependence on domestic animals. In nonagricultural areas like the Straits of Mackinac, with no urban market, access to meat from domestic animals was limited to the wealthy and the military. Most of the less wealthy British colonists at Michilimackinac modified their dependence on domestic animals. Written accounts indicate that they often relied upon wild animal resources, especially when they were effectively isolated for about six months each year (Bald 1938; Gates 1965; Quaife 1921, 1922, 1928). The food remains from a household in which relatively poor British foot soldiers resided in the early 1760s revealed just such a diet, although pork, beef, and chicken (*Gallus gallus*) were consumed also (Shapiro 1978).

Several wild plants and their products were utilized by people at Michilimackinac and described by 18th-century observers (Bald 1938; Gates 1965; Quaife 1921, 1922, 1928). Among those most commonly mentioned was the maple sugar made by French-Canadian and *metis* families at Michilimackinac and by local Ottawa and Ojibwa families. Many kinds of berries were used as food, and at least some were used medicinally.

Jewish Diet

Although limited in comparison to information on Ottawa, Ojibwa, French-Canadian, and British diets in Canada during the last half of the 18th century, some documentary and archaeological evidence for a contemporary Jewish diet is available. Marcus (1970:1227) notes that the "isolated country Jew" in 18th-century North America was nonobservant. Peter Kalm reported in 1748 that Jewish colonists in North America often ate pork when traveling, though at home they "commonly ate no pork" (Benson 1987:129–130). Although dietary restrictions and licensing of *shohets* (those who properly slaughtered animals and certified kosher meats according to Jewish law) are prominent in the records for 18th-century congregations in Philadelphia and New York (Marcus 1959:93, 192–194), no such regulations were recorded in the constitution or minute book for the Montreal congregation (Marcus 1959:105–114).

Using material from 17th- and 18th-century sites in Amsterdam, The Netherlands, some of which were located in an area of the city inhabited by Portuguese Jews, Ijzereef (1989) proposed a method for detecting differences between Jewishness and non-Jewishness in the faunal remains from individual households. He based his conclusions on the relative percentages of bone weights for each animal species identified. The main criterion he used to distinguish Jewish and non-Jewish households was the percentage of pig remains by weight in the refuse. On the basis of these percentages, he identified four groups: (1) Jewish households (no pig remains); (2) Jewish households (0–1% pig remains), with the small quantity possibly due to contamination from later layers; (3) nonkosher households, or Jews with non-Jewish residents living in the same house (1–5% pig remains); and (4) non-Jewish households (greater than 5%, usually about 15%, pig remains) (Ijzereef 1989).

PREDICTED DIETARY ASSEMBLAGES

Thus, the archaeological evidence for differences in diet among households at Michilimackinac from A.D. 1761–1781 could reflect both ethnicity and socioeconomic position. We might expect a British household to depend more heavily on domestic animals than a French-Canadian or Native American household. The British diet would be characterized also by use of deer or hare as the primary wild mammals and by predominance of chicken, "game birds," and passenger pigeons (*Ectopistes migratorius*).

We might expect that a French-Canadian diet at Michilimackinac would include a greater use of wild mammals and birds than a British one. We would expect an observant Jewish household to contain significantly fewer pig remains than other households and to depend less on wild animals in general. Ottawa or Ojibwa diets should include few, if any, domestic animals, and should reveal a heavy dependence on maize, with gourd/squash and beans also present in Ottawa deposits.

Socioeconomic differences, however, might cut across ethnic lines. We would expect the diets of both wealthy British and wealthy French-Canadian inhabitants to be more heavily dependent on domestic animals than would the diets of the middle or lower classes in either group. The cuts of meat from domestic mammals could be used to gauge socioeconomic position among both British and French-Canadian colonists. Wealthy households might be characterized by animal remains from *all* parts of the skeleton in their refuse. Poor households might be characterized by the presence of *only* the bony portions. Middle income households would contain the remains from both meaty and bony portions, although not from entire animals. The wealthy of both ethnic groups enjoyed a greater variety of foods, especially plant foods, than did the less wealthy. The diet for the less wealthy and poorer colonists would be expected to depend more heavily on wild resources among both the British and French-Canadians.

ARCHAEOLOGICAL EVIDENCE OF DIVERSITY IN DIET

With these expectations derived from documentary evidence, we now turn to an evaluation of the archaeological evidence (Tables 18-1, 18-2, and 18-3). The food remains from each household in this study will be compared with the predicted diets for the socioeconomic and ethnic groups at Michilimackinac.

Table 18-1. Animal Food Remains from Each Household, Percentage of Total Biomass[a]

Taxa	House F		House A-B	House C
	1760s	1770s	1770s	1765–1781
Domestic species				
Cow (*Bos taurus*)	–	13.4	3.4	17.1
Pig (*Sus scrofa*)	17.4	17.0	5.7	13.0
Sheep and caprines (*Ovis aries* and Caprinae)	–	0.5	3.4	4.2
Chicken (*Gallus gallus*)	0.1	–	12.4	0.3
Subtotals	17.5	30.9	24.9	34.6
Wild species				
Deer (*Odocoileus virginianus*)	–	–	–	1.4
Badger (*Taxidea taxus*)	0.1	–	–	–
Foxes (*Vulpes vulpes* and *Vulpes/Urocyon* spp.)	–	0.4	–	0.1
Beaver (*Castor canadensis*)	13.4	2.1	–	4.9
Squirrels (*Tamiasciurus hudsonicus* and *Glaucomys* spp.)	–	–	–	<0.1
Unidentifiable rodent	–	–	–	<0.1
Rabbits and hares (Leporidae)	0.5	1.4	0.8	0.9
Subtotals	14.0	3.9	0.8	7.3
Swan (*Cygnus* spp.)	–	–	3.0	0.2
Canada goose (*Branta canadensis*)	0.3	0.8	–	0.4
Other ducks and geese (Anatidae)	1.4	1.1	0.5	1.0
Hawks and falcons (Falconiformes)	0.6	0.1	–	<0.1
Grouse (Phasianidae)	<0.1	–	<0.1	0.1
Gulls and terns (Charadriiformes)	<0.1	–	0.1	0.1
Passenger pigeon (*Ectopistes migratorius*)	2.5	2.0	1.4	1.8
Small birds	<0.1	–	<0.1	<0.1
Subtotals	4.8	4.0	5.0	3.6
Lake sturgeon (*Acipenser fulvescens*)	0.6	0.7	0.6	0.6
Gars (*Lepisosteus* spp.)	–	0.1	–	<0.1
Longnose sucker (*Catostomus catostomus*)	–	–	0.1	<0.1
Pikes (Esocidae)	–	0.2	–	0.1
Trouts and whitefishes (Salmonidae)	4.2	3.8	3.7	2.8
Lake trout (*Salvelinus namaycush*)	1.3	1.7	5.8	2.3
Burbot (*Lota lota*)	<0.1	0.1	–	<0.1
Basses (*Micropterus* spp.)	0.1	0.2	0.1	<0.1
Yellow perch (*Perca flavescens*)	–	–	0.1	–
Walleye (*Stizostedion vitreum*)	0.2	0.5	0.3	0.4
Freshwater drum (*Aplodinotus grunniens*)	–	0.2	–	<0.1
Unidentifiable fish	6.1	6.5	12.7	9.2
Subtotals	12.5	14.0	23.4	15.4
Domestic/wild				
Dog family (Canidae)	–	–	–	0.3
Unidentifiable mammal	34.5	44.5	37.9	33.7
Unidentifiable bird	2.7	2.4	7.7	4.3
Subtotals	37.2	46.9	45.6	38.3
Total biomass[a]	12.698 kg	11.006 kg	9.998 kg	72.067 kg
Total fragments[a] (NISP)	3724	2043	5426	71,455

[a] Total includes food and nonfood remains for each household. NISP = number of identified specimens.

Table 18-2. Animal Food Remains from House C Yard: Change Through Time, Percentage of Total Biomass for Each Period

Taxa	1760s	1770s
Domestic species		
Cow	9.1	27.1
Pig	26.9	9.1
Chicken	0.2	0.2
Sheep and caprines	–	8.3
Subtotals	36.2	44.7
Wild species		
Deer	2.9	0.4
Beaver	10.4	0.4
Squirrels	0.1	-
Rabbits and hares	0.6	0.7
Subtotals	14.0	1.5
Swan	0.5	–
Canada goose	0.3	0.1
Other ducks and geese	1.0	0.3
Hawks and falcons	–	<0.1
Grouses	–	0.2
Gulls and terns	<0.1	–
Passenger pigeon	2.3	0.7
Small birds	<0.1	–
Subtotals	4.1	1.3
Lake sturgeon	1.4	0.2
Pikes	–	0.1
Trouts and whitefishes	2.9	1.2
Lake trout	4.2	1.6
Burbot	0.1	–
Walleye	0.5	0.1
Freshwater drum	–	<0.1
Unidentifiable fish	7.1	4.1
Subtotals	16.2	7.3
Indeterminate domestic/wild		
Dog family	1.4	–
Unidentifiable mammal	22.7	43.3
Unidentifiable bird	2.2	1.7
Subtotals	26.3	45.0
Total biomass[a]	16.635 kg	36.097 kg
Total fragments[a] (NISP)	4867	6103

[a] Total includes food and nonfood remains for each period. NISP = number of identified specimens.

House F, 1760s

The occupant(s) of House F during the 1760s relied less on domestic animals than any other household. No cattle remains were recovered, only pig and a small amount of chicken. Pig remains were from the meaty forelimb and hindlimb, but also included portions from the bony head, back, and trunk. These findings suggest occupant(s) of at least medium socioeconomic position.

Table 18-3. Plant Remains Present for Each Household

Remains	House F 1760s	House F 1770s	House C Interior	House C Exterior	House A-B 1770s
Cultivated					
Maize (*Zea mays*)	X	X	X	X	X
Gourd/squash (*Cucurbita pepo*)			X	X	X
Wheat (*Triticum aestivum*)			X		
Common bean (*Phaseolus vulgaris*)			X		
Edible Wild Plants					
Black/chokecherry (*Prunus serotina/ virginiana*)			X		
Wild cherry (*Prunus* spp.)			X		
Raspberry/blackberry (*Rubus* spp.)			X		X
Elderberry (*Sambucus* spp.)			X		
Blueberry (*Vaccinium* spp.)			X		
Serviceberry (*Amelanchier* spp.)			X		
Grape (*Vitis* spp.)			X		
Hawthorn (*Crataegus* spp.)			X		
Hazelnut (*Corylus* spp.)			X		
Wild, possibly edible					
Pin cherry (*Prunus pennsylvanica*)	X	X	X	X	X
Other					
Sumac (*Rhus* spp.)			X		
cf. Bouncing bet (cf. *Saponaria officinalis*)		X			
cf. Bedstraw (cf. *Galium* spp.)			X		

This suggestion is supported by the relatively wide variety of wild birds and fishes used. Of all the households, the 1760s household at House F was the most dependent on wild mammals, suggesting that it was not in the upper wealth group. In fact, beavers (*Castor canadensis*) accounted for most of the wild mammals in the diet at House F. This predominance suggests that the occupant or occupants were French-Canadian or *metis*, since British consumption of beaver meat is not indicated by the documents. It is possible that French-Canadians consumed beaver more than British colonists did for religious reasons. Peter Kalm noted as early as 1749 that the Pope classified beavers with fishes because the beaver "spends most of his time in the water" (Benson 1987:534). Beaver, then, could be eaten by Catholics on fast days. Maize and pin cherry (*Prunus pennsylvanica*) remains were found in this household as well. The predominant use of local wild resources in House F during the 1760s seems most like the French-Canadian diet for the 1715–1761 period. The presence of pork and chicken, plus the variety of wild species used, indicates at least medium socioeconomic position.

The material culture recovered from the yard of House F suggests that someone of medium socioeconomic position lived there (Heldman and Grange 1981). The dietary evidence also suggests occupants of medium wealth, and strongly suggests French-Canadian ethnicity as well. This suggestion is at variance with the British officer reported to be in the house in Lt. Perkins Magra's 1766 sketch of the fort (Magra 1766). House F was owned just prior to British occupation of Michilimackinac by M. Boye. Perhaps Boye or another French-Canadian lived in the house before and after the British officer rented it in 1766. Or perhaps Magra was mistaken in his identification of the occupant of House F. It is possible that a British officer did live there and chose to eat a French-Canadian diet rather than a British one, but this seems unlikely, as will be shown below.

House F, 1770s

All the households occupied in the 1770s contain evidence that mutton was consumed at Michilimackinac. The smallest percentage of mutton remains is found in the 1770s component at House F, these remains being from the bony lower leg and foot only. This household depended in addition on beef and pork. In fact, domestic animals contributed about one third of the meat diet for the household. The beef portions are from the bony lower leg and foot, the back, and the trunk. The pork cuts are from the meaty hindlimb and forelimb as well as from the head. The domestic animal remains suggest a medium socioeconomic position.

Compared to the other households, the occupant(s) of House F during the 1770s relied more on hare, although there was overall little dependence on wild mammals and birds. This finding suggests a British diet. Passenger pigeon as well as ducks and geese (Anatidae) were the primary birds consumed. A wide variety of fishes from deep lakes and shallower inland ponds and streams were used. The plants recovered were maize and pin cherry.

The occupancy of House F in the 1770s, then, appears to be different from that in the 1760s. Available documentary records do not address the occupancy of House F in the 1770s, but the material culture recovered from the yard suggests that a wealthy trader lived there (Heldman and Grange 1981). The predominance of domestic animals and the kinds of wild animals used suggest someone of medium to upper socioeconomic position and of British ethnicity.

House A-B (Privy), 1770s

The documents and artifact assemblage identify this house as one in which British officers of the King's Eighth Regiment lived between 1774 and 1781 (Heldman 1977; Heldman and Grange 1981). Because deposits from the privy were those least disturbed when the fort was destroyed in 1781, bones from the privy were chosen for use here. Domestic mammals were in general less important to the diet here than in any of the other 1770s households, although chicken was more important to the diet in House A-B than elsewhere. Beef and mutton were of equal importance and pork was slightly more important to the diet. Beef and mutton are represented only by bony lower leg and foot remains, but pork portions include the meaty shoulder and hindlimb. These remains suggest a medium socioeconomic position.

The only wild mammals used were hare, and these contributed less than 1% of the meat in the diet. The most important wild birds in the diet were swans (*Cygnus* sp.), ducks (Anatidae), grouses (Phasianidae), and passenger pigeons. All these wild species are mentioned in the British colonial cookbooks. Swans were associated with high-status households in Britain as early as the Middle Ages (Serjeantson 1989:2).

The variety of fishes used also corresponds to the variety evident in British cookbooks. The diet revealed by the privy remains was dominated by fish more than was the diet in any other household. Lake trout (*Salvelinus namaycush*) and whitefishes (Salmonidae) contributed nearly 10% of the meat in the diet of the occupant(s) of House A-B, and several species of fishes from inland streams and shallow lakes were used also. The heavy dependence on fish could reflect supplies provided to the officers by soldiers. There is abundant documentary evidence that soldiers engaged in fishing year-round as part of their duties and that they preserved large supplies of fish for long-term use.

The diet revealed in the privy deposits seems almost stereotypically British in its lack of dependence on wild mammals and in the importance of chickens, "game" birds, and fishes. The material culture and dietary evidence from the privy suggest that the occupants of House A-B in the 1770s were of medium to upper socioeconomic position. The privy also contained maize, gourd/squash, raspberry/blackberry (*Rubus* sp.), and pin cherry remains, this variety suggesting at least a medium socioeconomic position.

House C, 1765–1781

In 1765, two German-Jewish traders, Ezekiel Solomon and Gershon Levy, bought House C from its French owners (Halchin 1985). They lived in the house through 1767, after which time Solomon apparently lived there alone. Solomon was married in August 1769 in Montreal, and thereafter probably used the house at Michilimackinac only as a summer residence, until the settlement was abandoned in 1781 (Scott 1991).

Setting aside this documentary evidence for a moment, when the food remains from all the occupation deposits associated with House C (interior and exterior) are combined for the period 1765–1781, the diet suggests occupants of medium to upper socioeconomic position who possibly were British. Of all the households examined, the diet in House C was the most dependent on domestic animals. As can be seen in Table 18-1, mutton and lamb were consumed in greater amounts there than in any other household, and in portions both meaty (forelimb) and bony (skull and trunk). Chicken was consumed also. Beef and pork were consumed in cuts from all portions of the body.

The occupants of House C depended to a greater degree on wild mammals than did the 1770s occupants of Houses F and A-B. They utilized the greatest variety of wild animals and plants of any household. The variety of wild resources used, then, suggests a medium to upper socioeconomic position. On the basis of documentary evidence for British, French-Canadian, and Native American diets at Michilimackinac, we might conclude that House C was occupied by British colonists, due to the importance of domestic animals in the diet.

A different picture emerges, however, if we look at change through time in the diet for the occupants of House C. The yard deposits for this house can be separated into those from the 1760s and those from the 1770s. As can be seen in Table 18-2, there is an increase in the importance of domestic animals and a pronounced decrease in the importance of wild mammals and birds and fishes in the diet. Whereas pork was the most important domestic meat in the 1760s, beef was most important in the 1770s.

This shift was predicted by Ijzereef's distinction between kosher and nonkosher Jewish households in 18th-century Amsterdam. His use of bone weight percentages is comparable to the biomass percentages used here, since biomass is estimated from bone weight. To note again, in Ijzereef's non-Jewish households, pig usually comprised about 15% of the faunal assemblage by bone weight. Kosher households contained 0–1% pig remains, and nonkosher households contained 1–5% pig remains. In the House C assemblage, pork consumption decreased through time from 27% of the diet to 9% (Table 18-2). Following Ijzereef's categories, this suggests a non-Jewish occupant during both periods. The difference between 27% and 9%, however, is similar in scale to Ijzereef's difference between non-Jewish (15%) and nonkosher (1–5%) households.

As can be seen in Table 18-3, the overwhelming majority of plant species from House C were recovered from the interior of the house. There is some indication that the maize remains found inside House C resulted from hominy preparation. The variety of plant remains, far more than in any other household when the interior and exterior deposits are combined, might be regarded as evidence of Jewishness, although it is perhaps stronger evidence of high socioeconomic position.

FOOD PREFERENCES AND ETHNICITY

As with the households studied by Ijzereef in Amsterdam, the material culture from House C gave no indication of the Jewishness of the inhabitants. They appeared from the artifacts to be like other successful fur traders at the fort and, had the record of the land sale to Solomon and Levy in 1765 not been found, the interpretation of the household likely would be that they were successful fur traders. When the food remains from the interior and exterior deposits are combined, there is no indication of a kosher diet. Rather, wealth is what is indicated, since Solomon and Levy's food remains are the most varied of all households examined, and include beef, pork, mutton, lamb, and chicken, in addition to many wild mammals and birds. Yet, when we examine the temporal components from the yard, we see a shift through time toward what appears to be a more observant diet.

Although in the 1770s Ezekiel Solomon was a prominent member of the Jewish congregation in Montreal, he married a French-Canadian Protestant woman; their children were baptized; and the boys were not circumcised (Katz 1948; Lefebvre 1962). Solomon never completely eliminated pork from his diet. There is no question, however, that Solomon considered himself Jewish and that this perception was shared by others at Michilimackinac; these views are clear in the written records (for a similar case, see Stewart-Abernathy and Ruff 1989).

Also indicated in the records, and to some degree in the artifactual evidence, is what appears to be anti-Semitism, at least on the part of some of the British residents at Michilimackinac (Heldman 1986; Scott 1991:61). It might therefore be useful to consider how food can be used to emphasize or deemphasize one's ethnic and religious differences.

As a German-Jewish trader, Solomon was in a tiny minority among his British, French-Canadian, and *metis* neighbors at Michilimackinac. There may have only been two other German-Jewish traders in the community. His diet in the 1760s, when he was a poor trader newly arrived from Europe, resembled that of French-Canadians, the largest group of residents. Eating as everyone else did may have been a way of "fitting in" or at least deemphasizing that aspect that most distinguished him from them, his Jewishness. Later, when he was successful in the fur trade and could afford not only the more expensive meats from domestic animals but also the more expensive material culture, he was able to emphasize his "similarity" with other members of the community in those ways. He could afford, therefore, to risk being viewed as "different" in his food habits without endangering his ability to be successful in the community. Consequently, corresponding to Solomon's increase in wealth and his increased involvement with the Jewish community in Montreal, we see in the 1770s a dramatic decrease in his use of pork and virtual elimination of wild birds and mammals from his diet.

Other ethnic differences in the community are emphasized by the dietary evidence as well. By the mid-18th century, there was a long history of rivalry and animosity between Britain and France. The French-Canadians at Michilimackinac had only recently been "occupied" by British troops and colonists. Because French-Canadians were allowed by the peace treaty to remain, to retain ownership of their houses and property, and to continue practicing Catholicism (*Wisconsin Historical Collections* [1931]8:216), one can imagine that they might have used food, as they did language, dress, religion, and architectural style, to emphasize their differences from the British citizens in the community. The trader living in House F during the 1760s seems to have steadfastly maintained the diet that French-Canadians at Michilimackinac had chosen for the preceding 50 years, with a strong reliance on wild species; his consumption of wild mammals, especially beaver, is far greater than in any of the other households.

Just as determined not to change their diets were the British officers in House A-B. Their food remains suggest a British diet changed very little by the frontier location, as might be expected of elite officers recently posted to Michilimackinac. The trader who lived in House F in the 1770s also had a diet that seems more British than anything else, although his greater consumption of wild mammals indicates less adherence to stereotypical British foods than was seen with the officers.

CONCLUSION

When the Michilimackinac evidence is compared with diets at 18th-century sites in other North American colonies, these ethnic differences appear muted (Scott 1990). It seems that in many cases the geographic setting (frontier versus urban) and the ecological range and abundance of particular food resources affected the diet more than did ethnic food preferences. For example, there were, overall, greater differences in diet between British residents in eastern seaboard colonies (e.g., Pendery 1984; Reitz and Honerkamp 1983) and British settlers at Michilimackinac than there were between British and French colonists at Michilimackinac itself.

Yet when diet within a specific multiethnic community is examined, it appears that ethnic and socioeconomic food preferences were used to accentuate differences among groups in that community. Despite the limiting and "homogenizing" influence on available foods occasioned by the frontier setting, people identifying with specific ethnic groups at Michilimackinac between 1761 and 1781 did distinguish themselves from each other through their diets, through socioeconomic and ethnic differences in both food consumption and food-related activities.

This case study emphasizes the necessity of integrating faunal and botanical analyses into the earliest stages of socioeconomic and ethnicity studies. To include environmental data merely as supplemental appendices ignores the significant role that environmental archaeology can play in the reconstruction of past societies.

ACKNOWLEDGMENTS

This chapter is based on portions of my dissertation (Scott 1991); parts of the chapter have been presented previously in conference papers (Scott 1990, 1992). I thank Philip Armitage for the reference to F. Gerard Ijzereef's work, and Sylvia Scudder for suggesting I get in touch with Dr. Armitage. I thank the late Francis B. Stahl for generously drafting Figure 18-1. The research on which this chapter is based benefitted greatly from comments by Janet Spector, Christine Hastorf, Elizabeth Wing, Leonard Blake, and my husband, Donald Heldman. I thank Terrance Martin and R. Bruce McMillan for allowing me to use the comparative osteological collections at the Illinois State Museum, Springfield, and the Department of Anthropology at the University of Minnesota in Minneapolis for a travel grant enabling me to go to the Illinois State Museum. Finally, I thank Elizabeth Wing for many years of advice, constructive criticism, and support; I consider myself fortunate indeed to have learned zooarchaeology, and much more, from her.

REFERENCES

Bald, F. C. (ed.), 1938, *From Niagara to Mackinac in 1767* (J. Porteous, comp.), *Historical Bulletin* 2, Algonquin Club, Detroit, Michigan.

Benson, A. B. (ed.), 1987, *Peter Kalm's Travels in North America: The English Version of 1770*, Dover Publications, New York.

Blake, L. W., 1981, Floral Remains from the 1978–1979 Excavations along the Rue de la Babillarde, in: *Excavations at Fort Michilimackinac, 1978–1979: The Rue de la Babillarde* (D. P. Heldman and R. T. Grange, Jr., comps.), *Archaeological Completion Report* 3, Mackinac Island State Park Commission, Mackinac Island, Michigan, pp. 366–375.

Blake, L. W., 1982, Analysis of Plant Remains from the 1981 Excavations at Fort Michilimackinac, manuscript on file, Office of Archaeology, Mackinac Island State Park Commission, Mackinaw City, Michigan.

Blake, L. W., 1985, Analysis of Plant Remains from the 1983 Excavations at Fort Michilimackinac, manuscript on file, Office of Archaeology, Mackinac Island State Park Commission, Mackinaw City, Michigan.

Blake, L. W., 1987, Analysis of Plant Remains from the 1984 Excavations at Fort Michilimackinac, manuscript on file, Office of Archaeology, Mackinac Island State Park Commission, Mackinaw City, Michigan.

Blake, L. W., 1990, Analysis of Plant Remains from the 1978–1984 Excavations at Fort Michilimackinac, manuscript on file, Office of Archaeology, Mackinac Island State Park Commission, Mackinaw City, Michigan.

Cardinal, E. A., 1976, Faunal Remains, in: *Archaeological Excavations at the Marquette Mission Site, St. Ignace, Michigan, in 1972* (J. E. Fitting, ed.), *Michigan Archaeologist* 22(2–3):226–234.

Colburn, M. L., and Martin, T. J., 1985, *P-Flat Site, 47AS47, Manitou Island, Wisconsin: Faunal Remains from the 1984 Excavations*, Illinois State Museum Society, Archaeological Research Program, *Technical Report* 85–212–3, Springfield.

Crabtree, P. J., 1990, Zooarchaeology and Complex Societies: Some Uses of Faunal Analysis for the Study of Trade, Social Status, and Ethnicity, in: *Archaeological Method and Theory*, Volume 2 (M. B. Schiffer, ed.), University of Arizona Press, Tucson, pp. 155–205.

Feest, J. E., and Feest, C. F., 1978, Ottawa, in: *Handbook of North American Indians*, Volume 15 (W. C. Sturtevant, general ed.), (B. G. Trigger, ed.), Smithsonian Institution Press, Washington, DC, pp. 772–786.

Gates, C. M., 1965, The Narrative of Peter Pond, in: *Five Fur Traders of the Northwest* (C. M. Gates, ed.), Minnesota Historical Society, St. Paul, pp. 11–59.

Gerin-Lajoie, M., 1976, *Fort Michilimackinac in 1749, Lotbiniere's Plan and Description, Mackinac History*, Volume II, Leaflet 5, Mackinac Island State Park Commission, Mackinac Island, Michigan.

Halchin, J. Y., 1985, *Excavations at Fort Michilimackinac, 1983–1985: House C of the Southeast Row House, Archaeological Completion Report* 11, Mackinac Island State Park Commission, Mackinac Island, Michigan.

Heldman, D. P., 1977, *Excavations at Fort Michilimackinac, 1976: The Southeast and South Southeast Row Houses, Archaeological Completion Report* 1, Mackinac Island State Park Commission, Mackinac Island, Michigan.

Heldman, D. P., 1986, Michigan's First Jewish Settlers: A View from the Solomon-Levy Trading House at Fort Michilimackinac, 1765–1781, *Journal of New World Archaeology* 6(4):21–34.

Heldman, D. P., and Grange, R. T., Jr., 1981, *Excavations at Fort Michilimackinac, 1978–1979: The Rue de la Babillarde, Archaeological Completion Report* 3, Mackinac Island State Park Commission, Mackinac Island, Michigan.

Ijzereef, F. G., 1989, Social Differentiation from Animal Bone Studies, in: *Diet and Crafts in Towns* (D. Serjeantson and T. Waldron, eds.), *British Archaeological Reports International Series* (Oxford) 199:41–53.

Katz, I. I., 1948, Ezekiel Solomon: The First Jew in Michigan, *Michigan History* 32(3):247–256.

Lefebvre, J. J., 1962, Letter to Mr. John K. Adams re: E. Salomon & Elizabeth Dobois, April 12, 1962, Department of Civil Status and of Archives, Montreal.

Magra, P., Lt., 1766, Lt. Perkins Magra's 1766 Sketch of Michilimackinac, Original in the Clements Library, University of Michigan, Ann Arbor.

Marcus, J. R., 1959, *American Jewry-Documents-Eighteenth Century: Primarily Hitherto Unpublished Manuscripts*, Hebrew Union College Press, Cincinnati, Ohio.

Marcus, J. R., 1970, *The Colonial American Jew, 1492–1776*, Volumes I–III, Wayne State University Press, Detroit, Michigan.

Martin, T. J., 1981, Animal Remains from the Gros Cap Site: An Evaluation of Fish Scales versus Fish Bones in Assessing the Species Composition of an Archaeological Assemblage, *Michigan Archaeologist* 27(3–4):77–86.

McKibbin, J. (ed.), 1976 [1772], reprint of S. Carter, *The Frugal Colonial Housewife*, Dolphin Books, Garden City, New York.

Pendery, S. R., 1984, The Archaeology of Urban Foodways in Portsmouth, New Hampshire, in: *Foodways in the Northeast* (P. Benes and J. M. Benes, eds.), University Press, Boston, Massachusetts, pp. 9–27.

Quaife, M. (ed.), 1921, *Alexander Henry's Travels and Adventures*, Lakeside Press, Chicago, Illinois.

Quaife, M. (ed.), 1922, *John Long's Voyages and Travels in the Years 1768–1788*, Lakeside Press, Chicago, Illinois.

Quaife, M. (ed.), 1928, *The John Askin Papers*, Volume I: 1741–1795, Detroit Historical Society, Detroit, Michigan.

Reitz, E. J., and Honerkamp, N., 1983, British Colonial Subsistence Strategy on the Southeastern Coastal Plain, *Historical Archaeology* 17(2):4–26.

Reitz, E. J., and Scarry, C. M., 1985, *Reconstructing Historic Subsistence with an Example from Sixteenth Century Spanish Florida, The Society for Historical Archaeology Special Publication* 3:1–150.

Rogers, E. S., 1978, Southeastern Ojibwa, in: *Handbook of North American Indians*, Volume 15 (W. C. Sturtevant, general ed.), (B. G. Trigger, ed.), Smithsonian Institution Press, Washington, DC, pp. 760–771.

Schulz, P. D., and Gust, S. M., 1983, Faunal Remains and Social Status in 19th Century Sacramento, *Historical Archaeology* 17(1):44–53.

Scott, E. M., 1985, *French Subsistence at Fort Michilimackinac, 1715–1781: The Clergy and the Traders, Archaeological Completion Report* 9, Mackinac Island State Park Commission, Mackinac Island, Michigan.

Scott, E. M., 1990, Subsistence and Ethnicity in Colonial North America, paper presented at the 6th International Conference of the International Council for Archaeozoology, Washington, DC.

Scott, E. M., 1991, *"Such Diet as Befitted his Station as Clerk": The Archaeology of Subsistence and Cultural Diversity at Fort Michilimackinac, 1761–1781*, Ph.D. dissertation, Department of Anthropology, University of Minnesota, University Microfilms International, Ann Arbor.

Scott, E. M., 1992, At Home They "Commonly Ate No Pork": Observance of Jewish Dietary Restrictions at the Solomon-Levy Trading House at Fort Michilimackinac, 1761–1781, paper presented at the 15th Annual Conference of the Society for Ethnobiology, Washington, DC.

Serjeantson, D., 1989, Introduction, in: *Diets and Crafts in Towns* (D. Serjeantson and T. Waldron, eds.), *British Archaeological Reports International Series* (Oxford) 199:1–12.

Shapiro, G., 1978, Early British Subsistence Strategy at Michilimackinac: An Analysis of Faunal Remains from the 1977 Season, in: *Excavations at Fort Michilimackinac, 1977: House One of the South Southeast Row House* (D. P. Heldman, comp.), *Archaeological Completion Report 2*, Mackinac Island State Park Commission, Mackinac Island, Michigan, pp. 161–177.

Simmons, A., 1984 [1796], *The First American Cookbook, A Facsimile of American Cookery*, Dover Publications, New York.

Smith, B. A., 1984, The Use of Animal Resources by the 17th Century Inhabitants at the Mission of St. Ignace, manuscript on file, Museum, Michigan State University, East Lansing.

Stevens, S. K., Kent, D. H., and Woods, E. E. (eds.), 1941, *Travels in New France by J.C.B*, Pennsylvania Historical Commission, Harrisburg.

Stewart-Abernathy, L. C., and Ruff, B. L., 1989, A Good Man in Israel: Zooarchaeology and Assimilation in Antebellum Washington, Arkansas, *Historical Archaeology* 23(2):96–112.

Wing, E. S., and Brown, A. B., 1979, *Paleonutrition: Method and Theory in Prehistoric Foodways*, Academic Press, Orlando, Florida.

Wisconsin Historical Collections, 1931 [1855–1931], Collections of the Wisconsin Historical Society (R. G. Thwaites, ed.), State Historical Society of Wisconsin, Madison.

Chapter 19

Seasonal Slaughter Cycles and Urban Food Supply in the Colonial Chesapeake

DAVID B. LANDON

People have lived in cities for over five millennia, thus understanding the emergence, development, and characteristics of urban settlements is a core research area in archaeology. Archaeologists have traditionally linked urbanization and urbanism to environmental questions because the concentration of people that defines a city requires a large supply of food as well as social and economic mechanisms to bring it to the city and distribute it. Cities typically rely on systems of intensive agriculture and high levels of surplus production from intensive agriculture systems are either a prerequisite for urbanization or a quick by-product. Cities are permanent settlements with dense populations, complex social systems, productive specializations, and strong appetites for food, fuel, and other natural resources. In addition to depleting local resources, cities change the environment though continued building and discharge of large volumes of waste.

This case study investigates urban food supplies and distribution systems by comparing seasonal slaughter patterns of domestic animals at urban and rural sites in North America. The data are from sites in the states of Virginia and Maryland dating from the mid-17th century through the end of the 18th century (Figure 19-1). Incremental growth structures in the dental cementum of cattle (*Bos taurus*) teeth are examined on thin-sections of teeth to interpret the season of slaughter. In the New England (New Hampshire, Massachusetts,

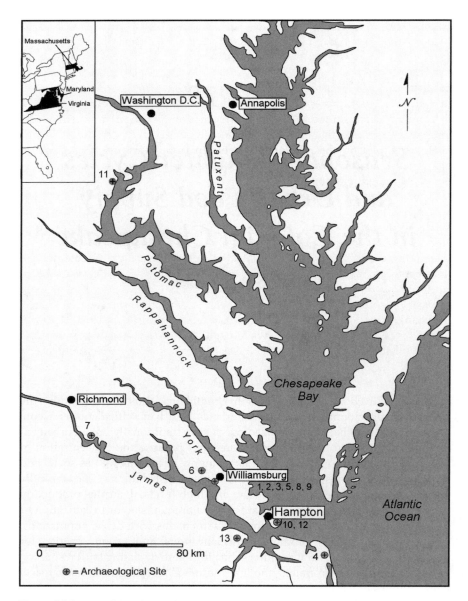

Figure 19-1. Map of the Chesapeake region showing the location of archaeological sites included in the study: (1) Kingsmill Slave Quarter, 1775–1800; (2) Draper, 1775–1800; (3) Anthony Hay, 1750–1800; (4) Virginia Beach, 1750–1800; (5) Firehouse, 1700–1740; (6) Rich Neck Slave Quarter, 1750–1775; (7) Curles Neck, 1750–1775; (8) Brush-Everard, 1750–1775; (9) Geddy, 1750–1762; (10) Settler's Landing, late 18th–early 19th century; (11) Mount Vernon, 1750–1775; (12) Hampton University, 1620–1660; and (13) Boothe Site, mid-late 18th century. Fort Criswell (1750s) is off the map to the west. Adapted from Walsh et al. (1997:Figure 2.2).

Connecticut, and Rhode Island) and Mid-Atlantic (New York, New Jersey, Pennsylvania, Delaware, Maryland, and Virginia) colonies, the slaughter of large domestic animals followed a seasonal pattern, based in part on broader rhythms of seasonal agricultural practice. Comparing these patterns at rural and urban sites provides insight into seasonal dietary variation, the nature of food distribution within cities and towns, and the links between rural production and urban consumption. This work focuses on just one component of the complex system of urban foodways. Nonetheless, this study demonstrates the merits of exploring the relationship between rural agricultural production and urban consumption through seasonal agriculture cycles and slaughter cycles for domestic animals.

URBANISM AND URBAN FOOD SUPPLY SYSTEMS

Placing this research in the context of urban environmental archaeology provides perspective on the range of issues and approaches archaeologists take to the study of urban assemblages (Hall and Kenward 1982, 1994). In a general sense, the research questions can be grouped around three interrelated topics: (1) changes in the local environment as a result of urbanization (Redman 1999:127–158; Rothschild and Balkwill 1993); (2) urban health and sanitation (Geismar and Janowitz 1993; Keene 1982; Mrozowski et al. 1989; Reinhard et al. 1986); and (3) urban foodways, including social or economic variation in urban residents' access to resources, as well as the character and development of food supply and distribution systems (Henry 1987; Landon 1997; Reitz 1986; Schulz and Gust 1983; Waateringe 1994; Zeder 1988). Archaeologists typically study these topics by integrating plant, animal, soil, and other archaeobiological data with the complex archaeological and historical records that frequently characterize urban sites (see O'Connor [2003] for a zooarchaeological approach). Often these studies benefit from a comparative perspective, either by examining changes in urban assemblages through time, or by contrasting assemblages from different sites.

The focus of this study is on the nature and development of urban food supply and distribution systems. Cities and towns generally have dense populations and a high degree of productive specialization. As centers of administration and specialized craft production, urban areas require a sufficiently developed agricultural hinterland to support non-farmer specialists, and a production and distribution system sufficient to bring necessary food supplies into the urban center. Furthermore, as these food supply systems develop, the diet of town residents can begin to diverge from that of local farmers. These issues are often the focus of zooarchaeological studies, which use characteristics of faunal assemblages to gain insight into urban-rural connections.

Several examples highlight the diverse approaches zooarchaeologists take towards investigating urban food supply systems. Zeder (1991), in her study

of the ancient Near Eastern site of Tal-e Malyan, Iran, studies zooarchaeological assemblages to understand the effects of the development of a specialized
urban economy on animal production and distribution. She examines changes
in taxonomic representation, animal ages, and butchery patterns across the
major periods of site development. In this case, urban development fostered
specialized meat distribution systems from the very beginning. Growing cen-
tralization of control over the distribution of animals within the city, however,
seems to have had little impact on the management of local herds. This apparently reflects the strong differences between sedentary urban consumers and
nomadic rural producers.

Maltby (1994) studies more recent zooarchaeological assemblages from
British sites to look at characteristics of urban foodways and the development of
urban-rural connections. In his study of the meat supply of Romano-British
towns he investigates the degree of urban self-sufficiency and the nature of town
demand on local production (Maltby 1994). He examines a variety of attributes
of the faunal assemblages, including taxonomic representation, animal ages as
determined by tooth wear, and animal size and morphology based on bone measurements. His work shows strong evidence for specialized urban butchers for
cattle, husbandry of different varieties of sheep (*Ovis aries*), and instances where
urban pork consumption apparently exceeded what was produced on local rural
farms. This work highlights the complexity of urban food supply systems, and
shows the value of broad comparative studies across multiple sites.

My own earlier work focused on delineating broad characteristics of the city
of Boston's urban food supply and distribution systems through comparative
analysis of urban and rural zooarchaeological assemblages in eastern
Massachusetts (Landon 1996). Although these data are from sties in New
England, and thus farther north, they provide useful background to the
Chesapeake research. In this study (Landon 1996), I compared taxonomic
representation, skeletal part representation, butchery patterns, and animal ages
at two sites in Boston and two rural farmsteads outside the city. Bostonians ate
more seafood and lamb and less pork, but in fundamental ways the patterns of
urban and rural assemblages were quite similar. Although meat and other foodstuffs came into the city from surrounding farms, this appears to be largely rural
surplus more than specialized production for urban markets.

Seasonal slaughter patterns of animals in the Massachusetts urban and rural
assemblages are important components of this research (Landon 1993, 1996).
Seasonal dietary variation is still little studied for sedentary populations, especially urban populations. As Bowen (1988) observes, archaeological interest in
seasonality is typically framed around non-sedentary populations, where the
emphasis is on interpreting seasonal variation in site occupation and patterns
of resource use. In many modern cities the food supply and distribution systems
are so developed that there is little variation in the seasonal availability of
foodstuffs, with the exception of some fresh fruits and vegetables and other
seasonal specialties. This lack of seasonal variation is a new phenomenon.

Earlier food supplies were probably much more strongly linked to seasonal agricultural cycles of planting and harvesting crops and the birth and growth of animals. Characterizing the seasonal agricultural cycle in the past provides insight into rural patterns of agricultural production. Assessing seasonality in urban areas demonstrates their similarity to rural patterns. Although an earlier study took this approach to urban and rural seasonal slaughter cycles in Massachusetts (Landon 1993, 1996), this case study expands upon this work by adding the perspective of seasonal variation in cattle slaughter in the Chesapeake Bay region of eastern Maryland and Virginia.

SITES AND SAMPLES

This study of urban and rural slaughter seasonality builds on Colonial Williamsburg's large-scale historical and zooarchaeological study of town provisioning systems in the 18th- and early 19th-century Chesapeake (Walsh et al. 1997). The Colonial Williamsburg study used surviving account books, probate records, and skeletal remains from archaeological sites to examine agricultural production, food supply and distribution, and urban consumption. Even though Chesapeake towns such as Williamsburg and Annapolis were not large cities, they were sufficiently large and dense to affect rural agricultural production. The zooarchaeological data play a central role in this discussion, providing direct evidence of the meat component of the diet, as well as of animal ages at death and husbandry strategies. The results show that beef provided most of the meat in the diet during this period. Moreover, detailed study of plantation account books suggests that the slaughter of cattle, like other domestic animals, often followed a distinct seasonal pattern. Documenting seasonal growth structures in cattle teeth from the zooarchaeological assemblages adds an additional component to this study of seasonality and allows urban and rural sites to be compared. This study includes cattle teeth from 14 different sites spanning the period A.D. 1620 to circa 1800 (Figure 19-1). Although specific aspects of the histories and environmental contexts of these sites vary, they are all located in the broadly similar Chesapeake region and embedded in a regional agricultural economy driven by tobacco (*Nicotiana* sp.) production. Looking at these sites in a very broad comparative perspective is thus a valid way to investigate aspects of the regional agricultural economy.

During the Colonial period (1607–1775), the Chesapeake Bay region developed an agricultural economy focused on the production of tobacco for export. Maize (*Zea mays*) became the staple grain in the rural diet, with wealthy planters growing some wheat (*Triticum* sp.) and other crops. The colonists raised cattle, pigs (*Sus scrofa*), and some sheep. With an abundance of land and a general shortage of labor, effort went first into tobacco production. Domestic stock received little care, and were generally fenced out of fields and left to forage for themselves rather than fenced into pastures (Anderson 2002). Despite the stereotype of pork as the primary meat, zooarchaeological assemblages show beef was the most important meat in the diet (Bowen 1996). Looking at slaughter patterns for cattle thus focuses on the central source of meat in the regional diet.

In the colonial Chesapeake cattle were raised primarily for meat and limited draft purposes, with virtually no emphasis on dairy products (Bowen 1994). Cattle typically grazed on woodlands and unimproved lands adjacent to tobacco and grain fields, foraging for their own grass and receiving very little in shelter or supplemental feed. This general lack of care continued through the winter, perhaps with the exception of occasionally feeding animals some left over maize stalks. Contemporary accounts describe cattle shivering in the cold and experiencing a hard, starving time during the winter (Anderson 2002; Bowen 2000:364–365). By the end of the 17th century, inattention to cattle's care, overgrazing, and depletion of the land was apparently damaging the health of cattle sufficiently that they became smaller in size (Anderson 2002), a conclusion supported by measurements of cattle bones from Annapolis (Reitz and Ruff 1994).

DENTAL CEMENTUM INCREMENT ANALYSIS

The tooth tissues (dentine, enamel, and cementum) of cattle all grow in incremental patterns that record aspects of life history (Hillson 1986:107–175; Klevezal 1996). Cementum growth increments, the focus of this study, are the best known, being widely used by wildlife biologists and anthropologists to estimate age and season of death across a broad spectrum of mammalian taxa (Beasley et al. 1992; Burke and Castanet 1995; Coy and Garshelis 1992; Coy et al. 1982; Gordon 1988; Grue and Jensen 1979; Hillson 1986:223–230; Kay 1974; Klevezal 1996; Klevezal and Shishlina 2001; Landon et al. 1998; Lieberman 1994; Pike-Tay 1991; Saxon and Higham 1968; Stallibrass 1982). Cementum grows on the exterior of the roots of mammal teeth, anchoring them to the periodontal ligament (Figure 19-2). Variations in chewing forces on teeth and in the seasonal growth rates of dental tissues leave visible growth lines in the cementum (Lieberman 1993, 1994). Not surprisingly, these structures are most visible in animals that experience pronounced seasonal cycles (such as hibernation) or live in environments with marked seasonal variation. As described above, the winter was undoubtedly a period of great dietary stress and slowed growth for the Chesapeake's cattle. It is this yearly period of dietary stress that helped create the cementum growth increments visible in the animals' teeth.

Growth increments in dental cement are observed under the microscope on sections of the teeth, and are visible on thick-sections, stained histological thin-sections, and ground thin-sections (see Beasley et al. 1992; Klevezal 1996:234–241; and Lieberman et al. 1990 for discussions of techniques). Archaeological teeth are typically examined in ground thin-sections, using methods that reinforce the tooth to help protect it during the sectioning process. Tooth samples were selected for this study following five major

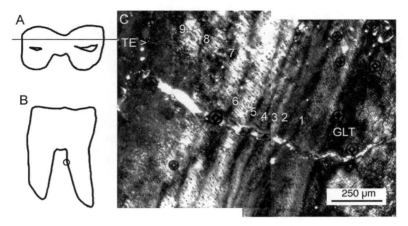

Figure 19-2. Sectioned M₁ from Kingsmill Slave Quarter: (A) occlusal view of tooth (anterior edge to the left) showing section cut line; (B) sketch of section showing circled location of cementum micrograph; and (C) composite digitized micrograph of cementum band under polarized light. TE is the exterior tooth edge; GLT is the Granular Layer of Tomes at the edge of the dentine; and the numbers mark the winter growth lines.

criteria: (1) minimize the destructive aspects of sectioning by selecting loose teeth where appropriate, rather than cutting up mandibles; (2) avoid sectioning teeth from the same animal; (3) include no teeth with pathologies; (4) selectively emphasize the lower fourth premolar, lower first molar, and lower second molar; and (5) avoid teeth with cracked, abraded, missing, or damaged roots, making certain that at least half of the anterior root is present.

This study is based on polarized light observations of calcified, unstained sections of the teeth. Viewed under these conditions, the generally narrower opaque bands represent winter growth and the wider translucent bands represent summer growth (Figure 19-2). Counting the bands allows the age of the animal to be identified and characteristics of the outermost band allows the season of death to be estimated. A full translucent outermost band was taken to represent a full summer's growth, interpreted as a fall kill, and a full opaque band was interpreted as a spring kill.

The majority of the teeth had significantly altered microstructure, and a general consideration of post-deposition alteration to tooth microstructure was necessary. Teeth are subject to post-depositional alteration that destroys incremental growth structures as microorganisms attack the tooth tissues, breaking down organic proteins with enzymes and dissolving tooth minerals with organic acids (Jackes et al. 2001; Nielsen-Marsh and Hedges 2000:1141). This microorganism attack is not apparent with an unaided visual inspection of the tooth, but is very clear on tooth thin-sections as a series of small tunnels through the tissue (Figure 19-3). Heavily damaged and stained areas

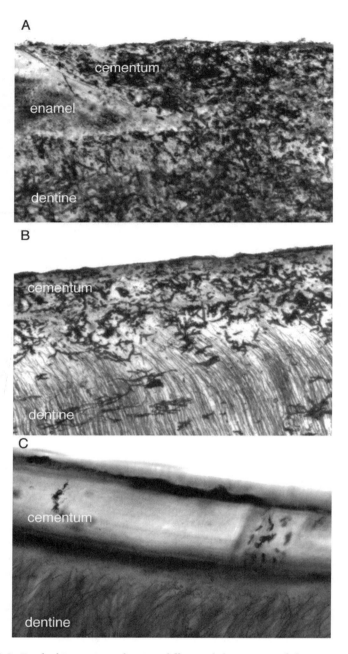

Figure 19-3. Tooth thin-sections showing differential destruction of the microstructure, with the microstructure of the top sample almost entirely destroyed and the microstructure of the bottom sample mostly intact. All images are transmitted light micrographs: (A) Kingsmill Slave Quarter, field width 1 mm; (B) Anthony Hay Site, field width 1 mm; and (C) Anthony Hay Site, field width 350 μm. Note that B and C are from the same site, suggesting that micro-environmental conditions influence microorganism attack.

appear light tan to brown and opaque in cross-polarized light. In extreme examples, where the microorganism damage is extensive, the microstructure of the entire cement band is altered to such an extent that analysis of the growth structures is difficult if not impossible. Understanding the patterns of microorganism attack helps with interpretation, as these alterations can be recognized in the sections. Sample selection can help overcome this problem, if samples are chosen from sites or specific site contexts where microorganism activity is suppressed, such as dry, waterlogged, or anaerobic environments.

This study includes teeth from temperate terrestrial sites, thus samples showed extensive micro-structural alteration. In these cases, two or more sections were typically made from the same tooth, boosting the chance of an intact section of cement being present and allowing the sample to be interpreted. Although this strategy had some success, it meant that the greatest effort was spent on sectioning the teeth that had the least intact micro-structure, and thus returned the least information on the age and season of death of the animal. Careful sample selection based on the environmental characteristics of specific site contexts is recommended for future analyses. This study excludes any tooth where the outermost band could not be properly interpreted. As a result, the starting sample of teeth from 79 cattle across 20 sites was reduced to a sample of teeth from 41 cattle across 14 sites. This is still an adequate sample for a seasonality study of this type. All but two of these sites date to the second half of the 18th century, strengthening inter-site comparison.

RESULTS

The season of death data indicate a highly-focused slaughter pattern emphasizing cold weather slaughter in the fall and winter (Figure 19-4). A smaller number of animals was killed in the spring and summer. It appears from this frequency distribution that the rural slaughter season was less concentrated and the urban slaughter season was more focused. It is possible that this is simply an artifact of sample size, with the larger sample appearing to be more diverse. Overall, the urban and rural patterns are markedly similar, and the difference in the proportional slaughter patterns is not statistically significant (chi-square $= 4.779$, $p = 0.5724$).

DISCUSSION

The focused seasonal slaughter of cattle broadly supports other zooarchaeological and historical indications of seasonal dietary patterns in colonial Chesapeake. Miller's (1984, 1988) analysis of zooarchaeological assemblages documents seasonal variation in the diet of 17th-century Chesapeake settlers by using short-term trash deposits from different times of the year to look at

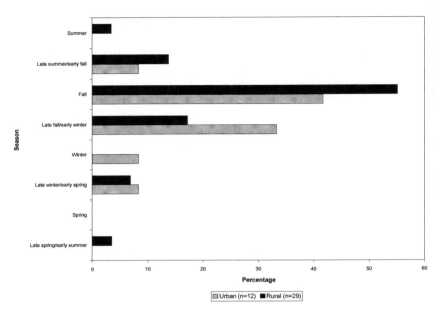

Figure 19-4. Urban and rural seasonal kill-off pattern for Chesapeake cattle (*Bos taurus*) based on the cementum analysis.

patterns of meat consumption. Beef and pork clearly dominated the meat diet, with wild animal food playing an important part in bridging seasonal dietary shortfalls. Late fall and winter were the primary slaughter seasons for swine and cattle, with much of this meat salted and/or smoked for consumption through the winter and into the spring. Beef and pork were major sources of winter and spring meat, augmented by small quantities of venison from white-tailed deer (*Odocoileus virginianus*), raccoon (*Procyon lotor*), and waterfowl. As the stores of preserved meat dwindled through the spring, the importance of wild foods grew, with a summer dietary emphasis on fish augmented by hunting or trapping wild mammals. In New England, historical studies show that the summer diet was augmented by the production and consumption of milk, butter, and cheese (McMahon 1989). Dairying was never as important in the Chesapeake region (Bowen 1994).

Cold weather slaughter of large domestic animals was based on both meat preservation practices and broader aspects of the seasonal agricultural and husbandry cycles. Cold weather kept meat from spoiling quickly, and cold storage areas in an unheated barn or outbuilding allowed some meat to be hung fresh (or even frozen) to start the winter. Meat was preserved by being packed in salt, brine, or smoked. Beef was most commonly eaten fresh or cured in brine and pork was most commonly salted and smoked. Households that could do so likely slaughtered and put up several animals each year, eating barreled and salted meat throughout the winter and into the spring. Animals could be fattened on maize stalks and some grain in the fall before being slaughtered. Given the limited amount of care provided to animals in the winter, many were likely in very poor shape by spring. As the season of death data show, few cattle were slaughtered during this season.

By the end of the 17th century the importance of wild animal meat in the diet had decreased dramatically, at least for the wealthy. Herds of cattle, pigs, and sheep had prospered, and by 1700 domestic animals apparently supplied over 90% of the meat in the diet (Bowen 1994; Miller 1988:191). Beef was clearly the most important meat in the diet, perhaps approaching two-thirds of the meat consumed (Bowen 1996:100). The regional agricultural economy was still strongly seasonal, tied to natural rhythms of planting, growth, and harvest, and the intertwined cycles of animals' birth, growth and slaughter. Although some wild foods were still seasonally important, such as anadromous fish, wealthy planters had a diet that varied seasonally primarily in the type of domestic meat consumed.

Larger Chesapeake planters ensured a continuous supply of meat, predominately fresh, by coordinating the varying optimal slaughter times of different types of domestic livestock (Walsh et al. 1997:179). Eighteenth-century account books document this seasonal pattern by showing the types and quantities of meat exchanged at different times of the year (Walsh et al. 1997:179–180, Figure 6.1). The exchange of hogs and pork was greatest from November through January, as pigs were slaughtered and pork was salted or smoked for consumption through the winter and into the spring. Beef and cattle exchanges peaked in October and January, reflecting cold weather slaughter, first of grass-fed cattle and then of cattle fattened on some extra grain. As beef and pork stocks decreased in the spring, first lamb and then veal became more important in the diet. The smallest numbers of meat exchanges were recorded during the months of June and July, and most of these were of lamb and veal. This pattern is broadly similar to that described by Miller (1984, 1988) for the early 17th century, except that lamb and veal have partially taken the place of wild meat in the spring and summer. This is likely a reflection of the success of animal husbandry in the region, with more people having access to increasing stores of beef and preserved pork that lasted longer through the year, and herds prosperous enough to cull out lambs and calves.

This study of cattle slaughter seasonality from cattle tooth cementum increments provides an additional archaeological perspective on seasonal dietary variation in the Chesapeake. It supports the patterns derived from other archaeological and historical data, and demonstrates the strongly focused cattle slaughter season in the fall and winter. Including archaeological specimens from a variety of sites establishes a regional pattern not limited to the seasonal scheduling decisions of wealthy planters, but broadly affecting people across the Chesapeake.

The similarity in the urban and rural patterns ties urban dwellers' food consumption to rural farmers' agricultural production, linking them through a similar agricultural cycle. Although it is likely that urban residents had greater access to imported foodstuffs, and potentially some seasonal specialty foods, the types of meats consumed were apparently similar in both rural and urban areas, varying with the seasons in similar fashions. This important point suggests the limits on the development of the market systems in the Chesapeake towns. One way around limitations on slaughtering large animals in warm weather is to distribute the meat, through neighbor, kinship, or market exchanges, so that all of the meat is consumed before it can spoil. Although larger towns with developed market systems and larger populations might have had fresh beef available all year, this appears not to have been the case. Neither the pull of urban demand for food nor the development of town markets was sufficient to change the rural seasonal slaughter cycle.

The cities and towns of the Chesapeake undoubtedly affected their surrounding agricultural hinterlands, but perhaps in ways that were more subtle than dramatic. Although towns likely helped encourage the development of market-oriented agricultural production, much of the Chesapeake agricultural economy was already built around market production of tobacco and grain for international markets (Walsh 1999:271). Animal husbandry was much less commercially oriented in the region, and the towns' meat supplies thus came from farms throughout the region, some relatively distant from the town centers. By the end of the 18th century a greater consumption of lamb and veal in towns is indicated, with a greater proportion of cattle under four years old, apparently animals fattened for market rather than just grass-fed (Walsh 1999:275). These differences are minor and the diet of urban and rural people appears to be fundamentally similar, both following patterns set by rural agricultural cycles.

Of course one problem with generalizing about regional diet and agricultural practices in this fashion is that it glosses over the significant differences in the lives of different members of Chesapeake society. Despite our understanding of the broad nature of seasonal variation in the meat supply, and the indications that this was a shared urban and rural phenomenon, it is not clear how different people in the colonies experienced seasonal dietary variation. Although wealthy planters carefully scheduled slaughter times to ensure a supply of fresh meat, many middling farmers undoubtedly felt thankful for their barrels of salt pork and maize, as the poorest struggled to stretch their maize through the lean end of winter. Although environmental archaeology studies have increased our understanding of slave diets in the Chesapeake (Crader 1984, 1990; Franklin 2001; McKee 1987), we still know significantly more about the diet of the wealthy than of the poor. Increasing our understanding of individual experiences within the broad parameters of regional patterns remains important.

Interestingly, the data on seasonal slaughter practices and dietary variation in the Chesapeake correspond closely with those from studies in New England, suggesting broad similarities across the Mid-Atlantic and northeastern American colonies for both urban and rural dwellers. Although probate inventories, account books, diaries, and zooarchaeological data provide different perspectives, they nonetheless are remarkably convergent (Bowen 1988; Derven 1984; Landon 1993; McMahon 1989; Miller 1988). Differences in the importance of dairying and the types and relative importance of the main crops undoubtedly created variations in regional diets, but within an overarching seasonal cycle that was fundamentally similar, especially in the seasonality of slaughtering and meat consumption. Some of this similarity likely arises from the broad pattern of the seasons and limitations on food preservation without refrigerators; slaughtering large cattle and pigs during cold weather helped keep meat from spoiling, the birth season of lambs and calves dictated when they were available, and the fall harvest generated a stock of grain and vegetables to be dried and stored for the winter. The changes that altered this system came in the 19th century, as industry created larger cites, and the growth of roads, canals, and ultimately the railroad transformed the way agricultural products are shipped and marketed.

CONCLUSION

Today the diet in the Chesapeake is basically "deseasonalized," with the tradition of lamb or ham for Easter dinner remaining one of the few reminders of early seasonal dietary practices. In the 17th and 18th centuries, colonial American cities and towns operated in a fundamentally pre-industrial, mercantilist mode, embedded in a culture of agriculture. Looking at seasonal slaughter patterns provides one window into this culture, and helps us recognize the close ties that existed in the past between town and country through urban food supply and distribution systems.

ACKNOWLEDGMENTS

A Postdoctoral Fellowship in the Archaeobiology Laboratory of the Department of Anthropology, Smithsonian Institution, supported this research. This study could not have been completed without the encouragement of Joanne Bowen, whose assistance is gratefully acknowledged. Thanks to Steve Atkins for help gathering the Virginia collections. Tim Gooding helped with the sectioning protocol and Tim Rose helped arrange lab facilities in the Smithsonian's Department of Mineral Science. Thanks especially to Melinda Zeder for all of her support of this work, and Andrea Shapiro for her assistance with interpreting tooth histology and diagenesis. Stephen Mrozowski and Heather Trigg both read and commented on a version of this paper. Any errors of fact or interpretation remain mine.

REFERENCES

Anderson, V. D., 2002, Animals into the Wilderness: The Development of Livestock Husbandry in the Seventeenth-Century Chesapeake, *William and Mary Quarterly* 59(2):377–408.

Beasley, M. J., Brown, W. A. B., and Legge, A. J., 1992, Incremental Banding in Dental Cementum: Methods of Preparation for Teeth from Archaeological Sites and for Modern Comparative Specimens, *International Journal of Osteoarchaeology* 2:37–50.

Bowen, J., 1988, Seasonality: An Agricultural Construct, in: *Documentary Archaeology in the New World* (M. C. Beaudry, ed.), Cambridge University Press, Cambridge, England, pp. 161–171.

Bowen, J., 1994, A Comparative Analysis of the New England and Chesapeake Herding Systems, in: *Historical Archaeology of the Chesapeake* (P. A. Shackel and B. J. Little, eds.), Smithsonian Institution Press, Washington, DC, pp. 155–167.

Bowen, J., 1996, Foodways in the Eighteenth-Century Chesapeake, in: *The Archaeology of Eighteenth-Century Virginia* (T. R. Reinhart, ed.), *Archaeological Society of Virginia Special Publication* 35, Richmond, pp. 87–130.

Bowen, J., 2000, The Chesapeake Landscape and the Ecology of Animal Husbandry, in: *Old and New Worlds* (G. Egan and R. Michael, eds.), Oxbow Books, Oxford, England, pp. 358–367.

Burke, A., and Castanet, J., 1995, Histological Observations of Cementum Growth in Horse Teeth and Their Application to Archaeology, *Journal of Archaeological Science* 22:479–493.

Coy, P. L., and Garshelis, D. L., 1992, Reconstructing Reproductive History of Black Bears from the Incremental Layering in Dental Cementum, *Canadian Journal of Zoology* 70:2150–2160.

Coy, J. P., Jones, R. T., and Turner, K. A., 1982, Absolute Ageing of Cattle From Tooth Sections and its Relevance to Archaeology, in: *Ageing and Sexing Animals from Archaeological Sites* (B. Wilson, C. Grigson, and S. Payne, eds.), *British Archaeological Reports British Series* (Oxford) 109:127–140.

Crader, D. C., 1984, The Zooarchaeology of the Storehouse and the Dry Well at Monticello, *American Antiquity* 49(3):542–558.

Crader, D. C., 1990, Slave Diet at Monticello, *American Antiquity* 55:690–717.

Derven, D. L., 1984, Wholesome, Toothsome, and Diverse: Eighteenth-Century Foodways in Deerfield, Massachusetts, in: *Foodways in the Northeast* (P. Benes, ed.), Dublin Seminar for New England Folklife Annual Proceedings 1982, Boston University Scholarly Publications, Boston, Massachusetts, pp. 47–63.

Franklin, M., 2001, The Archaeological Dimensions of Soul Food: Interpreting Race, Culture, and Afro-Virginian Identity, in: *Race and the Archaeology of Identity* (C. E. Orser Jr., ed.), Foundations of Archaeological Inquiry (J. M. Skibo, general ed.), University of Utah Press, Salt Lake City, pp. 88–107.

Geismar, J. H., and Janowitz, M. F. (eds.), 1993, *Health, Sanitation, and Foodways in Historical Archaeology*, *Historical Archaeology* 27(2):1–111.

Gordon, B. C., 1988, *Of Men and Reindeer Herds in French Magdalenian Prehistory*, British Archaeological Reports International Series (Oxford) 390.

Grue, H., and Jensen, B., 1979, Review of the Formation of Incremental Lines in Tooth Cementum of Terrestrial Mammals, *Danish Review of Game Biology* 11:1–48.

Hall, A. R., and Kenward, H. K. (eds.), 1982, *Environmental Archaeology in the Urban Context*, Council for British Archaeology Research Report 43, London.

Hall, A. R., and Kenward, H. K. (eds.), 1994, *Urban-Rural Connexions: Perspectives from Environmental Archaeology*, Symposia of the Association for Environmental Archaeology 12, Oxbow Monograph 47, Oxbow Books, Oxford, England.

Henry, S., 1987, A Chicken in Every Pot: The Urban Subsistence Pattern of Turn-of-the-century Phoenix, Arizona, in: *Living in Cities: Current Research in Urban Archaeology* (E. Staski, ed.), *Society for Historical Archaeology Special Publication* 5:19–28.

Hillson, S., 1986, *Teeth*, Cambridge University Press, Cambridge, England.

Jackes, M., Sherburne, R., Lubell, D., Barker, C., and Wayman, M., 2001, Destruction of Microstructure in Archaeological Bone: A Case Study from Portugal, *International Journal of Osteoarchaeology* 11:415–432.

Kay, M., 1974, Dental Annuli Age Determination on White-Tailed Deer from Archaeological Sites, *Plains Anthropologist* 19:224–227.

Keene, D. J., 1982, Rubbish in Medieval Towns, in: *Environmental Archaeology in the Urban Context* (A. R. Hall and H. K. Kenward, eds.), *Council for British Archaeology Research Report* 43, London, pp. 26–30.

Klevezal, G. A., 1996, *Recording Structures of Mammals*, revised and updated English edition (A. A. Balkema, ed.), Brookfield, Vermont.

Klevezal, G. A., and Shishlina, N. I., 2001, Assessment of the Season of Death of Ancient Human from Cementum Annual Layers, *Journal of Archaeological Science* 28:481–486.

Landon, D. B., 1993, Testing a Seasonal Slaughter Model for Colonial New England Using Tooth Cementum Increment Analysis, *Journal of Archaeological Science* 20:439–455.

Landon, D. B., 1996, *Feeding Colonial Boston. A Zooarchaeological Study, Historical Archaeology* 30(1):1–153

Landon, D. B., 1997, Interpreting Urban Food Supply and Distribution Systems from Faunal Assemblages: An Example from Colonial Massachusetts, *International Journal of Osteoarchaeology* 7:51–64.

Landon, D., Waite, C., Peterson, R., and Mech, L. D., 1998, Evaluation of Age Determination Techniques for Gray Wolves, *Journal of Wildlife Management* 62(2):674–682.

Lieberman, D. E., 1993, Life History Variables Preserved in Dental Cementum Microstructure, *Science* 261:1162–1164.

Lieberman, D. E., 1994, The Biological Basis for Seasonal Increments in Dental Cementum and Their Application to Archaeological Research, *Journal of Archaeological Science* 21:525–539.

Lieberman, D. E., Deacon, T. W., and Meadow, R. H., 1990, Computer Image Enhancement and Analysis of Cementum Increments as Applied to the Teeth of *Gazella gazella*, *Journal of Archaeological Science* 17:519–533.

Maltby, M., 1994, The Meat Supply in Roman Dorchester and Winchester, in: *Urban-Rural Connexions: Perspectives from Environmental Archaeology* (A. R. Hall and H. K. Kenward, eds.), Symposia of the Association for Environmental Archaeology 12, Oxbow Monograph 47, Oxbow Books, Oxford, England, pp. 85–102.

McKee, L., 1987, Delineating Ethnicity from the Garbage of Early Virginians: Faunal Remains from the Kingsmill Plantation Slave Quarter, *American Archaeology* 6(1):31–39.

McMahon, S. D., 1989, "All Things in Their Proper Season": Seasonal Rhythms of Diet in Nineteenth Century New England, *Agricultural History* 63(2):130–151.

Miller, H. M., 1984, *Colonization and Subsistence Change on the Seventeenth Century Chesapeake Frontier*, Ph.D. dissertation, Department of Anthropology, Michigan State University, East Lansing.

Miller, H. M., 1988, An Archaeological Perspective on the Evolution of Diet in the Colonial Chesapeake, 1620–1745, in: *Colonial Chesapeake Society* (L. G. Carr, P. D. Morgan, and J. B. Russo, eds.), University of North Carolina Press, Chapel Hill, pp. 176–199.

Mrozowski, S., Bell, E., Beaudry, M., Landon, D., and Kelso, G., 1989, Living on the Boott: Health and Well Being in a Boardinghouse Population, *World Archaeology* 21(2):298–319.

Nielsen-Marsh, C. M., and Hedges, R. E. M., 2000, Patterns of Diagenesis in Bone I: The Effects of Site Environments, *Journal of Archaeological Science* 27:1139–1150.

O'Connor, T. P., 2003, *The Analysis of Urban Animal Bone Assemblages, The Archaeology of York Principles and Methods 19/2*, Council for British Archaeology and York Archaeological Trust, York, England.

Pike-Tay, A., 1991, *Red Deer Hunting and the Upper Paleolithic of Southwestern France: A Study in Seasonality*, British Archaeological Reports International Series (Oxford) 569.

Redman, C. L., 1999, *Human Impact on Ancient Environments*, University of Arizona Press, Tucson.

Reinhard, K. J., Mrozowski, S. A., and Orloski, K. A., 1986, Privies, Pollen, Parasites and Seeds: A Biological Nexus in Historic Archaeology, *MASCA Journal* 4(1):31–36.

Reitz, E. J., 1986, Urban/Rural Contrasts in Vertebrate Fauna from the Southern Atlantic Coast, *Historical Archaeology* 20(2):47–58.

Reitz, E. J., and Ruff, B., 1994, Morphometric Data for Cattle from North America and the Caribbean Prior to the 1850s, *Journal of Archaeological Science* 21(5):699–713.

Rothschild, N. A., and Balkwill, D., 1993, The Meaning of Change in Urban Faunal Deposits, *Historical Archaeology* 27(2):71–89.

Saxon, A., and Higham, C., 1968, Identification and Interpretation of Growth Rings in the Secondary Dental Cementum of *Ovis aries* L., *Nature* 219:634–635.

Schulz, P. D., and Gust, S. M., 1983, Faunal Remains and Social Status, *Historical Archaeology* 17(1):43–53.

Stallibrass, S., 1982, The Use of Cement Layers for Absolute Ageing of Mammalian Teeth: A Selective Review of the Literature, with Suggestions for Further Studies and Alternative Applications, in: *Ageing and Sexing Animal Bones from Archaeological Sites* (B. Wilson, C. Grigson, and S. Payne, eds.), *British Archaeological Reports British Series* (Oxford) 109:109–126.

Waateringe, W. G.-v., 1994, The Menu of Different Classes in Dutch Medieval Society, in: *Urban-Rural Connexions: Perspectives from Environmental Archaeology* (A. R. Hall and H. K. Kenward, eds.), Symposia of the Association for Environmental Archaeology 12, Oxbow Monograph 47, Oxbow Books, Oxford, England, pp. 147–169.

Walsh, L. S., 1999, Feeding the Eighteenth-Century Town Folk, or, Whence the Beef? *Agricultural History* 73(3):267–280.

Walsh, L. S., Martin, A. S., and Bowen, J., 1997, *Provisioning Early American Towns. The Chesapeake: A Multidisciplinary Case Study*, National Endowment for the Humanities completion report, Colonial Williamsburg Foundation, Williamsburg, Virginia.

Zeder, M. A., 1988, Understanding Urban Process through the Study of Specialized Subsistence Economy in the Near East, *Journal of Anthropological Archaeology* 7:1–55.

Zeder, M. A., 1991, *Feeding Cities: Specialized Animal Economy in the Ancient Near East*, Smithsonian Institution Press, Washington, DC.

Chapter **20**

Crop Husbandry Practices in North America's Eastern Woodlands

C. Margaret Scarry

Considerable research has been devoted to documenting the existence of indigenous crops in North America's Eastern Woodlands and to understanding the timing of and reasons for their replacement by maize (*Zea mays*) and beans (*Phaseolus vulgaris*). Hence, it is now well-established that for millennia before native farmers adopted crops introduced from Mexico, they cultivated and domesticated a suite of starchy- and oily-seeded plants. In response to this evidence, we have modified our depictions of Archaic and Woodland lifeways and changed our questions about how and why maize became important.

We have given surprisingly little attention, however, to how native farmers raised and tended the indigenous crops. Most frequently, pre-maize husbandry is labeled horticulture or gardening. This creates images of small-scale, shifting, polycropped plots. In this case study, I argue that consideration of the characteristics of indigenous crops suggests different practices than those conveyed by the labels. Reshaping our understanding of pre-maize husbandry in turn has implications for modeling pre-Columbian economics, for understanding the changes that took place as maize came to dominate later farming systems, and for schemes which presume "horticulture" precedes "agriculture" in a neoevolutionary trajectory. To make my case, I first present an historical summary for the Eastern

Woodlands. Then I describe the practices of late pre-Columbian farmers who relied on maize and beans. Finally, I turn to practices I infer were used by earlier native farmers who grew the indigenous crops.

CULTURAL CHRONOLOGY AND CROP HISTORY

North America's Eastern Woodlands stretch from the Atlantic coast westward across the Mississippi to the edge of the Plains and from southern Canada to the Gulf of Mexico (Figure 20-1). This enormous territory encompasses considerable regional and temporal variation in topography, climate, and ecology. Needless to say, there were also significant temporal and regional differences in

Figure 20-1. Map of Eastern North America showing the locations of areas and sites numbered in the order they are mentioned in the text. (1) Hidatsa; (2) Natchez; (3) American Bottom; (4) Ozark Bluff Shelters including Edens Bluff and Whitney Bluff; (5) Russell Cave; (6) Ash Cave; (7) Newt Kash Hollow; and (8) Gypsy Joint.

the lifeways of the Native Americans who lived here. We know that the extent to which people relied on crops for food varied (Crawford and Smith 2003; Johannessen 1993; Lopinot 1992; Scarry 1993; Smith and Cowan 2003) and there are good reasons to believe that their farming practices varied as well. To start thinking about crop husbandry, however, I gloss over regional variation and focus on broad temporal differences in farming practices appropriate for indigenous versus introduced crops. Nonetheless, it is important to keep in mind that at any point in time people must have adjusted their practices to fit their local circumstances (e.g., food demand, labor pool, scheduling conflicts, weather, and soil conditions).

The relevant cultural histories and chronologies of crop domestication and introduction are well known to archaeologists who work in the Eastern Woodlands but, perhaps, less familiar to readers unfamiliar with the region. The cultural history outlined here follows the scheme used by most archaeologists who work in the Midwest and Southeast (Smith 1986; Steponaitis 1986). The periods are largely heuristic chronological divisions but they also chart broad changes in material culture as well as other aspects of native lifeways. Because material traits and considerations of social and economic organization are used in defining the cultural periods the temporal boundaries between them vary somewhat from region to region. While the nuances of dating are important for understanding cultural trajectories and interaction patterns, for the present purposes, the gross categories are satisfactory.

Insofar as we know, the ancestors of the Native Americans first entered the Eastern Woodlands sometime around 10,000 B.C. The Paleoindian period (circa 10,000–8,000 B.C.) spans the end of the last glacial period and the transition to the early Holocene. We believe that during this time, people were hunter-gatherers who lived in small, mobile bands. The Archaic period (circa 8,000–1,000 B.C.) spans the period when modern vegetation and river drainage patterns were established. While people were predominantly hunter-gatherers, there is evidence for increasing sedentism and local cultural differentiation throughout the Archaic. The Archaic period is generally divided into three sub-periods: the Early Archaic (8,000–6,000 B.C.), the Middle Archaic (6,000–4,000 B.C.) and the Late Archaic (4,000–800 B.C.). The latter period is noted not only for greater sedentism but also for the beginnings of indigenous crop husbandry (Fritz 1990; Smith and Cowan 2003; Yarnell 1993). The Woodland period (circa 800 B.C.–A.D. 1000) was originally distinguished from the preceding period on the basis of a technological innovation, specifically the development of pottery. We now know that some pottery was made in the Late Archaic period but the temporal boundary continues to be used. The Woodland period is generally characterized by further increases in sedentism, by more varied social and political organizations, and by economies that increasingly rely on the production and storage of crops (Johannessen 1993). Like the Archaic period, the Woodland period is generally divided into three sub-periods: the Early Woodland (1000 B.C.–A.D. 300), the Middle Woodland

(A.D. 350–650), and the Late Woodland (A.D. 650–1000). The Late pre-Columbian period spans the time from about A.D. 1000 to the arrival of Europeans in the middle of the 16th century. Throughout most of the Eastern Woodlands, the Late pre-Columbian period is characterized by considerable cultural complexity and by subsistence economies based on maize husbandry (Scarry and Scarry 2005). Two broad regional cultural patterns are defined for this period. In the Ohio Valley area, Fort Ancient peoples were farmers, who lived for much of the year in large nucleated villages (see Wagner, this volume) but whose societies were not hierarchically organized. Elsewhere in the Midwest and Southeast, the Mississippian peoples were farmers, who lived in hierarchical societies headed by hereditary elites.

Thanks to the work of Yarnell (1972, 1976, 1978, 1993), Asch (Asch and Asch 1977, 1985; Asch and Hart 2004), Asch Sidell (Asch and Asch 1978), Smith (1985a, b, 1992), Fritz (1986, 1987, 1990, 1994), Gremillion (1993b, 2004) and others who documented changes in seed morphology, and to a host of archaeologists and archaeobotanists who diligently collected, floated, and sorted thousands of soil samples, we have substantial evidence about the history of plant domestication in the Eastern Woodlands (for summaries see Asch and Hart 2004; Fritz 1990; Scarry and Yarnell 2006; Smith 2006; Smith and Cowan 2003). Based on current evidence a gourd or squash (*Cucurbita pepo* ssp. *ovifera*) was domesticated by 2500 B.C. (calibrated). This gourd/squash, which was similar to today's ornamental gourds, had oily seeds and a thin, hard rind suited to a variety of uses. Gourd/squash culti-vars with edible, thick, fleshy rinds (the ancestors to today's crookneck, acorn, and pattypan squashes) were developed later. Sunflower (*Helianthus annuus*) and its relative sumpweed (*Iva annua*) were domesticated by 2000 B.C. (calibrated). Through time, both cultigens show increases in seed size and (at least for sunflower) fruiting heads. The seeds of sunflower and sumpweed are excellent sources of fat and protein. Changes in seed shape and surface texture indicate that chenopod (*Chenopodium berlandieri*) was domesticated by 1500 B.C. (calibrated). Erect knotweed (*Polygonum erectum*) was cultivated by 900 B.C. but not clearly domesticated until about A.D. 1000. Both chenopod and knotweed produce small starchy seeds that are high in plant proteins. Maygrass (*Phalaris caroliniana*) and little barley (*Hordeum pussilum*) are consistently found outside their native ranges and in associa-tion with the other indigenous crops. We lack definitive evidence for domes-tication of these cereals but it is generally agreed that they were cultivated, probably as early as 2000 B.C.

Maize and, especially, beans were late-comers to Eastern fields. By the 15th century A.D., the "three sisters" (maize, beans, and squash) had displaced indigenous cultigens and formed an integrated suite in the fields, food, and thought of many Native Americans. Nonetheless, archaeological evidence indicates quite disparate histories for the three crops. Some squash cultivars were indigenous (see above) while pumpkins (*Cucurbita pepo* spp. *pepo*) and cushaw squashes (*C. argyrosperma*) spread from Mexico via the Southwest arriving in the Eastern Woodlands sometime after A.D. 1000. Ephemeral, but securely dated, evidence indicates that maize was present in the Eastern Woodlands by the beginning of the common era (Chapman and Crites 1987; Conard et al. 1984; Crawford et al. 1997; Ford 1987; Riley et al. 1994). This is centuries, and in some cases almost a millennium, before maize became prominent in native diets and farming strategies. Multiple lines of evidence point to increased production and consumption of maize beginning about A.D. 800 to 900. By A.D. 1200, maize was a staple food for people throughout most of the Eastern Woodlands. These changes are documented in the archaeological record by marked increases in the frequency and abundance of maize remains, by bone chemistry signatures that indicate increased consumption of C_4 plants, and by increases in the rates of dental caries (Bender et al. 1981; Buikstra et al. 1994; Crawford and Smith 2003; Fritz 1992; Hart and Asch Sidell 1996; Lynott et al. 1986; Simon 2000; Yarnell and Black 1985). Beans were the last of the "three sisters" to be adopted by native farmers. They seem to have reached the Eastern Woodlands around A.D. 1200 after which they spread rapidly (Hart et al. 2002; Hart and Scarry 1999).

HUSBANDRY PRACTICES USED FOR MAIZE, BEANS AND SQUASH

Early European explorers and colonists observed native fields and farming practices in various parts of the Eastern Woodlands. Their reports provide general characterizations of maize-based farming that can be projected back into the pre-Columbian period. We know from these accounts that maize-based farming was communal and gendered. In late winter or early spring, men created new fields by girdling trees and burning stumps and underbrush. They also burned the grass and weeds from existing fields to prepare them for planting. In some places, men trenched the fields for planting and helped with the harvest when crops were ripe. Women bore the primary responsibility for preparing fields, planting, weeding, and harvesting. They used digging sticks and hoes to create hills spaced at intervals of a meter or so in ranks across the fields. The women then planted the maize by placing four to ten grains in each hill. In some areas, they planted squash and beans between the ranks of maize. In others, they planted squash and beans in the hills so the vines could climb the maize stalks. Once the crops sprouted, the women thinned the plants, weeded the fields, and hoed the soil up around the plants to promote root development (Swanton 1946:306–308, 710–717; Wenhold 1936:13).

When the crops ripened, the women harvested and prepared them for storage. Maize and squash were plucked or cut from stalks and vines in the field. When beans were grown in small quantities, they too may have been picked from the vines. When the bean crop was large, however, it is more likely that, as among the historic Hidatsa of the Missouri River Valley, vines were allowed to dry before being pulled and threshed in the field (Wilson 1987).

The scale of production varied considerably among native communities. Schroeder (1999, 2001) and Baden and Beekman (2001) used 19th-century data to estimate field sizes and maize yields. They disagree on the methods used to derive these estimates, but their figures provide some sense of the order of magnitude of household farming activities. A family of five to six would cultivate a plot averaging a quarter hectare and ranging as high as two hectares when there were several adult women in the household. Yields varied depending on soil, maize variety, etc., but the average yield over the life of a field was 500 to 1000 kg/ha. This was sufficient to provide 25% to 50% of a family's caloric needs. These are probably reasonable estimates for those areas (and times) such as the upper Midwest, and the Northeast where people mainly grew crops to meet their household needs. In the Mississippian world, however, we know that in addition to tending household plots people worked in communal and/or chief's fields. The harvests from such fields were stored in community granaries and provided the foodstocks that underpinned the activities of the elite (Braund 1990; Hann 1986:144; Swanton 1946:286, 307, 309; Wenhold 1936). We get some indication of this larger scale of production from the accounts of the 16th-century Soto expedition in which the Spaniards describe places where they rode through leagues of fields and found sufficient grain to feed the army for months on end. While there may be some exaggeration, we know that the Spaniards survived for several years by appropriating the grain they found in native storehouses (Robertson 1993:72; Shelby 1993:194, 253).

In sum, native husbandry of maize, beans, and squash involved interplanting and labor intensive practices commonly associated with modest-scale production. In some parts of the Eastern Woodlands, however, these strategies were used for expansive large-scale production for which we lack modern parallels.

HUSBANDRY PRACTICES USED FOR INDIGENOUS CROPS

Native farmers shifted their production efforts from indigenous crops to maize centuries before Europeans arrived. Thus, we have only tantalizing glimpses in the written record of pre-maize husbandry practices (see Smith [1992] for a discussion of chenopod cultivation). Nonetheless, we can make some inferences about how people raised the indigenous crops. To do so, I draw on characteristics of the plants themselves, information about how crops with similar characteristics are grown today, and archaeological data.

These multiple lines of evidence suggest that native farmers used husbandry practices for the indigenous cultigens that differed from those subsequently used for maize, beans, and squash. In contrast to the "three sisters," chenopod, knotweed, maygrass, and little barley have quite small seeds. These small grains are more suited to broadcast sowing than they are to setting individual seeds (Smith and Cowan 2003). Yields from experimental harvesting of wild populations of the small grains indicate that dense stands would be required to make their cultivation and use worthwhile (Asch and Asch 1978; Munson 1984; Smith 1992). Broadcast sowing resulting in dense stands is also suggested by husbandry practices used for analogous plants (e.g., quinoa [*Chenopodium quinoa*], buckwheat [*Fagopyrum esculentum*], millet [*Setaria* spp.]) elsewhere in the world (Gade 1970; Halstead and Jones 1989). In the early 18th century, Le Page du Pratz (cited in Smith 1992) described the Natchez of Mississippi planting "*choupichoul*" (probably chenopod) which he compared to the "*belle-dame-sauvage*" of Europe. Le Page wrote that the women and children sowed the grain on dry sandbanks covering the grain with their feet but not tending it during the growing season (Gilmore 1931; Smith 1992). This method of planting seems more akin to broadcast sowing than to the hilling used for maize planting.

Though circumstantial, these considerations suggest that the small grains of the Eastern Woodlands were planted by broadcasting, either randomly or in rows, on lightly tilled soils rather than by dibbling. In this scenario, it is not implausible that plants such as chenopod, and perhaps knotweed, which set more seed when not overly crowded, were thinned once the seedlings emerged. This would leave some to grow to maturity and provide others for use as greens.

Besides sowing and spacing, the small grains probably differed from maize in the timing of their planting and harvesting. Wild stands of maygrass and little barley ripen in late spring; maygrass is ready for harvest in May, little barley generally ripens several weeks later (Asch and Asch 1985:189, 193). These crops may have been sown in the fall or early spring to produce a harvest when other plant staples were scarce. Though chenopod seeds ripen in the fall, they require cold to germinate. Thus, they too were probably planted in fall or early spring.

Squash, sunflower, and sumpweed may have been handled differently. Squashes and gourds with their sprawling vines were probably planted along field edges or other places where they had room to spread or climb. Initially, small-seeded sunflowers and sumpweeds may have been planted by broadcasting. As the seeds (and plants) enlarged under domestication, however, native women probably switched to planting them in hills set around the edges of fields (Wilson 1987). It seems worth considering whether hilling for the indigenous sunflowers, sumpweeds, and squashes provided the model (or at least set a precedent) for the subsequent adoption of this method of planting for maize and beans (see Gremillion [1993a] for a discussion of transferal of husbandry practices in the adoption of new crops).

Hoes of stone, bone, and shell have been recovered from numerous sites in the Eastern Woodlands, but there is no evidence for sickles or other blades suitable for harvesting crops by cutting bundles of stems. Large sunflower heads were probably cut from their stalks and gourds and squashes were likely picked individually (cf., Wilson 1987). Small sunflower and sumpweed heads as well as the small grains were probably harvested by hand stripping or by beating either uprooted or standing plants so that their seeds dropped onto hides or blankets or into waiting containers. Once harvested, the small grains had to be parched, threshed, and winnowed.

Native husbandry of indigenous crops is often described as gardening or horticulture with shifting, small-scale plots and polycropping implied by analogy to classic descriptions of tropical swidden systems. The use of slash-and-burn techniques for clearing plots, shifting cultivation with successional plants encouraged to grow in fallow plots, and a preference for clearing growth from old fields rather than carving new plots out of the forest are probably reasonable inferences. It also seems likely that useful plants were tolerated and tended when they "volunteered" in active plots and that when old fields were cleared useful shrubs (e.g., hazel [Corylus spp.]) and fruit trees (e.g., persimmon [Diospyros virginiana] and plum [Prunus spp.]) were left standing (Fritz 2000a; Hammett 1992; Scarry and Scarry 2005).

The implication of small-scale conveyed by the use of the terms "garden" and "horticulture" has been questioned, however, on several grounds. Johannessen (1993) argues that the massive quantities of small grains recovered from sites in the American Bottom indicate Late Woodland people in that area were farmers. Based on the characteristics of the plants themselves, Asch and Asch (1985) contend that they are more appropriately raised in fields. Given evidence for diverse cropping schemes over time and space, there probably is no "one size fits all" answer to the question of scale (cf., Fritz 2000b). In some times and in some places, the plots in which Native Americans raised indigenous crops may have been quite modest. At other times and in other places they were probably expansive.

There is surprisingly little attention given to the question of whether the indigenous cultigens were polycropped. The stereotypical horticultural plot has tubers, grains, shrubs, and trees intermingled in a manner that "apes the jungle" in diversity and structure-replicating ground cover, understory, and forest canopy (e.g., Conklin 1954; Geertz 1963). Since Native American farmers interplanted maize, beans, and squash, the unstated assumption seems to be that the small grains and oil seeds were raised in a similar fashion (but see Asch and Asch 1978:334). But, if the indigenous crops were broadcast and grown in dense stands, then the riotous swidden plot is not the appropriate model. More likely, plots were planted with a single cultigen or, as in some places in Latin America, crops may have been planted in zones within plots for ease of tending and harvesting (Beckerman 1983).

Another intriguing possibility can be suggested by analogy not to tropical swidden systems but to traditional cereal production in Europe and the Mediterranean. Wheat (*Triticum* spp.) and barley (*Hordeum vulgare*), which have similar growth habits, ripening schedules, processing requirements, and cooking properties, were often grown as maslins (Halstead and Jones 1989). That is, a mixture of wheat and barley seeds was broadcast and resulted in fields and harvests in which the cereals were intermingled. Perhaps, the spring ripening grains (maygrass and little barley) or the fall ripening seeds (chenopod and knotweed) were grown as maslins. There is some archaeological evidence from the Eastern Woodlands to support these suggestions. Some caches of seedstock in rockshelters and some pits filled with seeds apparently burned during parching contain nearly pure assemblages of a single species. For example, a woven bag full of chenopod seeds was found at Edens Bluff, Arkansas and caches of chenopod seeds were found at Russell Cave, Alabama and Ash Cave, Ohio (Figure 20-1; Fritz 1986; Smith 1985a, b). Likewise, sheaves of maygrass stalks with seed heads attached were found in several Ozark Bluff shelters in Arkansas and at Newt Kash Hollow in Kentucky (Figure 20-1; Cowan 1978; Fritz 1986; Gilmore 1931; Jones 1936). But other caches and parching accidents contain mixtures of small grains. Gourds containing a mixture of chenopod and erect knotweed seeds were found at Whitney Bluff, Arkansas and pits containing thousands of these same seeds were found at Gypsy Joint, Missouri (Figure 20-1; Fritz 1986, 1997; Smith and Wetterstrom 1978). Finding both pure and mixed caches suggests that at times people grew the various crops in separate fields or at least in separate patches within a field, while at other times they grew them as maslins. In neither instance, would the plots resemble the horticultural plots of tropical Asia or Oceania.

The above speculations concern people's husbandry of indigenous crops as it may have existed prior to the widespread adoption of maize in the Eastern Woodlands. What about later when people grew small grains along with maize? In areas such as the American Bottom (Figure 20-1; Lopinot 1992), where substantial quantities of both maize and indigenous crops were grown, it seems likely that they were raised in separate fields. Or in the case of the spring ripening small grains and fall ripening maize, people may have double cropped; planting maize after they harvested maygrass or little barley. In areas such as Ohio (Wagner 1987; Wymer 1993), where planting of small grains dwindled with the adoption of maize, small crops of spring-ripening grains may have been raised in separate plots or harvested from fields before maize was planted. Likewise, small crops of chenopod and knotweed, which grow over the summer and do not produce well when crowded or shaded, may have been planted around the edges of fields (as sunflower was in historic Hidatsa villages [Wilson 1987]) where they would not be overgrown or overshadowed by maize.

To summarize, I argue that husbandry of indigenous crops employed broadcast sowing of single crops or maslins resulting in dense stands of plants more akin to a field of wheat than one of maize. Crops such as squash and sunflowers may have been planted in separate plots or around the edges of "grain" fields. I also suggest that planting of some indigenous crops took place in the fall or early spring either to provide a late spring harvest or because the plants require cold to germinate.

DISCUSSION

The husbandry practices I envision for indigenous crops are quite different from those commonly implied by the terms horticulture and gardening. Moreover, the shift to maize production involved changes in practice that run counter to those presumed to accompany a switch from small-scale horticulture to large-scale agriculture. Replacing indigenous crops with maize, beans, and squash involved shifting from planting single crops or maslins to intercropping at least three and sometime more plants. It involved shifting from broadcast sowing densely planted plots to dibbling carefully spaced hills. It involved shifting from fields of unknown (most likely variable) sizes to sometimes very extensive "fields." And, it involved shifting from harvesting by uprooting or handstripping and threshing to harvesting by plucking and shucking. Neither the indigenous crops nor the "three sisters" yield well in shade. People may have selectively left fruit trees when they cleared their fields, but the canopy would have been largely removed. Both cropping regimes might be described as swidden in the sense that clearing was accomplished by slashing and burning and fertility was maintained by bush-fallowing and permitting "natural" regeneration in old fields. Nonetheless, the changes in husbandry practices do not fit comfortably in trajectories derived from combining Boserup's (1965) models with neoevolutionary cultural categories. It is misleading to contrast Woodland tribes harvesting crops from horticultural plots to later Mississippian chiefdoms harvesting maize from agricultural fields.

The differences in husbandry practices also have implications for understanding what happened when native farmers shifted from growing indigenous crops to raising maize, beans, and squash. It was not simply a matter of people recognizing the greater productivity of maize, acquiring a superior cultivar, or responding to increased social demands for food. Longstanding farming practices had to be replaced and along with them the ecological knowledge that provided clues as to when to plant, etc. had to be reevaluated. Moreover, the indigenous crops and maize placed different demands on household labor (and especially on women's labor). The rhythm and tempo of work differed throughout the season-from planting, to cultivating, to harvesting, to processing, to cooking. In other words, habitual ways of doing things had to be altered. Tasks had to be reallocated among people and over the course of the growing season. The realignment of work schedules must have had effects that rippled throughout people's daily lives.

CONCLUSION

I offer these thoughts with full recognition that I am glossing over tremendous variability and have only skimmed the surface. Nonetheless, it should be evident that attention to native husbandry practices is critical if we are to understand the lifeways in which production of indigenous crops were embedded and the changes that ensued when the "three sisters" displaced them.

ACKNOWLEDGMENTS

My inspiration for this chapter came from working on two articles. The first article, focusing on Mississippian maize production and storage (Scarry and Scarry 2005), was originally written for the symposium that resulted in the first edition of this book. I subsequently revived and revised this paper for an issue of *World Archaeology* devoted to "garden agriculture." I thank Betsy Reitz for the invitation to participate in the symposium that set things in (slow) motion and Marika van der Veen for the invitation to contribute to *World Archaeology* that brought the article to fruition. The second article was a chapter on the domestication of plants in the Eastern Woodlands (Scarry and Yarnell 2006) for the environmental volume of *The Handbook of North American Indians*. Dick Yarnell wrote several versions of this chapter over a period of several decades. When the volume was resurrected in 2003, Dick generously asked me to be his co-author. I updated his histories of the indigenous crops and added a new section on husbandry practices. The final drafts of both articles were written in January and February of 2005. Thinking and writing about pre-maize and maize husbandry practices back-to-back made me "see" new things. The result is this chapter. Besides these direct influences, I owe a debt to the many Eastern Woodlands archaeobotanists whose high-caliber work provided the evidence I draw on here. Bruce Smith read a draft of this paper and offered comments. John Scarry acted as my sounding board, read drafts, and prepared the map for Figure 20-1. I thank everyone who made this chapter possible but take sole responsibility for the ways I have put together the data and interpretations offered here.

REFERENCES

Asch, D. L., and Asch, N. B., 1977, Chenopod as Cultigen: A Reevaluation of Some Prehistoric Collections from Eastern North America, *Midcontinental Journal of Archaeology* 2:3–45.

Asch, D. L, and Asch, N. B., 1985, Prehistoric Plant Cultivation in West-central Illinois, in: *Prehistoric Food Production in North America* (R. I. Ford, ed.), Anthropological Papers 75, University of Michigan Museum of Anthropology, Ann Arbor, pp. 149–203.

Asch, D. L., and Hart, J. P., 2004, Crop Domestication in Prehistoric Eastern North America, *Encyclopedia of Plant and Crop Science*, Marcel Dekker, New York, pp. 314–319.

Asch, N. B., and Asch, D. L., 1978, The Economic Potential of *Iva annua* and its Prehistoric Importance in the Lower Illinois Valley, in: *The Nature and Status of Ethnobotany* (R. I. Ford, ed.), Anthropological Papers 67, University of Michigan Museum of Anthropology, Ann Arbor, pp. 300–341.

Baden, W. M., and Beekman, C. S., 2001, Culture and Agriculture: A Comment on Sissel Schroeder, Maize Productivity in the Eastern Woodlands and Great Plains of North America, *American Antiquity* 66(3):505–516.

Beckerman, S. J., 1983, Does the Swidden Ape the Jungle?, *Human Ecology* 11:1–12.

Bender, M. M., Baerreis, D. A., and Steventon, R. L., 1981, Further Light on Carbon Isotopes and Hopewell Agriculture, *American Antiquity* 46:346–353.

Boserup, E., 1965, *The Conditions of Agricultural Growth: The Economics of Agrarian Change Under Population Pressure*, Aldine, Chicago, Illinois.

Braund, K. E. H., 1990, Guardians of Tradition and Handmaidens to Change: Women's Roles in Creek Economic and Social Life During the Eighteenth Century, *American Indian Quarterly* 14: 239–258.

Buikstra, J. E., Rose, J. C., and Milner, G. R., 1994, A Carbon Isotopic Perspective on Dietary Variation in Late Prehistoric Western Illinois, in: *Agricultural Origins and Development in the Midcontinent* (W. Green, ed.), Report 19, Office of the State Archaeologist, University of Iowa, Iowa City, pp. 155–170.

Chapman, J., and Crites, G. D., 1987, Evidence for Early Maize (*Zea mays*) from the Icehouse Bottom Site, Tennessee, *American Antiquity* 52:352–354.

Conard, N., Asch, D. L., Asch, N. B., Elmore, D., Gove, H., Rubin, M., Brown, J., Wiant, M., Farnsworth, K., and Cook T., 1984, Prehistoric Horticulture in Illinois: Accelerator Radiocarbon Dating of the Evidence, *Nature* 308:443–446.

Conklin, H. C., 1954, An Ethno-ecological Approach to Shifting Agriculture, *Transactions of the New York Academy of Sciences* 17:133–142.

Cowan, C. W., 1978, The Prehistoric Use and Distribution of Maygrass in Eastern North America: Cultural and Phytogeographical Implications, in: *The Nature and Status of Ethnobotany* (R. I. Ford, ed.), Anthropological Papers 67, University of Michigan Museum of Anthropology, Ann Arbor, pp. 263–288.

Crawford, G. W., and Smith, D. G., 2003, Paleoethnobotany in the Northeast, in: *People and Plants in Ancient Eastern North America* (P. Minnis, ed.), Smithsonian Institution Press, Washington, DC, pp. 172–257.

Crawford, G. W., Smith, D. G., and Bowyer, V. E., 1997, Dating the Entry of Corn (*Zea mays*) into the Lower Great Lakes Region, *American Antiquity* 62:112–119.

Ford, R. I., 1987, Dating Early Maize in the Eastern United States, paper presented at the 10th Annual Meeting of the Society for Ethnobiology, Gainesville, Florida.

Fritz, G. J., 1986, *Prehistoric Ozark Agriculture: The University of Arkansas Rockshelter Collections*, Ph.D. dissertation, University of North Carolina at Chapel Hill, University Microfilms, Ann Arbor, Michigan.

Fritz, G. J., 1987, The Trajectory of Knotweed Domestication in Prehistoric Eastern North America, paper presented at the 10th Annual Meeting of the Society of Ethnobiology, Gainesville, Florida.

Fritz, G. J., 1990, Multiple Pathways to Farming in Precontact Eastern North America, *Journal of World Prehistory* 4:387–435.

Fritz, G. J., 1992, 'Newer,' 'Better' Maize and the Mississippian Emergence: A Critique of Prime Mover Explanations, in: *Late Prehistoric Agriculture* (W. Woods, ed.), Studies in Illinois Archaeology 8, Illinois Historic Preservation Agency, Springfield, pp. 19–43.

Fritz, G. J., 1994, Precolumbian *Cucurbita argyrosperma* ssp *argyrosperma* (Cucurbitaceae) in the Eastern Woodlands of North America, *Economic Botany* 48:280–292.

Fritz, G. J., 1997, A Three-thousand-year-old Cache of Crop Seeds from Marble Bluff, Arkansas, in: *Plants, People, and Landscapes: Studies in Paleoethnobotany* (K. Gremillion, ed.), University of Alabama Press, Tuscaloosa, pp. 42–62.

Fritz, G. J., 2000a, Levels of Native Biodiversity in Eastern North America, in: *Biodiversity and Native America* (P. Minnis and W. Elisens, eds.), University of Oklahoma Press, Norman, pp. 223–247.

Fritz, G. J., 2000b, Native Farming Systems and Ecosystem in the Mississippi River Valley, in: *Imperfect Balance: Landscape Transformations in the Precolumbian Americas* (D. Lentz, ed.), Columbia University Press, New York, pp. 225–250.

Gade, D. W., 1970, Ethnobiology of Canihua (*Chenopodium pallidicuale*), Rustic Seed Crop of the Altiplano, *Economic Botany* 24(1):55–61.

Geertz, C., 1963, *Agricultural Involution*, University of California Press, Berkeley.

Gilmore, M. R., 1931, Vegetal Remains of the Ozark Bluff-dweller Culture, *Papers of the Michigan Academy of Science, Arts and Letters* 14:83–102.

Gremillion, K. J., 1993a, Adoption of Old World Crops and Processes of Culture Change in the Historic Southeast, *Southeastern Archaeology* 12:15–20.

Gremillion, K. J., 1993b, The Evolution of Seed Morphology in Domesticated *Chenopodium*: An Archaeological Case Study, *Journal of Ethnobiology* 13:149–169.

Gremillion, K. J., 2004, Seed Processing and the Origins of Food Production in Eastern North America, *American Antiquity* 69:215–233.

Halstead, P. W., and Jones, G. E. M., 1989, Agrarian Ecology in the Greek Islands: Time Stress, Scale and Risk, *The Journal of Hellenic Studies* 109:41–55.

Hammett, J. E., 1992, Ethnohistory of Aboriginal Landscapes in the Southeastern United States, *Southern Indian Studies* 41:1–50.

Hann, J. H., 1986, Translation of Alonso de Leturiondo's Memorial to the King of Spain, *Florida Archaeology* 2:165–225.

Hart, J. P., Asch, D. L., Scarry, C. M., and Crawford, G. W., 2002, The Age of the Common Bean (*Phaseolus vulgaris* L.) in the Northern Eastern Woodlands of North America, *Antiquity* 76:377–383.

Hart, J. P., and Asch Sidell, N. B., 1996, Prehistoric Agricultural Systems in the West Branch of the Susquehanna River Basin, A.D. 800 to A.D. 1350, *Northeast Anthropology* 52:1–30.

Hart, J. P., and Scarry, C. M., 1999, The Age of Common Beans (*Phaseolus vulgaris*) in the Northeastern United States, *American Antiquity* 64:653–658.

Johannessen, S., 1993, Farmers of the Late Woodland, in: *Foraging and Farming in the Eastern Woodlands* (C. M. Scarry, ed.), University Press of Florida, Gainesville, pp. 57–77.

Jones, V. H., 1936, The Vegetal Remains of Newt Kash Hollow Shelter, in: *Rock Shelters in Menifee County, Kentucky* (W. Webb and W. Funkhauser, ed.), Reports in Archaeology and Anthropology 3, University of Kentucky, Lexington, pp.147–167.

Lopinot, N. H., 1992, Spatial and Temporal Variability in Mississippian Subsistence: The Archaeobotanical Record, in: *Late Prehistoric Agriculture* (W. Woods, ed.), Studies in Illinois Archaeology 8, Illinois Historic Preservation Agency, Springfield, pp. 44–94.

Lynott, M. J., Boutton, T. W., Price, J. E., and Nelson, D. E., 1986, Stable Carbon Isotopic Evidence for Maize Agriculture in Southeast Missouri and Northeast Arkansas, *American Antiquity* 51:51–65.

Munson, P. J., 1984, *Experiments and Observations on Aboriginal Wild Plant Food Utilization in Eastern North American*, Indiana Historical Society, Indianapolis.

Riley, T. J., Walz, G., Bareis, C., Fortier, A., and Parker, K., 1994, Accelerator Mass Spectrometry (AMS) Dates Confirm Early *Zea mays* in the Mississippi River Valley, *American Antiquity* 59:490–498.

Robertson, J. A. (trans.), 1993, The Account by a Gentleman from Elvas, in: *The De Soto Chronicles,* Volume I (L. Clayton, V. Knight, and E. Moore, eds.), University of Alabama Press, Tuscaloosa, pp. 19–219.

Scarry, C. M., 1993, Variability in Mississippian Crop Production Strategies, in: *Foraging and Farming in the Eastern Woodlands* (C. M. Scarry, ed.), University Press of Florida, Gainesville, pp.78–90.

Scarry, C. M., and Scarry, J. F., 2005, Native American "Garden Agriculture" in Southeastern North America, *World Archaeology* 37(2):259–274.

Scarry C. M., and Yarnell, R. A., 2006, Domestication of Plants in the East, in: *Handbook of North American Indians: Environment, Origins, and Population*, Volume 3 (W. C. Sturtevant, general ed.), (D. H. Ubelaker, ed.), Smithsonian Institution Press, Washington, DC, pp. 428–236.

Schroeder, S., 1999, Maize Productivity in the Eastern Woodlands and Great Plains of North America, *American Antiquity* 64(3):499–516.

Schroeder, S., 2001, Understanding Variation in Prehistoric Agricultural Productivity: The Importance of Distinguishing among Potential, Available, and Consumptive Yields, *American Antiquity* 66:517–525.

Shelby, C., 1993, La Florida by Garcilaso de la Vega, the Inca, in: *The De Soto Chronicles,* Volume II (L. Clayton, V. Knight, and E. Moore, eds.), University of Alabama Press, Tuscaloosa, pp. 25–560.

Simon, M. L., 2000, Regional Variations in Plant Use Strategies in the Midwest during the Late Woodland, in: *Late Woodland Societies* (T. Emerson, D. McElrath, and A. Fortier, eds.), University of Nebraska Press, Lincoln, pp. 37–75.

Smith, B. D., 1985a, *Chenopodium berlandieri* ssp. *jonesianum*: Evidence for a Hopewellian Domesticate from Ash Cave, Ohio, *Southeastern Archaeology* 4:107–133.

Smith, B. D., 1985b, The Role of *Chenopodium* as a Domesticate in Pre-maize Garden Systems of the Eastern United States, *Southeastern Archaeology* 4:51–72.

Smith, B. D., 1986, The Archaeology of the Southeastern United States: From Dalton to DeSoto, 10,500 B.P. to 500 B.P., *Advances in World Archaeology* 5:1–92.

Smith, B. D., 1992, *Rivers of Change: Essays on Early Agriculture in Eastern North America*, Smithsonian Institution Press, Washington, DC.

Smith, B. D., 2006, Eastern North America as an Independent Center of Plant Domestication, *Proceedings of the National Academy of Sciences* 103(33):12223–12228.

Smith, B. D., and Cowan, C. W., 2003, Domesticated Crop Plants and the Evolution of Food Production Economies in Eastern North America, in: *People and Plants in Ancient Eastern North America* (P. Minnis, ed.), Smithsonian Institution Press, Washington, DC, pp. 105–125.

Smith, B. D., and Wetterstrom, W., 1978, Energy-capture Analysis, in: *Prehistoric Patterns of Human Behavior: A Case Study in the Mississippi Valley* (B. D. Smith, ed.), Academic Press, New York, pp. 99–115.

Steponaitis, V., 1986, Prehistoric Archaeology in the Southeastern United States 1970–1985, *Annual Review of Anthropology* 15:363–404.

Swanton, J. R., 1946, *The Indians of the Southeastern United States*, Bureau of American Ethnology Bulletin 137, Government Printing Office, Washington, DC.

Wagner, G. E., 1987, *Uses of Plants by the Fort Ancient Indians*, Ph.D. dissertation, Washington University, St. Louis, Missouri.

Wenhold, L. L., 1936, A 17th century Letter of Gabriel Díaz Vara Calderón, Bishop of Cuba, Describing the Indians and Indian Missions of Florida, *Smithsonian Miscellaneous Collections* 95(16), Government Printing Office, Washington, DC.

Wilson, G. L., 1987 [1917], *Buffalo Bird Woman's Garden*, Minnesota Historical Society Press, St. Paul and Minnesota.

Wymer, D. A., 1993, Cultural Change and Subsistence: Middle Woodland-Late Woodland Transition in the Mid-Ohio Valley, in: *Foraging and Farming in the Eastern Woodlands* (C. M. Scarry, ed.), University Press of Florida, Gainesville, pp. 138–156.

Yarnell, R. A., 1972, *Iva annua* var. *macrocarpa*: Extinct American Cultigen?, *American Anthropologist* 74:335–341.

Yarnell, R. A., 1976, Early Plant Husbandry in Eastern North America, in: *Culture Change and Continuity* (C. Cleland, ed.), Academic Press, New York, pp. 265–273.

Yarnell, R. A., 1978, Domestication of Sunflower and Sumpweed in Eastern North America, in: *The Nature and Status of Ethnobotany* (R. I. Ford, ed.), Anthropological Papers 67, University of Michigan Museum of Anthropology, Ann Arbor, pp. 289–299.

Yarnell, R. A., 1993, The Importance of Native Crops during the Late Archaic and Woodland. in: *Foraging and Farming in the Eastern Woodlands* (C. M. Scarry, ed.), University Press of Florida, Gainesville, pp. 13–26.

Yarnell, R. A., and Black, M. J., 1985, Temporal Trends Indicated by a Survey of Archaic and Woodland Plant Food Remains from Southeastern North America, *Southeastern Archaeology* 4:93–106.

Chapter **21**

Pets and Camp Followers in the West Indies

Elizabeth S. Wing

The many different relationships people have with animals are of abiding interest. As we know from present-day experiences, these relationships range from revering animals to reviling them. Revered animals often possess bold and admired characteristics such as depicted by lions and stags on heraldic shields, eagles on currency and other national symbols, and ravens on Northwest Coast family totems. At the other end of the spectrum are feared and hated animals. Serpents may personify evil in paradise and dogs (*Canis familiaris*) may be considered unclean. In other cultures, snakes may be worshiped and dogs kept as beloved pets. Such attributions are related to cultural perceptions of the world.

One of the most profound relationships between people and animals is that of domestication. For many years it was assumed that animal domestication, broadly practiced on other continents, was restricted to the Andes in the Americas. Evidence from the Caribbean suggests that the process of domestication was more complex than usually described and some aspects of this relationship were more widespread in the Americas than previously thought. This study examines evidence for two domestic animals that were brought into the West Indies during pre-Columbian times. They, and two managed animals, may be thought of as pets though not necessarily in the way we think of pets today. Other animals may have moved around with people but more as camp followers than intentional introductions.

RELATIONSHIPS WITH ANIMALS

A whole range of relationships lies between the two extremes of reverence and fear, including indifference. Most animals with which we share the earth are and were considered unimportant by most people. Animals that are small in size, rarely encountered, or of little economic use rarely are appreciated. However, exceptions exist. For example, the work of bees (Meliponidae) was well known to the Maya and their descendants. Bees held a sacred place in the household that went beyond their contributions as pollinators or providers of honey (Schwarz 1948).

We are most familiar with animals that are a source of meat, whether obtained by herding, hunting, fishing, or gathering. Through the ages hunting has focused on larger herbivores such as white-tailed deer (*Odocoileus virginianus*), fishing on near-shore species from marine, estuarine, and fresh waters, and gathering on mollusks and the eggs of birds and reptiles. To be successful in these pursuits requires a great fund of knowledge about the prey species. This knowledge is based on generations of accumulated observations and techniques devised to ensure success. Some resources peak in abundance during parts of the year due to migrations or breeding congregations. People doubtless incorporated such life history information into plans for hunting or fishing schedules in the past as they do today. This concept can be extended to planting and harvesting schedules and regulated hunting and fishing seasons. One way to resolve seasonal abundance and scarcity is to hold some wild animals in captivity until they are slaughtered and eaten.

Some of the captured animals are not eaten but are tamed and kept as pets; adding another dimension to the life histories of these species (Serpell 1986:48–58). Pet-keeping is widespread today and has deep roots in human history. Young individuals of social animals may be the easiest to tame. Usually women and children are most involved in the care of young animals, a practice that becomes part of the social structure of the human community. As the pet grows older, it may become more dominant and harder to manage.

The selection of individuals which are submissive, exhibit juvenile characteristics, and whose breeding can be controlled is part of the process leading to domestication. Modern domestic animals are easily recognized. The coat or plumage patterns of domestic animals are distinctive and differ from wild color patterns that offer protective camouflage, for example, they may be white with or without bold marking. Many physical characteristics are different in domestic animals. The clearest differences are seen in dogs with large droopy ears and curly tails. Such ears and tails cannot be raised to signal dominance. Many of these characteristics, so obvious in contemporary domestic animals, are not preserved in most archaeological deposits.

Archaeological evidence for domestic animals includes size and proportions of skeletal and dental elements which are different from those of the wild progenitor, the archaeological context and associated remains, and the presence of the species beyond the range of the wild ancestor. Human selection of submissive animals with juvenile characteristics is accompanied by changes in overall size and proportions; domestic animals are usually smaller than the wild progenitor. Another clue to the domestic status of the animal is the context in which its remains are found. For example, one of the oldest dogs is a puppy found between the hands of a women buried at Ein Mallaha, Isreal (Davis 1987:145–147). The burial of horses (*Equus caballus*) with tack shows these animals were used to pull wagons or carry burdens (Davis 1987:164–166). The remains of an animal at a site outside the present range of the species may be interpreted as evidence for the introduction of the animal by people to the location or that the former range of the species once was larger. The ranges of animals change as climates and other aspects of landscapes change. One can determine the former range of an animal by its absence in earlier archaeological or paleontological deposits in the area.

The condition of animals that lay somewhere between tamed or managed is important for understanding the processes that led to full domestication. How long did the process of domestication take for different species? Undoubtedly genetic modification of domestic animals took different lengths of time depending on the breeding capacities and generation lengths of the animals and the degree to which they were isolated from wild relatives. Under what circumstances were animals domesticated? Can managed animals satisfy the need for access to animals, their meat, and products without the effort of maintaining them in domestication?

Incipient domestication and maintenance of tame animals in the West Indies are the subject of this case study (Figure 21-1). The animals of particular concern are agouti (*Dasyprocta leporina*); Jamaican hutia (*Geocapromys brownii*); Bahamian hutia (*Geocapromys ingrahami*); Greater Antillean hutia (*Isolobodon portoricensis*); and Lesser Antillean rice rats (tribe Oryzomyini). Comparisons are made between these animals and the remains of fully domestic animals such as dogs and guinea pigs (*Cavia porcellus*). Examples of these animals were recovered from a series of sites in the West Indian archipelago (Newsom and Wing 2004). This study focuses on the identified remains of these animals, their relative abundance, and the contexts in which they were found in order to detect evidence for efficient hunting, management, taming, or domestication of these animals.

MATERIALS AND METHODS

The faunal samples were excavated from a series of islands that extend the length of the Caribbean archipelago, otherwise known as the West Indies (Figure 21-1). This island chain includes small isolated islands and large islands divided into the Lesser Antilles, the Virgin Islands, the Greater Antilles, and the Bahamas. Many of the islands are within view of one another.

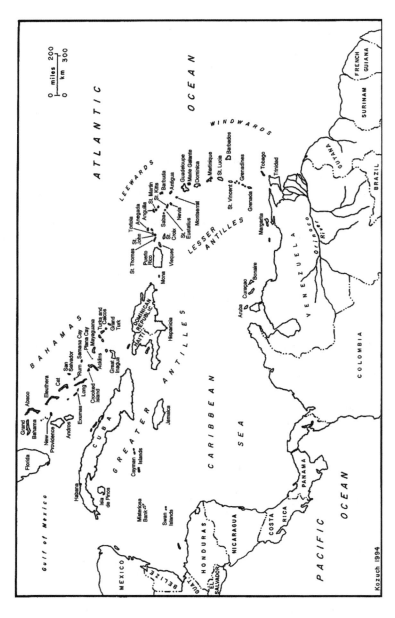

Figure 21-1. Map of study area. Jost van Dyke is a small island just west of Tortola.

The Lesser Antilles lie between the South American mainland in the south and the Anegada Passage in the north. Most of the smaller islands are in the Lesser Antilles. The Anegada Passage, approximately 90 km wide, is the largest gap between the islands and separates the Lesser Antilles from the Virgin Islands between St. Croix and Saba. The Virgin Islands are a cluster of small islands between the Anegada Passage and Puerto Rico. Most of the largest islands are in the northern part of the chain and are known as the Greater Antilles. The Bahamas and Turks and Caicos extend north of the Greater Antilles to immediately east of southern Florida. The smallest island from which faunal samples were studied is Saba (13 km^2). At the other extreme is Hispaniola, on which the countries of Haiti and the Dominican Republic are located. Hispaniola is 76,484 km^2. Unfortunately, data from the largest island, Cuba, are not available.

The faunal samples were recovered primarily by sieving the midden material through fine-gauge screens (at least as fine as 1.5 mm). Most mammals are large enough that their fragmentary remains will be caught in larger screen sizes (3 mm); however, finer gauged sieves are necessary to recover most of the fishes and to place mammals in the context of the entire suite of animals used at the site. Generally, the contents of all faunal samples from chronologically-secure contexts were identified. This allowed the relative abundance of the key animals in each faunal collection to be assessed. This is important because tamed and wild animals may have been disposed of in different ways. Animal remains from two human burial sites, Silver Sands (Barbados) and Sorcé (Vieques), are considered in order to investigate this possibility.

The relative abundance of the key animals in different sites and the contexts in which they were found were examined in order to understand the role of these animals. During the original study, the sizes of skeletal elements were measured and postmortem modifications were described. Skeletal or dental modifications that might have occurred during the life of the animal were described and quantified. Estimates of the individual animal's age at death were made based on tooth eruption, dental wear stages, and long bone epiphyseal fusion (Reitz and Wing 1999:178–185). The contexts in which the remains were found include the island's location in the West Indian archipelago, the age of the deposit, associated animals, and whether the deposit was a midden or burial. The animal remains include all vertebrate taxa and are quantified by estimating the Minimum Numbers of Individuals (MNI *sensu* White; Reitz and Wing 1999:194–200). The resulting MNI estimates are expressed here as percentages to facilitate comparison of the results among deposits. The taxonomic protocol follows Morgan and Woods (1986) except in the case of dogs, which follows the ruling of the International Commission on Zoological Nomenclature (Gentry et al. 2004).

The data for this case study are from well-dated sites occupied primarily after 500 B.C. Though considerable variation exists in the cultural chronology within the West Indies, I use broad cultural time periods based primarily on the work of Rouse and Wilson (Rouse 1992; Rouse and Faber-Morse 1999; Wilson 2001). The earliest time period for which we have animal remains is the Archaic period (also called the Ortoiroid). Traditionally it is understood that the Ortoiroid peoples migrated from South America into the Antilles around 2000 B.C. Their economy was based on fishing and gathering. They did not practice horticulture or produce pottery. A second migration from South America occurred between 500 and 300 B.C. and brought to the Lesser Antilles, Virgin Islands, and Puerto Rico people who lived in permanent settlements, practiced horticulture, and produced pottery. This Ceramic period culture is generally called Saladoid. Around A.D. 600, island societies are marked by increased social and political complexity. Populations increased and people expanded into Cuba, Hispaniola, Jamaica, and the Bahamas. These people were sedentary agriculturalists who hunted, fished, and gathered marine resources. This period is known as post-Saladoid or Ostionoid in the Virgin Islands and Greater Antilles and Troumassoid in the Lesser Antilles. Exploration and colonization by Europeans, which began with Columbus' voyage into the Caribbean in A.D. 1492, changed the hemisphere dramatically.

RESULTS

The faunal data from the West Indies are summarized in Tables 21-1 through 21-6. In Tables 21-1, 21-3, and 21-5 the data are summarized in approximate geographical order. The cultural affiliation of each deposit, the sample sizes expressed in terms of MNI, and references are also given. The abundance of each of the seven mammals which are the focus of this study is presented as MNI and percent MNI among the total vertebrate individuals identified in each deposit in Tables 21-2, 21-4, and 21-6. The presence, absence, and relative abundance of the key mammals in West Indian sites reveal several trends in the use of these animals.

Rice Rats (Tribe Oryzomyini)

Endemic rice rats are present in all Lesser Antillean collections reported here. In addition to those sites listed on Tables 21-1 and 21-2, rice rats are reported for collections from St. Lucia and Marie Galante (Newsom and Wing 2004; Wing and Reitz 1982). At least two genera and several species of rice rats are described for the West Indies. These generally fall into two size classes: large and small. The large form is sometimes referred to as the genus *Megalomys* spp. or as unde-scribed species B. The small species is sometimes referred to the genus *Oryzomys* sp. or as undescribed species A. A rice rat known as *Oryzomys antillarum* is identified at many Jamaican sites. I will refer to these animals by

Table 21-1. Sites Surveyed, Time Period of Occupation, Sample Size (MNI), and References for Sites in the Lesser Antilles

Islands and sites	Time periods	MNI	Reference
St. Martin			
Hope Estate unit 10 zone 3	late Saladoid	89	Newsom and Wing (2004:Table B.8)
Hope Estate unit 16 zone 18	early Saladoid	70	Newsom and Wing (2004:Table B.8)
Saba			
Kelbey's Ridge 2	late post-Saladoid	211	Wing (1996:279)
Kelbey's Ridge 1	early post-Saladoid	202	Wing (1996:273)
Spring Bay 1 and 3	middle post-Saladoid	111	Wing (1996:274–275)
St. Eustatius			
Golden Rock	Saladoid	525	Klift (1992)
St. Kitts			
Sugar Factory Pier	late Saladoid	541	Wing and Scudder (1980)
Cayon	early Saladoid	109	Wing and Scudder (1980)
Nevis			
Sulphur Ghaut JO-2	post-Saladoid	172	Newsom and Wing (2004:Table B.6)
Indian Castle GE-1	post-Saladoid	125	Newsom and Wing (2004:Table B.6)
Hichmans' GE-5	Saladoid	64	Newsom and Wing (2004:Table B.5)
Hichmans' Shell Heap GE-6	Archaic	148	Newsom and Wing (2004:Table 6.1)
Montserrat			
Trants	Saladoid	97	Steadman et al. (1984)
Antigua			
Indian Creek excavation 2	Saladoid	235	Wing (1999:Table 4)
Jolly Beach	Archaic	126	Newsom and Wing (2004:Table 6.1)
Grenada			
Pearls	Saladoid	41	Newsom and Wing (2004:Table B.1)
Barbados			
Silver Sands	post-Saladoid	454	Newsom and Wing (2004:Table B.3)
Chancery Lane	Saladoid	67	Newsom and Wing (2004:Table B.3)
Heywoods	Archaic	26	Newsom and Wing (2004:Table 6.10)

Table 21-2. Occurrence and Relative Abundance of Key Mammals in the Lesser Antilles

Islands and sites	Rice rat		Agouti		Dog		Guinea pig	
	MNI	%	MNI	%	MNI	%	MNI	%
St. Martin								
Hope Estate unit 10 zone 3	25	28.1	–	–	–	–	–	–
Hope Estate unit 16 zone 18	2	2.9	–	–	–	–	–	–
Saba								
Kelbey's Ridge 2	48	22.8	–	–	–	–	–	–
Kelbey's Ridge 1	10	4.9	–	–	–	–	–	–
Spring Bay 1 and 3	10	9.0	2	1.8	–	–	–	–
St. Eustatius								
Golden Rock	66	12.6	2	0.4	–	–	–	–
St. Kitts								
Sugar Factory Pier	77	14.0	15	3.0	1	0.2	–	–
Cayon	29	27.0	10	9.0	–	–	–	–
Nevis								
Sulpher Ghaut JO-2	9	5.2	2	1.2	–	–	–	–
Indian Castle GE-1	8	6.4	1	0.8	–	–	–	–
Hichmans' GE-5	19	29.7	2	3.1	–	–	–	–
Hichmans' Shell Heap GE-6	7	4.7	–	–	–	–	–	–
Monserrat								
Trants	5	5.2	–	–	2	2.1	–	–
Antigua								
Indian Creek excavation 2	26	11.1	10	4.3	–	–	1	0.4
Jolly Beach	11	10.3	–	–	–	–	–	–
Grenada								
Pearls	2	4.9	4	9.8	–	–	–	–
Barbados								
Silver Sands	22	4.8	–	–	15	3.3	–	–
Chancery Lane	2	3.0	–	–	2	3.0	–	–
Heywoods	3	11.5	–	–	–	–	–	–

their more inclusive name, Oryzomyini, until the systematics of the West Indian rice rats are revised. Rice rats of only one size are found at most sites; although Trants (Montserrat) has at least two sizes and possibly an intermediate-sized third species (Reitz 1994; Steadman et al. 1984).

Rice rats appear to be more abundant in the later deposits of some sites, such as Hope Estate (St. Martin), Kelbey's Ridge (Saba), and Hichmans' Shell Heap (Nevis) than in the earlier deposits at these same sites (Table 21-2). They also are more abundant at sites located farther from the coast. For example, Hope Estate is located 2 km from the nearest shore and at 50 m elevation and the early Ceramic deposit at Cayon (St. Kitts) is located 1.7 km from the nearest shore at approximately 130 m elevation (Goodwin 1979:188). All West Indian rice rats became extinct during the late 1800s and early 1900s (Westermann 1953).

Table 21-3. Sites Surveyed, Time Period of Occupation, Sample Size (MNI), and References for Sites in the Greater Antilles and Virgin Islands

Islands and sites	Time periods	MNI	References
Greater Antilles: Jamaica			
Bellevue	Ostionoid	126	Newsom and Wing (2004:Table 7.9)
Rodney House	Ostionoid	150	Scudder (1991)
Cinnamon Hill	Ostionoid	162	Newsom and Wing (2004:Table 7.9)
White Marl	Ostionoid	713	Newsom and Wing (2004:Table C.3)
Greater Antilles, Hispaniola: Haiti			
En Bas Saline	contact	75	Wing (2001)
Greater Antilles: Puerto Rico			
Maisabel	late Saladoid	73	deFrance (1988)
Maisabel	early Saladoid	80	deFrance (1988)
El Bronce	contact	119	Reitz (1985)
Greater Antilles: Vieques			
Lujan midden A and B	Ostionoid	199	Newsom and Wing (2004:Table C.1)
Sorcé YTA	late Saladoid	132	Narganes Storde (1982)
Sorcé Z	early Saladoid	380	Narganes Storde (1982)
Virgin Islands: St. Thomas			
Tutu	late Ostionoid	151	Newsom and Wing (2004:Table C.1)
Tutu	early Saladoid	121	Newsom and Wing (2004:Table C.1)
Virgin Islands: St. John			
Trunk Bay	Ostionoid	178	Newsom and Wing (2004:Table C.1)
Cinnamon Bay	Ostionoid	147	Newsom and Wing (2004:Table C.1)
Virgin Islands: Jost van Dyke			
Cape Wright	Ostionoid	99	Newsom and Wing (2004:Table C.1)

Agouti (*Dasyprocta leporina*)

The introduced agouti, which is native to South America, occurs in small numbers at sites on most of the Lesser Antillean islands reported here. The aguoti is absent, however, from sites on Barbados and from the Archaic deposits of Jolly Beach (Antigua) and Hichmans' Shell Heap (Nevis) (Table 21-2). Agouti remains were identified in other archaeological collections that do not appear in Table 21-2. Specifically, agouti is reported in Trants (Montserrat) collections other than the ones reported here (Steadman et al. 1984) and from sites other than Hope Estate on St. Martin (Nokkert 1995). An agouti was reported from Sorcé (Vieques), but this find cannot be verified (Woods 1996). The intentional burial of this animal on St. Kitts offers further insight into the role of agouti in human life (Goodwin 1976). Agouti are more abundant in the collection from the Pearls site (Grenada), which is close to South American, than in collections farther from South America. European plantation owners initiated a second introduction of agouti to provide meat for their slaves (Vietmeyer 1991). Animals from either pre-Columbian or post-Columbian contexts, or both,

Table 21-4. Occurrence and Relative Abundance of Key Mammals in the Greater Antilles and Virgin Islands

Islands and sites	Rice rat		Jamaican hutia		Antillean hutia		Agouti		Dog		Guinea pig	
	MNI	%	MNI	%	MNI	%	MNI	%	MNI	%	MNI	%
Greater Antilles: Jamaica												
Bellevue	–	–	103	81.7	–	–	–	–	–	–	–	–
Rodney House	1	0.7	28	18.7	–	–	–	–	–	–	–	–
Cinnamon Hill	–	–	16	9.9	–	–	–	–	–	–	–	–
White Marl	14	2.0	357	50.1	–	–	–	–	5	0.7	–	–
Greater Antilles, Hispaniola: Haiti												
En Bas Saline	–	–	–	–	2	2.7	–	–	–	–	–	–
Greater Antilles: Puerto Rico												
Maisabel, late	–	–	–	–	1	1.4	–	–	–	–	–	–
Maisabel, early	–	–	–	–	1	1.3	–	–	–	–	–	–
El Bronce	–	–	–	–	20	17.4	–	–	1	0.9	–	–
Greater Antilles: Vieques												
Lujan midden A and B	–	–	–	–	18	9.0	–	–	1	0.5	2	1.0
Sorcé YTA	–	–	–	–	6	4.5	–	–	3	2.3	–	–
Sorcé Z	–	–	–	–	1	0.3	–	–	19	5.0	–	–
Virgin Islands: St. Thomas												
Tutu, late	–	–	–	–	5	3.3	–	–	–	–	–	–
Tutu, early	–	–	–	–	3	2.3	–	–	–	–	–	–
Virgin Islands: St. John												
Trunk Bay	–	–	–	–	2	1.1	–	–	–	–	–	–
Cinnamon Bay	–	–	–	–	3	2.0	–	–	–	–	–	–
Virgin Islands: Jost van Dyke												
Cape Wright	–	–	–	–	5	5.1	–	–	–	–	–	–

Table 21-5. Sites Surveyed, Time Period of Occupation, Sample Size (MNI), and References for Sites in the Bahamas and Turks and Caicos

Islands and sites	Time periods	MNI	References
Bahamas: San Salvador			
Long Bay (SS-9)	contact	143	Newsom and Wing (2004:Table D.5)
Palmetto Grove (SS-2)	Ostionoid	570	Newsom and Wing (2004:Table D.5)
Bahamas: Crooked Island			
CK-14	late pre-Columbian	185	Newsom and Wing (2004:Table D.4)
Turks and Caicos			
MC-6	Ostionoid	139	Wing and Scudder (1983)
MC-12	Ostionoid	127	Wing and Scudder (1983)

became naturalized throughout most of the Lesser Antilles and still are found on some islands such as Dominica (Westermann 1953).

Jamaican Hutia (*Geocapromys brownii*); Bahamian Hutia (*Geocapromys ingrahami*); and Greater Antillean Hutia (*Isolobodon portoricensis*)

The Jamaican hutia is endemic to the island for which it is named and has not been found outside of Jamaica. The remains of Jamaican hutia increase in abundance the farther inland the site is from the sea (Table 21-4). This can be seen by comparing their presence at Cinnamon Hill (on the shore) with data from Rodney House (circa 1 km from the coast), White Marl (5.6 km from the coast), and the Bellevue site (10 km from the coast). The Jamaican hutia constitutes almost 82% of the vertebrate individuals in the Bellevue collection. The Jamaican hutia survives today in remote parts of Jamaica but is vulnerable to hunters, especially those with dogs (Wilkins 2001).

The endemic Bahamian hutia is identified from sites in the Bahamas and in the Turks and Caicos but is rare (Table 21-6). Today, the Bahamian hutia has

Table 21-6. Occurrence and Relative Abundance of Key Mammals in Bahamas and Caicos Sites

Islands and sites	Bahamian hutia		Dog	
	MNI	%	MNI	%
Bahamas: San Salvador				
Long Bay	1	0.7	–	–
Palmetto Grove	1	0.2	–	–
Bahamas: Crooked Island				
CK-14	6	3.2	–	–
Turks and Caicos				
MC-6	1	0.7	–	–
MC-12	–	–	1	0.8

been extirpated from the rest of the Bahamas. It still lives on Plana Cay, Bahamas, where it is very vulnerable to predators and environmental change.

The Greater Antillean hutia is believed to have originated on Hispaniola but is found in archaeological collections from Hispaniola, Puerto Rico, and the Virgin Islands (Morgan and Woods 1986) (Table 21-4). Antillean hutia remains are present even in some of the eastern-most Virgin Islands, such as the Paraquita Bay (Tortola) and the Cape Wright (Jost van Dyke) sites (Wing 2000). Antillean hutias have not been identified from preceramic deposits in Puerto Rico or the Virgin Islands: they are absent at Krum Bay (1680–1530 B.C.) and Main Street (A.D. 180) on St. Thomas; Puerto Ferro (1900–790 B.C.) on Vieques; and Maruca (2890–395 B.C.) on Puerto Rico (Narganes Storde 1991, Newsom and Wing 2004; Reitz 1989). Antillean hutias are present in collections from St. Croix but have not been identified either from the Bahamas or the Lesser Antilles (Flemming and MacPhee 1999). They are most abundant (17.4% of the vertebrate individuals) at El Bronce (Puerto Rico), 7.3 km inland from the coast (Reitz 1985). No evidence of burials or ritual use of the Antillean hutia is known.

Dog (*Canis familiaris*)

In the West Indies, domestic dogs are most often associated with human burials; middens contain relatively few dog remains. In addition to Trants (Montserrat) and Chancery Lane (Barbados), dogs have been identified from the Grande Anse site (St. Lucia) (Wing and Reitz 1982). At least 15 (MNI) dog burials are associated with human burials at Silver Sands (Barbados) and at least 22 (MNI) dogs were identified from Sorcé (Vieques) (Wing 1991a, b). Other reports of dogs from burials and special deposits are not listed in Tables 21-2 and 21-4. These include one from Martinique, four specimens from the Dominican Republic (Hispaniola), three from Cuba, two from Puerto Rico, and four from Jamaica (Lawrence 1977).

Though burials are the most common contexts in which dogs are found, worked dog specimens, especially teeth, also are reported. A necklace made of more than 4,000 mammal canines, the majority from dogs, is believed to have been associated with a human burial near Cabo San Rafael (Hispaniola), though the necklace was not excavated scientifically (Rímoli 1977). Modified dog teeth also are reported from Barbados (Wing 1991b). The mutilation of the fourth lower premolar is characteristic of many dog mandibles. The crowns of these teeth were broken off during life. In many cases the roots of the teeth remained in the jaw and the alveolus healed over the broken tooth (Wing 1991a, b). This type of tooth mutilation was observed in 30% of the 26 lower jaws examined throughout the West Indies. The gap in the tooth row produced by the mutilation may have provided a place for tying a restraint that acted as a muzzle to control the dog's predation on other animals valued by people.

Guinea Pig (*Cavia porcellus*)

Guinea pigs were domesticated in the central Andes (Wing 1986) and their remains are rare in the West Indies. When they are found, it is in late pre-Columbian or early post-Columbian deposits. In addition to Indian Creek (Antigua) and Lujan (Puerto Rico), their remains are found in collections from a late deposit (Santa Barbara) on the island of Curaçao in the ABC Islands off the coast of Venezuela and Mill Reef (Antigua) (Newsom and Wing 2004). The faunal sample from the contact site of Finca Valencia (NCS-1, Puerto Rico), in which guinea pigs represent 11% of the individuals, contains an usually large quantity of these animals (Quitmyer and Kozuch 1996). Spanish documents describe guinea pigs on Hispaniola (Oviedo 1526, translation by Stoudemire 1959).

PETS AND CAMP FOLLOWERS

The location of each site and the cultural period of each deposit are keys to the presence and relative abundance of these animals. These species fall into three groups: (1) those endemic to the islands where they were used; (2) those introduced from the South American mainland to the West Indies or from large islands to small ones; and (3) those domesticated somewhere other than the West Indies. All archaeological data are fragmentary and additional samples will serve to either strengthen the interpretations presented here or modify them.

Both rice rats and the Jamaican hutia occur in Pleistocene and archaeological deposits, indicating they are part of the endemic fauna of the West Indies. They share some characteristics in their distribution patterns. Their remains are present in all sites within their range, rice rats in the Lesser Antilles and Jamaican hutia on Jamaica. They increase in abundance the farther inland the site is located, as shown by the abundance of rice rats at Hope Estate (St. Martin) and of Jamaican hutia at Bellevue (Jamaica). This suggests that at locations farther from marine resources, small terrestrial animals were intensively exploited.

The question is, of course, were these animals wild and hunted or maintained in captivity as tame animals readily available as sources of meat? Wilkins (2001) studied the wild hutia populations on Jamaica as well as a captive population. She doubts that their reproductive and nutritional requirements could be met in captivity. The abundance of hutia in the Bellevue (Jamaica) collection indicates that wild individuals could be successfully caught in large numbers. This success may be due to the assistance of dogs in hunting.

Rice rats, endemic to the Lesser Antilles, are relatively abundant in early deposits such as those from Heywoods (Barbados), Jolly Beach (Antigua), and

Cayon (St. Kitts) (Table 21-2). They are even more abundant in the later deposits at these sites and at sites located farther inland such as Hope Estate (St. Martin). Newsom found evidence for an increase in secondary woods in the later deposits of Hope Estate (Newsom and Wing 2004). This suggests forest clearing and expanded second growth forest, which may have provided more favorable habitats for rice rats. Increased garden products, especially if stored, would have been another food source for rice rats and kept them close to human habitation. People living farther inland may have engaged in a form of garden hunting, which would have been particularly attractive to rice rats (Linares 1976; see Neusius, this volume).

Animals that are absent from paleontological or early archaeological deposits but then appear consistently at sites after 500 B.C. are interpreted as evidence of human introduction. The introduction of live breeding animals would permit extended use of them and encourage introductions to other neighboring islands. The West Indies have two sources of diverse, large to moderately-large land animals: the islands adjacent to the circum-Caribbean mainland and the West Indies themselves. South America is the closest mainland and the region from which all but the earliest human populations migrated into the West Indies. Another source of animals is the large Greater Antillean islands such as Cuba and Hispaniola with their more diverse fauna. The agouti, endemic to the South American mainland, and the Greater Antillean hutia, originally from Hispaniola, are examples of introductions from these two sources.

Proximity to the source of diverse animal species may have influenced the success of introductions. For example, South American mammals such as the armadillo (*Dasypus novemcinctus*) and opossum (*Didelphis marsupialis*) are found on Grenada and St. Lucia. They were probably introduced to those islands by colonists during the Ceramic period (Wing and Reitz 1982); but armadillos and opossums are not found farther up the island chain. Likewise, collections from islands closest to South America appear to have more agouti remains compared to those farther north along the archipelago (Tables 21-2, 21-4, and 21-6). This is reflected in the relatively greater abundance of agouti remains in the southern Lesser Antilles compared to the northern Lesser Antilles. Agouti, which are rare in the northern Lesser Antilles, are virtually absent in the Greater Antilles, Virgin Islands, and Bahamas. The agouti doubtless was transported from island to island in a tamed state during the Ceramic period. The shorter the transportation distance the more successful the introduction may have been.

The agouti is associated with people wherever it occurs in the West Indies. According to Gade (1977:326), "at a household scale, the agouti is a candidate for domestication because of its high food value, its year-round breeding in captivity, and its ability to thrive on kitchen refuse." Vietmeyer (1991:199) reports that "Among the best known of all animals of the American tropics, agoutis (*Dasyprocta* species) are prolific rabbit- or hare-sized rodents that are probably easily farmed. They are valued for food and are hunted throughout most of their range." Agoutis are attracted to edible refuse or intentional handouts of food; thus they tame relatively easily. They are a choice meat in Central and South America.

In some respects the Greater Antillean hutia is similar to the agouti in that the hutia was a medium-sized rodent, probably weighing 2 to 3 kg. Using other members of the family as a guide, the Antillean hutia probably produced a few precocial young at a time. This hutia evolved on Hispaniola and is absent from preceramic deposits elsewhere; suggesting it was a Ceramic-period introduction beyond Hispaniola. Hutia remains are present in all Ceramic period sites in Puerto Rico and the Virgin Islands for which we have data, even at remote Jost van Dyke. Once it was introduced it became widespread and was intensively used particularly at inland sites such as El Bronce (Puerto Rico). Hutias have not been found in sites on the Bahamas, Turks and Caicos, or in the Lesser Antilles.

Observations made between A.D. 1536 and 1546 by the Spanish chronicler Oviedo described "animals called hutia, four-footed, and resembling a rabbit, but smaller sized, smaller ears and rat-tailed" (Miller 1929:12). Amerindians used dogs to catch hutia and other animals (Miller 1929:15). Large corrals were constructed out of poles standing close together and bound with intertwining creepers to confine hutias at Vega Real (Hispaniola) so that the meat, which was relished, would always be available (Lovén 1935:437–438). Such methods of capture and control may have been widespread in managing the Antillean hutia as well as other animals.

The two fully domestic animals introduced into the West Indies are dogs and guinea pigs. Both came with traditions of care and use and were kept in the houses (Miller 1929:14–15). Dogs were described by Oviedo from his observation between 1536 and 1546 as small animals that did not bark even under provocation. Their coats were varied in color and the fur characteristics were usually "between silky and short-haired" (Miller 1929:15). They had "ears lively and alert like those of wolves" (Miller 1929:15). No agreement exists among chroniclers about how these native dogs were used. "They are small, Las Casas said, of the size of a lap dog, and were eaten, –indeed, were kept to be eaten"(Sauer 1966:59). Las Casas reported that dogs being eaten may be a consequence of conditions of food shortages referred to as the "starving time of 1494" when Spaniards "ate them up" (Sauer 1966:59). "Oviedo thought they were used in hunting" (Sauer 1966:59). No clear case of dogs used for food is evident in the sites reported here. Occasional fragments of dog bones or teeth are present at many sites but dog remains are only abundant in burial contexts at these sites. In burials they are usually moderately complete and show no evidence of butchering scars or burning. Both dog and human bone is occasionally found in midden contexts probably due to site formation processes unrelated to food uses. The association of dogs with human burials suggests some ritual treatment, as does the modification of dog canines (Rímoli 1977).

Guinea pigs are the other fully domestic species of the West Indies. Guinea pigs were a late introduction into the West Indies. They occur sporadically throughout the islands but always in the latest deposits. Guinea pigs were called cori, which probably has the same root as cuy by which name they are called in Peru where they were originally domesticated. Oviedo describes them as multi-colored animals that were kept in the house and fed grass and some cassava (*Manihot esculenta*) to fatten them (Miller 1929:14). Guinea pigs continue to be used in curing and divination rituals in Peru today although we cannot detect evidence for any use other than as food in the West Indies.

SUMMARY

The distribution of these three groups of animals, the endemic, the introduced, and the domestic, is quite different. These differences suggest different degrees of control and manipulation.

The endemic mammals, rice rats and Jamaican hutia, were clearly important local resources. Landscape changes and site locations influenced the intensity of their exploitation. Rice rats were probably not maintained in captivity. Modification of the environment by forest clearing and planting house gardens may have attracted rice rats making them more accessible to hunters. There is no evidence that the Jamaican hutia was maintained in captivity or introduced to other islands, although hunting strategies were developed to catch them in large numbers. This may have depended on the hunting assistance of dogs.

The introduced mammals, the agouti and Antillean hutia, clearly were managed. This allowed successful introductions to distant locations. An aspect of this management must have been control and support, at least during sea voyages. This implies that food was supplied to them during such voyages. Whether they were kept in cages or corrals at each new island cannot be determined. Even if they were confined, survival to old age would not be expected if they were mainly used as food. The single burial of an agouti on St. Kitts indicates use of this animal beyond simply as meat (Goodwin 1976). Such a ritual role was apparently rare, though it may have been practiced outside of the borders of normal site excavation. The virtual absence of agouti remains north of the Anegada Passage suggests there were obstacles to supporting them in captivity. A similar obstacle may have impeded the transportation of Antillean hutias, which reached remote parts of the Virgin Islands but were not taken north to the Bahamas or south into the Lesser Antilles. This apparent obstacle may not have been so much the inability to transport the animal as a cultural barrier to the exchange. Though exchange of goods has been described among these islands, live animals are harder to transport and maintain. Any animal introduced to an isolated island would be expected to change over time through interbreeding within a small, isolated population.

Some changes also would occur through human selection. Such size changes are reported among the remains of Greater Antillean hutia from a cave in western Puerto Rico (Reynolds et al. 1953).

The two fully-domesticated animals, the dog and guinea pig, differ in their distribution from these other animals. Dogs were clearly widespread throughout the Caribbean and probably were not routinely used for food. However, they may have played an important role as hunting assistants, as is frequently mentioned in the chronicles. Their place in ritual associated with death was probably of paramount importance. Guinea pigs came into the West Indies so close to the disruption caused by European colonization that we do not have enough information about their use other than for food. We have no explanation for their spotty distribution.

CONCLUSION

People have manipulated the distribution of plants and animals for millennia. The human hand in the changed distributions of organisms may have been accidental or intentional allowing continued use of an animal or its products. Many of these introductions did not culminate in full-blown domestication but were nevertheless important as cultural components. Such a relationship is not restricted to the West Indies. One of the earliest introductions, which occurred between 19,000 to 10,000 years ago, was of the cuscus (*Phalanger orientalis*) transported to the island of New Ireland in Melanesia probably from the island of New Britain (Flannery and White 1991). One or another species of rat of the genus *Rattus* have hitched a ride with people wherever they have gone. Some of these animals were large and their remains are incorporated with other food remains particularly in middens on many Pacific Islands. An example within a mainland setting is the well-organized trade of macaws (*Ara macao*) that originated in Mexico and extended throughout the southwestern United States during pre-Columbian times. The macaws were reared in breeding pens with nest boxes, examples of which were excavated from the site of Casas Grandes in Chihuahua, Mexico (Hargrave 1970; Minnis et al. 1993). Some of these animals probably enriched the lives and contributed to the economies of the people who maintained them, much as the agouti and the hutia presumably did in the West Indies. None became domesticated in the strict sense though they might eventually have become domesticated had this trajectory not been interrupted in 1492.

We can recognize in the West Indies a continuum in human and animal relationships: (1) wild animals that were rarely caught; (2) animals that may be called camp followers by their association with human habitation and garden cultivation; (3) animals that are well known and reliably caught; (4) animals that are maintained in captivity and moved around; and (5) fully domestic animals. Animals kept in captivity doubtless had importance in the social fabric of their captors beyond the potential for domestication. The experiment that might ultimately have resulted in domestication of the hutia and the agouti was cut short by the events that followed 1492 (Crosby 1986). Both were undoubtedly maintained in captivity for a while after 1492. However, Europeans came into the Caribbean with an array of well-adapted domestic animals: dogs, cats (*Felis catus*), horses and donkeys (*Equus* spp.), pigs (*Sus scrofa*), cattle (*Bos taurus*), sheep (*Ovis aries*), goats (*Capra hircus*), and chickens (*Gallus gallus*). They were accompanied by commensal rats (*Rattus* spp.) and house mice (*Mus musculus*). These domestic animals and camp followers overwhelmed the West Indian fauna and Amerindian cultures.

ACKNOWLEDGMENTS

This research could not have been possible without the help of many people and support from a number of organizations. The work was initiated with the support of the National Science Foundation BNS 8903377. The NSF grant grew out of a paper written with the intention of summarizing the status of zooarchaeological work in the Caribbean by 1989 (Wing 1989). Zooarchaeological research and recovery methods have advanced greatly during the past two decades. Many new faunal samples have been recovered using improved techniques and form the basis of this research. The new work could not have been done without funding from the Virgin Islands Division of Archaeology and Historic Preservation, L'Association Archéologique St. Martin, LAW Environmental and Engineering Caribe, funds from Peter Drewett and Samuel Wilson. I am most grateful to the archaeologists who entrusted faunal sample to us for study. This research would not have been possible without the careful recovery of these samples or without the help of the many colleagues who granted permission to share their data. Those who worked on the identification and analysis of many samples are Susan deFrance, Laura Kozuch, Irv Quitmyer, Sylvia Scudder, and Elizabeth Reitz. I benefitted immensely from their expertise and keen insights. I am also most grateful to Elizabeth Reitz for editing this chapter. Naturally any mistakes or misinterpretations are my responsibility alone.

REFERENCES

Crosby, A. W., 1986, *Ecological Imperialism: The Biological Expansion of Europe, 900–1900*, University of Cambridge Press, Cambridge, England.

Davis, S. J. M., 1987, *The Archaeology of Animals*, Yale University Press, New Haven.

deFrance, S. D., 1988, *Zooarchaeological Investigations of Subsistence Strategies at the Maisabel Site, Puerto Rico*, M.A. thesis, Department of Anthropology, University of Florida, Gainesville.

Flannery, T. F., and White, J. P., 1991, Animal Translocation, *National Geographic Research and Exploration* 7(1):96–113.

Flemming, C., and MacPhee, R. D. E., 1999, Redetermination of Holotype of *Isolobodon portoricensis* (Rodentia, Capromyidae), with Notes on Recent Mammalian Extinctions in Puerto Rico, *American Museum Novitates* 3278:1–11.

Gade, D. W., 1977, Animal/Man Relationships of Neotropical Vertebrate Fauna in Amazonia, *National Geographic Society Research Reports 1977 Projects* 18:321–326.

Gentry, A., Clutton-Brock, J., and Groves, C. P., 2004, The Naming of Wild Animal Species and Their Domestic Derivatives, *Journal of Archaeological Science* 31:645–651.

Goodwin, R. C., 1976, Archeological Investigations on St. Kitts, West Indies: A Progress Report, in: *Public Works*, Government of St. Kitts, Nevis, and Anguilla, Basseterre, St. Kitts, pp. 32–58.

Goodwin, R. C., 1979, *Prehistoric Cultural Ecology of St. Kitts, West Indies: A Case Study in Island Archeology*, PhD dissertation, Department of Anthropology, Arizona State University, Tempe.

Hargrave, L. L., 1970, Mexican Macaws: Comparative Osteology and Survey of Remains from the Southwest, *Anthropological Papers of the University of Arizona* 20:1–67, Tucson.

Klift, H. M. van der, 1992, Faunal Remains of Golden Rock, in: *The Archaeology of St. Eustatius: The Golden Rock Site* (A. H. Versteeg and K. Schinkel, eds.), *St. Eustatius Historical Foundation and Foundation for Scientific Research in the Caribbean* 131, Amsterdam, The Netherlands, pp. 74–83.

Lawrence, B., 1977, Dogs from the Dominican Republic, *Cuadernos del Cendia, Centro Dominicano de Investigaciones Antropológicas* 168(8):3–19.

Linares, O. F., 1976, "Garden Hunting" in the American Tropics, *Human Ecology* 4(4):331–349.

Lovén, S., 1935, *Origins of the Tainan Culture, West Indies*, Elanders Bokfryckeri Akfiebolag, Göteborg, Sweden.

Miller, G. S., Jr., 1929, Mammals Eaten by Indians, Owls, and Spaniards in the Coastal Region of the Dominican Republic, *Smithsonian Miscellaneous Collections* 82(5):1–16.

Minnis, P. E., Whalen, M. E., Kelley, J. H., and Stewert, J. D., 1993, Prehistoric Macaw Breeding in the North American Southwest, *American Antiquity* 58(2):270–276.

Morgan, G. S., and Woods, C. A., 1986, Extinctions and the Zoogeography of West Indian Land Mammals, *Biological Journal of the Linnean Society* 28:167–203.

Narganes Storde, Y. M., 1982, *Vertebrate Remains from Sorcé, Vieques, Puerto Rico*, M.A. thesis, Department of Anthropology, University of Georgia, Athens.

Narganes Storde, Y. M., 1991, Los Restos Faunísticos del Sitio de Puerto Ferro, Vieques, Puerto Rico, in: *Proceedings of the Fourteenth Congress of the International Association for Caribbean Archaeology* (A. Cummins and P. King, eds.), Barbados Museum and Historical Society, Barbados, pp. 94–114.

Newsom, L. A., and Wing, E. S., 2004, *On Land and Sea: Native American Uses of Biological Resources in the West Indies*, University of Alabama Press, Tuscaloosa.

Nokkert, M., 1995, *Prehistoric Faunal Exploitation on St. Martin: An Analysis of Vertebrate Faunal Material from the Pre-Ceramic Norman Estate and the Cedrosan Saladoid Anse des Peres Sites*, M.A. thesis, University of Leiden, Leiden, The Netherlands.

Oviedo, Gonzalo F. de (translation by S. A. Stoudemire, 1959), 1526, *Natural History of the West Indies*, University of North Carolina Press, Chapel Hill.

Quitmyer, I. R., and Kozuch, L., 1996, Phase II Zooarchaeology at Finca Valencia (NCS-1) and Site NCS-4, Northwest Puerto Rico with a Note on the Guinea Pig Remains, manuscript on file, Florida Museum of Natural History, University of Florida, Gainesville.

Reitz, E. J., 1985, Vertebrate Fauna from El Bronce Archaeological Site, Puerto Rico, in: *Archaeological Data Recovery at El Bronce, Puerto Rico, Final Report Phase 2* (L. S. Robinson, E. R. Lundberg and J. B. Walker, eds.), Report to the U.S. Corp of Engineers, Jacksonville District, Jacksonville, Florida.

Reitz, E. J., 1989, Vertebrate Fauna from Krum Bay, Virgin Islands, in: *Preceramic Procurement Patterns at Krum Bay, Virgin Islands*, by E. R. Lundberg, PhD. dissertation, University of Illinois, Urbana, University Microfilms, Ann Arbor, pp. 374–389.

Reitz, E. J., 1994, Archaeology of Trants, Montserrat. Part 2. Vertebrate Fauna, *Annals of Carnegie Museum* 63(4):297–317.

Reitz, E. J., and Wing, E. S., 1999, *Zooarchaeology*, 1st ed., University of Cambridge Press, Cambridge, England.

Reynolds, T. E., Koopman, K. F., and Williams, E. E., 1953, A Cave Faunule from Western Puerto Rico with a Discussion of the Genus *Isolobodon*, *Breviora* 12:1–8.

Rímoli, R. O., 1977, Nuevas Citas Para Mamíferos Precolombinos en el Hispaniola, *Cuadernos del Cendia, Centro Dominicano de Investigaciones Antropológicas* 259(5):3–15.

Rouse, I., 1992, *The Tainos: Rise and Decline of the People who Greeted Columbus*, Yale University Press, New Haven, Connecticut.

Rouse, I., and Faber-Morse, B., 1999, Excavations at the Indian Creek site, Antigua, West Indies, *Yale University Publications in Anthropology 82*, New Haven, Connecticut.

Sauer, C. O., 1966, *The Early Spanish Main*, University of California Press, Berkeley.

Schwarz, H. F., 1948, Stingless Bees (Meliponidae) of the Western Hemisphere, *Bulletin of the American Museum of Natural History* 90:1–546.

Scudder, S. J., 1991, Early Arawak Subsistence Strategies on the South Coast of Jamaica, in: *Proceedings of the Thirteenth International Congress for Caribbean Archaeology* (E. N. Ayubi and J. B. Haviser, eds.), Report of Archaeological-Anthropological Institute of the Netherlands Antilles 9:297–312.

Serpell, J., 1986, *In the Company of Animals*, Basil Blackwell, Oxford, England.

Steadman, D. W., Watters, D. R., Reitz, E. J., and Pregill, G. K., 1984, Vertebrates from Archaeological Sites on Montserrat, West Indies, *Annals of Carnegie Museum* 53:1–29.

Vietmeyer, N. D. (ed.), 1991, *Microlivestock: Little Known Small Animals with a Promising Economic Future*, National Research Council, National Academy Press, Washington, DC.

Westermann, J. H., 1953, Nature Preservation in the Caribbean, *Foundation for Scientific Research in Surinam and the Netherlands Antilles* 9, Utrecht, The Netherlands.

Wilkins, L., 2001, Impact of Hunting on Jamaican Hutia (*Geocapromys brownii*) Populations: Evidence from Zooarchaeology and Hunter Surveys, in: *Biogeography of the West Indies: Patterns and Perspectives*, 2nd ed. (C. A. Woods and F. Sergile, eds.), CRC Press, Boca Raton, Florida, pp. 529–545.

Wilson, S. M., 2001, The Prehistory and Early History of the Caribbean, in: *Biogeography of the West Indies: Patterns and Perspectives*, 2nd ed. (C. A. Woods and F. Sergile, eds.), CRC Press, Boca Raton, Florida, pp. 519–527.

Wing, E. S., 1986, Domestication of Andean Mammals, in: *High Altitude Tropical Biogeography* (F. Vuillemier and M. Monasterio, eds.), Oxford University Press, Oxford, England, pp. 262–264.

Wing, E. S., 1989, Human Exploitation of Animal Resources in the Caribbean, in: *Biogeography of the West Indies*, 1st ed. (C. A. Woods, ed.), Sandhill Crane Press, Gainesville, Florida, pp. 137–152.

Wing, E. S., 1991a, Dog Remains from the Sorcé Site on Vieques Island, Puerto Rico, in: *Beamers, Bobwhites and Blue-points: Tributes to the Career of Paul W. Parmalee* (J. Purdue, W. Klippel, and B. Styles, eds.), Illinois State Museum, Springfield, pp. 379–386.

Wing, E. S., 1991b, Economy and Subsistence 1: Faunal Remains, in: *Prehistoric Barbados* (P. L. Drewett, ed.), Archetype Publications, Institute of Archaeology, University of London, London, pp. 134–152.

Wing, E. S., 1996, Vertebrate Remains Excavated from the Sites of Spring Bay and Kelbey's Ridge, Saba, Netherlands West Indies, in: *In Search of the Native Population of Pre-Columbian Saba (400–1450 A.D.)* Part Two, M. L. P. Hoogland, PhD. dissertation, Rijkuniversiteit, Leiden, The Netherlands, pp. 261–279.

Wing, E. S., 1999, Animal Remains from the Indian Creek Site, Antigua, in: *Excavations at the Indian Creek Site, Antigua, West Indies* (I. Rouse and B. Faber Morse, eds.), *Yale University Publication in Anthropology 82*, New Haven, Connecticut, pp. 51–66.

Wing, E. S., 2000, Economy and Subsistence 1: Animal Remains on Sites on Barbados and Tortola, in: *Prehistoric Settlements in the Caribbean: Fieldwork in Barbados, Tortola, and the Cayman Islands* (P. L. Drewett, ed.), Archetype Publications, Institute of Archaeology, University of London, pp. 147–153.

Wing, E. S., 2001, Native American Use of Animals in the Caribbean, in: *Biogeography of the West Indies: Patterns and Perspectives*, 2nd ed. (C. A. Woods and F. Sergile, eds.), CRC Press, Boca Raton, Florida, pp. 418–518.

Wing, E. S., and Reitz, E. J., 1982, Prehistoric Fishing Economies of the Caribbean, *New World Archaeology* 5(2):13–32.

Wing, E. S., and Scudder, S. J., 1980, Use of Animals by the Prehistoric Inhabitants on St. Kitts, West Indies, in: *Proceedings of the 8th International Congress for the Study of Pre-Columbian Cultures of the Lesser Antilles* (S. W. Lewenstein, ed.), *Arizona State University Anthropological Research Papers* 22:237–245.

Wing, E. S., and Scudder, S. J., 1983, Animal Exploitation by Prehistoric People Living on a Tropical Marine Edge, in: *Animals and Archaeology: 2. Shell Middens, Fishes and Birds* (C. Grigson and J. Clutton-Brock, eds.), *British Archaeological Reports International Series* (Oxford) 183:197–210.

Woods, C. A., 1996, The Land Mammals of Puerto Rico and the Virgin Islands. The Scientific Survey of Puerto Rico and the Virgin Islands, *Annals of the New York Academy of Sciences* 776:131–149.

Index of Common
and Scientific Names

Page references to figures and tables are italicized.

427

III. BIRDS

IV. AMPHIBIANS AND REPTILES

VII. OTHER

Subject Index

Page references to figures and tables are italicized.

447

Printed in the United States of America